SCOPE 28

Environmental Consequences of Nuclear War

Volume I

Physical and Atmospheric Effects

Second Edition

Scientific Committee on Problems of the Environment

SCOPE

Executive Committee, elected 10 June 1988

Officers

President: Professor F. di Castri, CEPE/CNRS, Centre L. Emberger, Route de Mende, BP 5051, 34033 Montpellier Cedex—France.

Vice-President: Academician M. V. Ivanov, Institute of Microbiology, USSR Academy of Sciences, Prospekt 60-letija Oktjabrja 7, 117811, Moscow—USSR

Vice-President: Professor C. R. Krishna Murti, Scientific Commission for Continuing Studies on Effects of Bhopal Gas Leakage on Life Systems, Cabinet Secretariat, 2nd floor, Sardar Patel Bhavan, New Delhi 110 001—India.

Treasurer: Doctor T. E. Lovejoy, Smithsonian Institution, Washington, DC 20560–USA.

Secretary-General: Professor J. W. B. Stewart, Saskatchewan Institute of Pedology, University of Saskatchewan, Saskatoon, S7N 0W0 Saskatchewan—Canada.

Members

Professor M. O. Andreae (I.U.G.G. representative), Max-Planck-Institut für Chemie, Postfach 3060, D-6500 Mainz—FRG.

Professor M. A. Ayyad, Faculty of Science, Alexandria University, Moharram Bey, Alexandria—Egypt.

Professor R. Herrera (I.U.B.S. representative), Centro de Ecologia y Ciencias Ambientales (IVIC), Carretera Panamericana km. 11, Apartado 21827, Caracas—Venezuela.

Professor M. Kecskés, Department of Microbiology, University of Agricultural Sciences, Pater K. ucta 1, 2103 Gödöllö—Hungary.

Professor R. O. Slatyer, School of Biological Sciences, Australian National University, P.O. Box 475, Canberra, ACT 2601—Australia.

Editor-in-Chief

Dr. R. E. Munn, Institute of Environmental Studies, University of Toronto, Toronto, Canada M5S 1A4.

SCOPE 28

Environmental Consequences
of Nuclear War

Volume I. Physical and Atmospheric Effects
Second Edition

A. Barrie Pittock
Commonwealth Scientific and Industrial Research Organization
Division of Atmospheric Research

Thomas P. Ackerman
Pennsylvania State University

Paul J. Crutzen
Max-Planck-Institut für Chemie

Michael C. MacCracken
Lawrence Livermore National Laboratory

Charles S. Shapiro
San Francisco State University and
Lawrence Livermore National Laboratory

Richard P. Turco
University of California

In collaboration with:

V. Aleksandrov, P. Connell, G. Golitsyn,
T. Harvey, S. Kang, K. Peterson, and G. Tripoli

Published on behalf of the
Scientific Committee on Problems of the Environment (SCOPE)
of the
International Council of Scientific Unions (ICSU)
by

JOHN WILEY & SONS

Chichester · New York · Brisbane · Toronto · Singapore

Copyright © 1985, 1989 by the
Scientific Committee on Problems of the Environment (SCOPE)

Library of Congress Cataloging in Publication Data:
Environmental consequences of nuclear war.—2nd ed.
    p. cm.—(SCOPE ; 28)
    Includes bibliographical references.
    Contents: v. 1. Physical and atmospheric affects / A. Barrie
Pittock—v. 2. Ecological and agricultural effects / Mark A.
Harwell and Thomas C. Hutchinson.
    ISBN 0 471 92469 5 (v. 1)—ISBN 0 471 92471 7 (v. 2)
    1. Nuclear warfare—Environmental aspects. I. Pittock, A.
Barrie, II. Harwell, Mark A. III. Hutchinson, T.C.
(Thomas, C.); IV. International Council of Scientific
Unions. Scientific Committee on Problems of the Environment.
V. Series: SCOPE (Series) ; 28.
QH545.N83E58 1989
574.5'222—dc20
                                                89-22712
                                                CIP

British Library Cataloguing in Publication Data:
Environmental consequences of nuclear war—2nd ed—
(SCOPE;28)
Vol. 1, physical and atmospheric effects
1. Environment. Effects of nuclear warfare
I. Pittock, A. Barrie II. International Council
of Scientific Unions, *Scientific Committee on
problems of the Environment* III. Series
333.7'1

ISBN 0 471 92469 5

Printed and bound in Great Britain by Biddles Ltd, Guildford

SCOPE 37: Biological Invasions: A Global Perspective 1989, 528pp

SCOPE 38: Ecotoxicology and Climate, 1989, 392pp

SCOPE 39: Evolution of the Global Biogeochemical Sulphur Cycle, 1989, 241pp

SCOPE 40: Methods for Assessing and Reducing Injury from Chemical Accidents, 1989, 303pp

Funds to meet SCOPE expenses are provided by contributions from SCOPE National Committees, an annual subvention from ICSU (and through ICSU, from UNESCO), subventions from the French Ministère de l'Environnement, contracts with UN Bodies, particularly UNEP, and grants from Foundations and industrial enterprises.

International Council of Scientific Unions (ICSU)
Scientific Committee on Problems of the Environment (SCOPE)

SCOPE is one of a number of committees established by a non-governmental group of scientific organizations, the International Council of Scientific Unions (ICSU). The membership of ICSU includes representatives from 74 National Academies of Science, 20 International Unions and 26 other bodies called Scientific Associates. To cover multidisciplinary activities which include the interests of several unions, ICSU has established 10 scientific committees, of which SCOPE is one. Currently, representatives of 35 member countries and 20 international scientific bodies participate in the work of SCOPE, which directs particular attention to the needs of developing countries. SCOPE was established in 1969 in response to the environmental concerns emerging at that time; ICSU recognized that many of these concerns required scientific inputs spanning several disciplines and ICSU Unions. SCOPE's first task was to prepare a report on Global Environmental Monitoring (SCOPE 1, 1971) for the UN Stockholm Conference on the Human Environment.

The mandate of SCOPE is to assemble, review, and assess the information available on man-made environmental changes and the effects of these changes on man; to assess and evaluate the methodologies of measurement of environmental parameters; to provide an intelligence service on current research; and by the recruitment of the best available scientific information and constructive thinking to establish itself as a corpus of informed advice for the benefit of centres of fundamental research and of organizations and agencies operationally engaged in studies of the environment.

SCOPE is governed by a General Assembly, which meets every three years. Between such meetings its activities are directed by the Executive Committee.

R. E. Munn
Editor-in-Chief
SCOPE Publications

Executive Secretary: V. Plocq

Secretariat: 51 Bld de Montmorency
75016 PARIS

Acknowledgements

This study of the global environmental consequences of nuclear war was conducted with the help and cooperation of hundreds of scientists from many countries. The authors wish to express deep appreciation to the many scientists who donated their time, knowledge, and skills to make possible this cooperative international effort.

The SCOPE-ENUWAR steering committee, chaired by Sir Frederick Warner, skilfully guided this Project through its many stages. The members of the Steering Committee are listed in the Foreword. Particular appreciation and acknowledgment are given to Sir Frederick Warner, Thomas Malone, and Gilbert White for their very active role throughout the Project, particularly in the concluding months in helping with the preparation for publication.

Mark Harwell played a central role in the entire ENUWAR project by participating in all of the Volume I workshops and providing critical review of the written material. He unselfishly devoted considerable energy and much time in a volunteer capacity, providing essential scientific guidance to the entire Project.

This book was written with the collaboration of Vladimir Aleksandrov, Peter Connell, George Golitsyn, Ted Harvey, Sang-Wook Kang, Kendall Peterson and Greg Tripoli. These people contributed to this publication by extensive review of working drafts and by the preparation of significant sections of some of the chapters. It is with considerable concern that we learned of the unexplained disappearance in Madrid of our colleague, Vladimir Aleksandrov, in March 1985, following a conference where he reported on his studies contributing to this Project.

An extensive reviewing process was an essential part of the SCOPE-ENUWAR project. At each of the workshops, at different stages of the drafting of this volume, and at the final synthesis workshop in Essex, United Kingdom, an attempt was made to incorporate the views and critical analyses of reviewers from many nations, organizations, and disciplines. From this large group, special thanks are given here to those who played major roles in this extensive review process, were major participants in the workshops, or contributed in other important ways. They each made important contributions to the quality and integrity of this volume. This group includes Keith Bigg, Alan Cairnie, Robert Cess, William Cotton, Jürgen Hahn, Peter Hobbs, Julius London, Robert Malone, George Mulholland, Edward

Patterson, Joyce Penner, Larry Radke, Joseph Rotblat, Steve Schneider, Tony Slingo, John Walton and Manuel Wik. Many others who also contributed in a significant way are listed at the end of this volume (Appendix 3). To those who may have been missed in the rush to prepare this volume, our apologies—be assured of our appreciation.

The authors would like to express special appreciation to Thomas Ackerman, who served as coordinating editor during the final months of this Project. From his work on disparate inputs emerged a cohesive draft that more clearly presented both results and uncertainties. Final editing and production of the report were carried out under the direction of Michael MacCracken. Without his unrelenting effort, this report could not have been completed on schedule.

Ms. Ann Freeman served as the administrative staff of the ENUWAR project, arranging for workshops, travel, and secretarial support to the program, keeping the minutes, issuing the newsletters, and assisting in other diverse, but important, ways. Michel Verstraete provided technical support to the project by compiling the glossary (included as Appendix 2) and index.

The authors express special appreciation to Ms. Floy Worden, who transformed the various contributions into an integrated and printable text, interpreting the authors' many comments, each time regenerating corrected versions in a very short time. Her many hours of dedicated work, and the work of other secretaries at Lawrence Livermore National Laboratory, are greatly appreciated.

The authors also thank the institutions with which they are primarily affiliated: Commonwealth Scientific and Industrial Research Organization, Division of Atmospheric Research; National Aeronautics and Space Administration, Ames Research Center; Max-Planck-Institut für Chemie; Lawrence Livermore National Laboratory; San Francisco State University; and R and D Associates. These institutions provided support for us in many ways; however, this does not represent an endorsement by these institutions or their sponsoring agencies of the findings expressed in this report.

Finally, while many individuals have been mentioned as playing important roles in the evolution of this volume, most of them worked only with parts of the study and are not responsible for its overall content. The six authors alone accept the responsibility for accurate reporting of current scientific results and for any scientific judgments presented herein.

A. B. Pittock
T. P. Ackerman
P. J. Crutzen
M. C. MacCracken
C. S. Shapiro
R. P. Turco

Authors' Note

The term 'nuclear winter' has been applied in some previous analyses of the environmental consequences of a nuclear exchange to describe the multitude of possible effects. In its original usage, this term envisaged the combination of darkened skies, subfreezing temperatures, and extensive toxic and radioactive pollution that, to a more or less severe degree, might follow a nuclear war. The phrase has since, however, come to be associated primarily with the most severe possibilities. Although it is a convenient metaphor for use in describing the generic consequences, we have chosen to avoid use of the term 'nuclear winter' in this study because it does not, in a strict scientific sense, properly portray the range, complexity, and dependencies of the potential global scale environmental consequences of a nuclear war. By this choice, we are not suggesting that the environmental effects of a major nuclear exchange would be inconsequential; to the contrary, we find that they would be substantial and significant.

Contents

Foreword

Beginning in the summer of 1982, approximately 300 scientists from more than 30 countries and a wide range of disciplines, under the auspices of the International Council of Scientific Unions (ICSU), joined in a deliberative effort to appraise the state of knowledge of the possible environmental consequences of nuclear war. Although it has been recognized since the first nuclear explosions over Hiroshima and Nagasaki in 1945 that multiple detonations could cause massive destruction on people and their culture, the effects of life support systems of air, water, and soil and on organisms received relatively little emphasis in public discussion.

In the mid-1970s, attention began to turn to the whole range of consequences that might be expected to follow a large-scale exchange of nuclear weapons. This reflected a growing recognition of the immense number and yield of thermonuclear devices in the arsenals of the nuclear powers. The renewed activities also reflected concern with effects beyond the direct destruction of cities and human life. While interest still centered on the well-studied issues of direct blast, thermal effects, and radioactive fallout from ground and air bursts, scientists began to consider the large-scale consequences (e.g., from possible global depletion of ozone and from perturbations to the atmosphere). This concern was manifested in studies of information that had accumulated from the detonations at Hiroshima and Nagasaki and the subsequent series of nuclear tests, and with extrapolation of these data to situations in which the current nuclear arsenal might be used. Among the analyses were those by the U.S. Senate Committee on Foreign Relations (1975), the U.S. National Academy of Sciences (1975), the Office of Technology Assessment of the U.S. Congress (1979), the United Nations Environmental Programme (1979), the United Nations (1980a), and A. Katz (1982).

In 1982, several organizations and individual scientists launched new examinations of anticipated global effects, including those of the American Association for the Advancement of Science, the U.S. National Academy of Sciences, and the World Health Organization. Appraisals commissioned by the Royal Swedish Academy of Sciences published in Ambio in April 1982 were particularly influential. A paper in that issue by P. Crutzen and J. Birks had been intended to deal with possible effects on the stratospheric ozone layer and regional air quality. While it did suggest that ozone changes might be of significance, the new suggestion was that smoke and soot generated by large urban and forest fires might cause reductions in light at the Earth's surface, inducing profound changes in weather. These suggestions stimulated a new round of research and appraisal around the world. Not since the 1960s, when agitation about the consequences of delayed radioactive fallout from bomb tests in the 1950s resulted in the signing in 1963 of the Treaty Banning Nuclear Weapon Tests in the Atmosphere, in Outer Space, and Under Water, had as much thoughtful attention been marshalled by scientists and citizens.

At its General Assembly in Ottawa in June 1982, the Scientific Committee on Problems of the Environment (SCOPE)—one of the ten Scientific Committees of the International Council of Scientific Unions (ICSU)—concluded that 'the risk of nuclear warfare overshadows all other hazards to humanity and its habitat' and asked its Executive Committee to consider what further action might be appropriate for SCOPE. In September 1982, the General Assembly of ICSU passed the following resolution:

Recognizing the need for public understanding of the possible consequences of the nuclear arms race and the scientific competence that can be mobilized by ICSU to make an assessment of the biological, medical and physical effects of the large-scale use of nuclear weapons.

Urges the Executive Board to appoint a special committee to study these effects and to prepare a report for wide dissemination that would be an unemotional, nonpolitical, authoritative and readily understandable statement of the effects of nuclear war, even a limited one, on human beings and on other parts of the biosphere.

Accordingly, a Steering Committee for the SCOPE-ENUWAR (Environmental Consequences of Nuclear War) study was established, with responsibility to initiate the study requested by ICSU and to oversee the selection and recruitment of participants. A SCOPE-ENUWAR coordinating office was established at the University of Essex. From the outset it was agreed that the report would not deal explicitly with questions of public policy, but would focus on scientific knowledge of physical effects and biological response. International aspects of the direct medical effects have already been dealt with explicitly by the World Health Organization, and thus are not taken up in this study.

The SCOPE-ENUWAR process involved the active collaboration of scientists, bringing together the insights and skills of numerous disciplines. Preparatory workshops were held in London and Stockholm, and major workshops were convened in New Delhi, Leningrad, Paris, Hiroshima and Tokyo, Delft, Toronto, Caracas, Melbourne, and finally at the University of Essex in an attempt to arrive at a consensus. Smaller groups gathered in a variety of other places, chiefly in connection with meetings of International Scientific Unions. Meanwhile, new findings were becoming available as noted in appropriate parts of this report, and further studies of likely effects were published (Turco et al., 1983a; Ehrlich et al., 1983; Aleksandrov and Stenchikov, 1983; Openshaw et al., 1983; World Health Organization, 1983; Covey et al., 1984; London and White, 1984; United Nations, 1984; Harwell, 1984; National Research Council, 1985; The Royal Society of Canada, 1985; The Royal Society of New Zealand, 1985).

Support for the project came from individual donations of time and from organizational grants. The Steering Committee is particularly grateful to those who committed the extensive time and effort to prepare the two volumes reporting these important scientific results. Barrie Pittock, Thomas Ackerman, Paul Crutzen, Michael MacCracken, Charles Shapiro, and Richard Turco have been responsible for preparation of the volume on physical and atmospheric effects. Mark Harwell, Thomas Hutchinson, Wendell Cropper, Jr., Christine Harwell and Herbert Grover have played the major role in preparing the volume on ecological and agricultural effects. Both sets of authors were assisted by many colleagues, listed elsewhere in these volumes, who collaborated with them and generously gave of their time to participate in discussion, analysis, writing, and review. It was very much a cooperative, voluntary effort.

The collaboration among these scientists was made possible by financial contributions covering the costs of travel, assistance by post-doctoral fellows, workshop arrangements, and secretarial support. Initial grants making possible the planning of the project came from the SCOPE Executive Committee, using contributions from its 36 member academies of science, and from ICSU. The Royal Society of London hosted the preliminary and concluding workshops and funded the SCOPE-ENUWAR office. Other workshops were hosted by the Royal Swedish Academy of Sciences, the Indian National Science Academy, the Academy of Sciences of the U.S.S.R., la Maison de la Chimie of France, the T.N.O. Institute of Applied Geosciences of the Netherlands, the Australian Academy of Science jointly with the Royal Society of New Zealand, the United Nations University and the Venezuelan Institute of Scientific Investigation. Major grants for travel and other expenses were provided by the Carnegie Corporation of New York, The General Service Foundation, The Andrew W. Mellon Foundation, the W. Alton Jones Foundation, The MacArthur Foundation and The Rockefeller Brothers Fund.

Recognizing that the issues dealt with in this report transcend science and technology and involve moral and ethical issues, SCOPE-ENUWAR co-sponsored an *ad hoc* meeting of scientists and scholars of ethics and morality at the Rockefeller Conference and Study Centre, Bellagio, Italy, in November 1984. The conference took note of the preliminary findings that a significant nuclear exchange could lead to an unprecedented climatic perturbation, killing crops and threatening countries distant from the target areas with mass starvation. A statement called for the development of more effective cooperative efforts for dealing with common interests and problems and urged collaboration between science and religion in the '... quest for a just and peaceful world' (*Bulletin of Atomic Scientists*, April 1985, pp. 49–50).

The Steering Committee has elected to publish the results of the SCOPE-ENUWAR studies in two volumes. The first volume deals with the physical aspects of the environmental impact of a nuclear war. The second volume addresses the biological impacts, principally the ecological and agricultural effects. As further background for the reader, each volume includes the Executive Summary of the companion volume, with its explanation of findings and research recommendations, as an appendix. In addition, the Committee has commissioned a less technical account intended for wide international distribution to fulfill the ICSU request for a '... readily understandable statement of the effects of nuclear war'. It is anticipated that this third volume will be translated into several languages.

The two volumes present a general consensus among the scientists concerned with the study. There is not unanimity on all points, but a concentrated effort has been made to describe those remaining points at issue. These unresolved issues suggest research that should be pursued in order to reduce the present degree of uncertainty. The report should be regarded as the first attempt by an international scientific group to bring together what is known, and what must still be learned, about the possible global environmental effects of nuclear war. It should not be the last. It should be taken as a point of departure rather than as a completed investigation.

A recurring issue in the recent discussion of the long-term, global environmental consequences of a nuclear war has been the degree to which uncertainties preclude a conclusion regarding the plausibility of severe effects. These uncertainties are of two kinds: (1) those resulting from the nature of human actions (e.g., number of weapons, yields, targets, height of detonation, time of conflict, accidents resulting from technological failure, societal response to an outbreak of hostilities); and (2) those resulting from an incomplete state of knowledge concerning physical and biological processes and the limited ability to simulate them faithfully by mathematical models.

Clearly, the specific circumstances of a large-scale nuclear war cannot be predicted with confidence, and the history of past wars reminds us that even carefully planned military actions rarely develop as expected. Thus, detailed scenarios of possible nuclear exchanges must remain highly speculative. Wherever practicable, as a basis for estimating environmental effects, the report considers specific ranges of physical parameters and responses—such as a given mass of smoke injected into the atmosphere, or the occurrence of a freezing episode—that are consistent with the detailed technical analyses, yet are not peculiar to any specific war scenario. In the absence of a nuclear war, many of the specific effects will continue to be in doubt.

Although uncertainties associated with knowledge of physical and biological processes could be substantially reduced by further research, some of these uncertainties are bound to remain large for many years, as explained in the report.

The report does not attempt to provide a single estimate of the likely consequences for humans and their societies of the physical and biological changes projected to be possible after a nuclear war. One reason is that the combinations of possible environmental perturbations are so large and the varieties of environmental and human systems are so numerous and complex that it would be an impossible task to look with detail into all of the ways in which those perturbations might result in an impact. Further, the environmental disruptions and dislocations from nuclear war would be of a magnitude for which there is no precedent. Our present interdependent, highly organized world has never experienced anything approaching the annihilation of people, structures, resources, and disruption of communications that would accompany a major exchange, even if severe climatic and environmental disturbances were not to follow it. The latter could aggravate the consequences profoundly. How the environmental perturbations which would occur at unprecedented scales and intensities would affect the functioning of human society is a highly uncertain subject requiring concerted research and evaluation. Nevertheless, whatever the uncertainties, there can be no doubt that there is a considerable probability a major nuclear war could gravely disrupt the global environment and world society. All possible effects do not have the same probability of occurrence. Sharpening these probabilities is a matter for a continuing research agenda.

The bases for these statements are to be found in the report, along with references to supporting or relevant information. From them we draw the following general conclusions:

1. Multiple nuclear detonations would result in considerable direct physical effects from blast, thermal radiation, and local fallout. The latter would be particularly important if substantial numbers of surface bursts were to occur since lethal levels of radiation from local fallout would extend hundreds of kilometers downwind of detonations.

2. There is substantial reason to believe that a major nuclear war could lead to large-scale climatic perturbations involving drastic reductions in light levels and temperatures over large regions within days and changes in precipitation patterns for periods of days, weeks, months, or longer. Episodes of short term, sharply depressed temperatures could also

produce serious impacts—particularly if they occur during critical periods within the growing season. There is no reason to assert confidently that there would be no effects of this character and, despite uncertainties in our understanding, it would be a grave error to ignore these potential environmental effects. Any consideration of a post-nuclear-war world would have to consider the consequences of the *totality* of physical effects. The biological effects then follow.

3. The systems that currently support the vast majority of humans on Earth (specifically, agricultural production and distribution systems) are exceedingly vulnerable to the types of perturbations associated with climatic effects and societal disruptions. Should those systems be disrupted on a regional or global scale, large numbers of human fatalities associated with insufficient food supplies would be inevitable. Damage to the food distribution and agricultural infrastructure alone, (i.e., without any climatic perturbations) would put a large portion of the Earth's population in jeopardy of a drastic reduction in food availability.

4. Other indirect effects from nuclear war could individually and in combination be serious. These include disruptions of communications, power distribution, and societal systems on an unprecedented scale. In addition, potential physical effects include reduction in stratospheric ozone and, after any smoke had cleared, associated enhancement of ultraviolet radiation; significant global-scale radioactive fallout; and localized areas of toxic levels of air and water pollution.

5. Therefore, the indirect effects on populations of a large-scale nuclear war, particularly the climatic effects caused by smoke, could be potentially more consequential globally than the direct effects, and *the risks of unprecedented consequences are great for noncombatant and combatant countries alike.*

A new perspective on the possible consequences of nuclear war that takes into account these findings is clearly indicated. In these circumstances, it would be prudent for the world scientific community to continue research on the entire range of possible effects, with close interaction between biologists and physical scientists. It would be appropriate for an international group of scientists to reappraise those findings periodically and to report its appraisal to governments and citizen groups. Increased attention is urgently required to develop a better understanding of potential societal responses to nuclear war in order to frame new global perspectives on the large-scale, environmental consequences. This task is a special challenge to social scientists.

In arriving at these conclusions, we have been moderate in several respects. We have tried to state and examine all challenges to theories about environmental effects of nuclear war, to minimize speculative positions and to factor valid criticisms into discussions and conclusions. Uncertainties in the projections could either reduce or enhance the estimated effects in specific cases. Nevertheless, as representatives of the world scientific community drawn together in this study, we conclude that many of the serious global environmental effects are sufficiently probable to require widespread concern. Because of the possibility of a tragedy of an unprecedented dimension, any disposition to minimize or ignore the widespread environmental effects of a nuclear war would be a fundamental disservice to the future of global civilization.

SCOPE-ENUWAR Steering Committee

Sir Frederick Warner, University of Essex, U.K., *Chairman*
J. Bénard, Ecole Supérieure de Chimie, Paris, France
S. K. D. Bergström, Karolinska Institutet, Stockholm, Sweden
P. J. Crutzen, Max-Planck-Institut für Chemie, Mainz, F.R.G.
T. F. Malone, (ICSU Representative) St. Joseph College, U.S.A.
M. K. G. Menon, Planning Commission, New Delhi, India
M. Nagai, United Nations University, Tokyo, Japan
G. K. Skryabin, Akademia Nauk, Moscow, U.S.S.R.
G. F. White, University of Colorado, U.S.A.

Preface to Second Edition

Early in 1988, SCOPE-ENUWAR, an international effort to provide authoritative estimates of the environmental consequences of nuclear war, completed a five-year study involving unprecedented scientific cooperation. The SCOPE-ENUWAR project began with a workshop in Stockholm in November 1983 and concluded, after a series of meetings around the world, with a workshop in Moscow in March 1988.

The completion of SCOPE's study on the Environmental Consequences of Nuclear War (ENUWAR) marks a major milestone in international scientific cooperation in the tradition of the International Geophysical Year, heralding also the developing International Geosphere-Biosphere Programme. Based on research and findings through 1985, the SCOPE-ENUWAR project published its initial findings confirming that a major nuclear war could lead to climatic and other environmental impacts which could threaten much of the world's population with starvation. Since publication of the summary findings in the first edition of SCOPE-28 early in 1986, additional workshops have been held in London, Bangkok, Geneva, and Moscow. This preface to the second edition summarizes the work accomplished between the two editions, and describes the present state of understanding and the gaps in knowledge that remain to be filled.

While the formal SCOPE-ENUWAR program has now been concluded, ENUWAR's active coordination of activities for a wide field of research will go forward under a variety of auspices. It is expected that our conclusions will need to be reassessed from time to time, particularly if fundamental changes occur in the range or scope of plausible scenarios for a nuclear exchange. The recent discussions on arms limitation pose an initial example. The agreement on the Intermediate-Range Nuclear Forces (INF) Treaty calls for destruction of the delivery systems of four percent of the weaponry; initial analyses suggest that such modest changes would not significantly affect our conclusions. An agreement to reduce the arsenals by fifty percent, however, may require a careful re-examination of some assumptions and findings, but there seems little likelihood that substantial reductions in smoke emissions or consequent environmental perturbations would occur, so long as the bombing of targets in urban areas and consequent burning of cities and major industrial installations remains as an integral element of nuclear war strategy.

SMOKE SOURCES AND PROPERTIES

Fuel Loading and Smoke Emissions

The quantities, or inventories, of flammable materials located within potential nuclear target zones, the amount of smoke aerosol (and particularly the amount of highly carbonaceous sooty aerosols) generated when such materials burn, and the overlap of nuclear thermal ignition patterns with fuel distributions are the major factors that determine the production of smoke in a nuclear conflict.

Recent studies of fuel inventories (Penner, 1986b; Turco et al., 1987; Small et al., 1989) confirm that the total quantities of flammable materials may be lower than the earliest estimates of Crutzen and Birks, (1982), Turco et al., (1983a) and Crutzen et al. (1984) by a factor of 2 to 3. These reductions have generally occurred in inventories of materials that are not large contributors to the very sooty, black smoke that is of critical importance.

In those critical fuel categories that are strongly soot-producing, including petroleum and related products, asphalt roofing, and plastics, the inventory estimates have remained fairly constant and now seem to be better established. The world's primary and secondary stores of petroleum and related products are estimated to total about 800 to 1,500 teragrams (1 Tg equals 10^{12} grams, or 1 million metric tons); of asphalt roofing, 250 to 500 Tg; and of plastics, 340 to 460 Tg (see p. 44, this volume). The smoke emission factor for many of these materials has been measured in the laboratory under varying conditions of combustion (e.g., Einfeld et al., 1987). In these experiments the approximate emission factors are ten percent for asphalt, three to nine percent for large (about 150 square-meter) JP-4 pools, three to eleven percent for a variety of plastics, and ten percent for rubber. These smokes contain a large fraction of soot (70–90 percent) which is somewhat larger than assumed before (see pp. 48–49, this volume).

Of order 90 percent of urban and rural fuels are cellulosic, or wood based. Estimates of the global inventory of lumber and related products range from 6,000 to 13,000 Tg (Penner, 1986b; Turco et al., 1987). In freely ventilated combustion, these fuels generally produce very little smoke (Dod et al., 1989). Large-scale test fires in dry vegetation give smoke emission factors in the range of one to three percent, although some larger vegetation fires show smoke emissions as large as seven percent (Radke et al., 1988). Small-scale laboratory vegetation fires generate much less smoke during flaming combustion (less than one percent) than during smoldering combustion (up to ten percent) (Golitsyn et al., 1988a; Patterson et al., 1986). Only about 10 percent of such smoke mass is soot (i.e., 0.1 to 0.3 percent of the original fuel is emitted as smoke). An important recent finding, however, is that construction lumber in 100-kilogram cribs burning under conditions of restricted ventilation, similar to what might occur in large wooden structures, generates one to three percent of black sooty smoke (Dod et al., 1989). A new detailed analysis of possible smoke emissions caused by a full-scale attack on the United States estimates production of about 40 Tg of sooty smoke (Small et al., 1989); an equivalent amount would likely be generated as a result of European and Soviet fires. An independent Soviet estimate of the total smoke generated under a base-line nuclear exchange scenario corresponding to that of the U.S. National Research Council (NRC, 1985) is 70 to 90 Tg (Petryanov-Sokolov et al., 1988). Given the many factors going into these estimates, the plausible range for smoke emissions about these baseline values is probably about a factor of 3 or 4 (Turco et al., 1987); such estimates remain generally consistent with information presented in pages 40–46 of this volume, indicating somewhat less total mass of smoke than estimated earlier, but a commensurate increase in the sooty component.

Optical Properties

From the standpoint of climatic impact, the most important component of smoke is the amount of soot. Sooty particles are characterized by their high elemental carbon content, typically 90 to 95 percent of the particle mass. For fresh soot, the absorption of light at visible wavelengths has been found to be quite invariant, with a typical absorption cross section of about 8 to 10 square meters per gram. For typical particles of blackish smoke, which are only partly soot, the absorptivity generally lies in the range of 5 to 9 square meters per gram of smoke (Mulholland et al., 1988; Patterson et al., 1986; Anikin and Shukurov, 1988). Observations indicating that the absorptivity of soot is relatively independent of the particle source (or size) are consistent with a newly developed theory of light absorption by soot-like fractal aggregates (Berry and Percival, 1986), and with micrographs showing that the soot particles have an irregular, but snowflake-like structure. Recent measurements of light absorption by sooty smoke from a large petroleum pool fire plume indicate that the absorptivity decreases slowly with increasing wavelength such that the soot remains a strong absorber throughout the solar spectrum (approximately 0.35 to 1.0 μm). There have been no new measurements to report on regarding the longer wavelength (infrared) absorptivity of smoke.

Nucleation Properties

Recent laboratory studies of a variety of smokes to determine their ability to serve as cloud condensation nuclei (CCN) reveal that about 25 to 75 percent of wood smoke particles, but only about one percent of JP-4 soot particles, can serve as CCN at early times. This range in smoke CCN characteristics

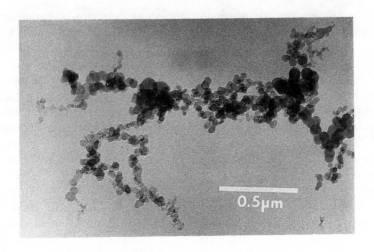

Figure 1. Typical aggregated particles of black soot – an effective absorber of sunlight – as generated by burning wood during a laboratory simulation. (Photo: George Mulholland)

introduces an important uncertainty, especially for smokes from mixed fuels. These CCN fractions are observed under conditions involving exposure to one percent supersaturation after aging in holding bags for 30 minutes, but the fractions are generally not significantly larger even after five hours of aging. The laboratory results for wood smoke are also consistent with extensive data collected from nine large-scale field burns. In these cases, the concentrations of CCN in the rising fire columns ranged from 100,000 to 1 million per cubic centimeter—two to three orders of magnitude greater than natural background values (Hallett et al., 1989).

Because of their high soot content the CCN activity of particles generated from liquid organic fuels (such as crude oil and kerosene) is of great importance. Hansen and Novakov (1988) find that CCN form fairly readily in crude oil fires, but direct measurements of CCN abundances in a large JP-4 fuel pool fire plume show relatively low CCN generation (Hallett et al., 1989). A similar result was found earlier in a soot plume generated by burning heavy diesel fuel. New laboratory studies of soot aggregates formed by aging dense sooty smoke for up to 16 hours indicate that exposure to supersaturations of as much as 3 percent cause only small variations in optical properties, which was expected given the particle sizes in the experiment (Mulholland, et al., 1988). In a series of experiments, soot aggregates exposed to very high supersaturations of 90 to 130 percent showed a negligible decrease in absorptivity (from 6.9 to 6.7 square meters per gram) and a moderate decrease in overall light attenuation (by 23 to 45 percent for soots aged for periods of 0.5 to 4 hours) (Colbeck et al., 1989; Mulholland et al., 1988).

Soot-Ozone Reaction

Several years ago it was proposed that soot particles might decompose upon reacting with ozone in the atmosphere. However, a new laboratory study demonstrates that the reaction is too slow, even under the most favorable conditions, to reduce significantly the atmospheric chemistry of injected soot particles. In a perturbed atmospheric environment, soot chemistry is controlled by ozone decomposition and soot lifetimes are projected to range from about one year to decades (Stephens et al., 1986), with the longer lifetimes being the more likely (Ackerman et al., 1988b). Accordingly, the residence time and removal of soot generated in nuclear war would be controlled by dynamical mixing and by microphysical processes yet to be definitively evaluated.

FIRE PLUME PROCESSING OF SMOKE

Injection

Many recent investigations indicate that the heat generated by urban fires could carry smoke up into the middle-to-upper troposphere (5–10 km) and, for the most intense phase of combustion, into the lowest few kilometers of the stratosphere (that is, up to about 15 kilometers in altitude at mid-latitudes) (Small et al., 1989; Tripoli and Kang, 1987; Marcus et al., 1987; Heikes et al., 1987; Bradley, 1987; and Small and Heikes, 1988). Lofting of smoke to such altitudes seems certain to assure that a substantial fraction of the smoke would remain aloft for long periods, months to years, if not removed by precipitation during its rise in the plume that is generated. Observations of smoke plumes from large-scale fires and wildfires (Radke et al., 1988) confirm that smoke can be injected well above the boundary layer, even when the areas aflame are as small as one square kilometer. The altitudes of smoke lofting generated by these and more intense fires have also been found to be consistent with a simplified theory for thermal plume rise in a stratified atmosphere (Golitsyn et al., 1988b).

Scavenging

The most recent assessments of smoke scavenging indicate that, of the microphysical mechanisms now being treated, direct nucleation is the primary mechanism by which smoke particles are incorporated into cloud water (Bradley, 1986; Tripoli and Kang, 1987; Edwards and Penner, 1987). Accordingly, the fraction of smoke particles incorporated into water droplets (that is, scavenged) condensing in plumes over large fires (in a humid environment) should roughly equal the CCN fraction of the smoke. Given the available data, it appears that less than half, perhaps significantly less, of the aerosol particles would be scavenged; because of re-evaporation of water droplets, one calculation indicates that nucleation would remove at most a few percent of the smoke particles by precipitation reaching the ground (Tripoli and Kang, 1987); nonetheless passage of smoke particles through the nucleation, drop coalescence, and evaporation process may alter the particle size and number distribution, optical coefficients, and other characteristics, generally making the particles less optically efficient and more likely to be removed in later precipitation systems.

Evaluation of the full range of possible scavenging mechanisms has not yet been thoroughly explored. The cloud electrification process, for example, has yet to be carefully evaluated, and this might increase the scavenging rate substantially (Penner and Molenkamp, 1989; Harvey and Edwards, 1988). For example, in one recent numerical simulation of a rising firestorm cloud, the half-lives of charged interstitial aerosol (0.01 μ – 1.0 μm) assuming an electrical field of 1000 volts/cm (not uncommon in thunderstorm clouds, and possible but as yet unmeasured, in large fires) were less than one minute (Harvey and Edwards, 1988), which would make electrical capture the dominant scavenging process.

As an additional complication, smoke particles generated from cellulosic fuels and by smoldering combustion could boost CCN concentrations in fire plumes to more than 100,000 per cubic centimeter (Hallett et al., 1989). Such an enormous enhancement of CCN (by a factor of a thousand or more) could tend to overseed the condensation cloud with small water droplets, possibly changing cloud optical properties. The precipitation efficiency of overseeded clouds is generally low (if they do not involve the ice phase), because the small droplets cannot efficiently coalesce into the size range required for precipitation (Pruppacher and Klett, 1978), resulting in a haze or fog rather than a precipitating cumulus cloud (Crutzen and Birks, 1982). However, many of the fires would be expected to form ice-clouds that would precipitate (Penner and Molenkamp, 1989). It is not yet certain, however, whether the drops containing smoke would end up on the ground; certainly, the Hiroshima 'black rain' indicated that some smoke would be removed, but the quantity is still highly uncertain.

Dispersion

The regional or local dispersion of large smoke plumes has recently been investigated through mathematical simulation (Westphal and Toon, 1988; Westphal et al., 1988; Giorgi, 1989), wildfire plume sampling (Hegg et al.,

1987), and historical data analysis (Robock, 1988b; Veltishchev, et al., 1988; Robock, 1988a). During the first several days of smoke dispersion, particle coagulation, washout, and transformation by condensation may be significant processes. Sophisticated model calculations, coupled with satellite photography of forest fire smoke, show that smoke can be transported rapidly over hemispheric distances and that significant quantities of smoke can be washed out in frontal systems during the first few days, if the smoke particles are good CCN. Often, however, smoke from only relatively small, isolated zones of fire can spread hazes that partially obscure continent-sized regions within a few days, indicating that large fractions of the smoke must be remaining aloft.

The surface cooling caused by extended smoke layers has been detected through observation and objective meteorological analyses (Robock, 1988b; Veltischev, 1988; Robock, 1988a). Smoke from a series of large fires in Alberta, Canada, in 1982, for example, appear to have reduced the average daytime maximum surface temperatures in the north-central United States by 1.5° to 4° C. Similarly, smoke from the great Siberian forest fires of 1915 apparently depressed average daytime temperatures by 2° to 5° C over large areas of Siberia (Veltishchev et al., 1988). More recently, the massive Chinese wildfires of May 1987 generated smoke that may have reduced daytime temperatures over Alaska by 2° to 6° C. In each of these cases, the minimum nighttime temperatures were not significantly reduced. Such reductions in the temperatures are consistent with theoretical expectations and calculations by Pittock et al. (1988, 1989). Urban smokes, which are much blacker than wildfire smokes and are generally lofted to higher altitudes in the atmosphere, would be expected to produce substantially stronger daytime cooling.

In September 1987 a series of large wildfires erupted in southern Oregon and northern California. During one period, a dense smoke pall remained in the Klamath River valley for three weeks. Through this period the town of Happy Camp, California experienced daytime temperatures more than 15° C below normal (Robock, 1988a). At roughly the same time, a large coherent plume of smoke moved southward from the Oregon fires along the coast of California. An aircraft team tracking this smoke cloud found indications of coagulation which doubled the original particle sizes over a period of 44 hours (Radke et al., 1988). The specific scattering coefficient of the smoke also decreased by roughly 25 percent over this time. After two days the smoke properties had been effectively 'frozen out' by dilution of the plume, suppressing further aging.

A separate investigation of the climatic effects of some 50 dust storms in Tadzhikistan, in south-central Asian USSR, indicated that maximum daytime surface temperatures were depressed by as much as 10° to 12° C (Golitsyn and Shukurov, 1987). Moreover, pronounced reductions in the yields of

Figure 2. The 17,000-foot smoke plume and capping cloud generated by a large, 1-square-kilometer test fire set in an Ontario, Canada, forest on August 28, 1987. (Photo: L. Radke)

a variety of crops were associated with the occurrence of the dust storms (Shukurov and Golitsyn, 1988).

GLOBAL-SCALE CLIMATE IMPACTS

Recent calculations using sophisticated three dimensional general circulation models (GCMs) have incorporated improved treatments of many important processes. Additional studies have been done to refine estimates of the sharp acute-phase perturbations in climate that would be expected to occur primarily in the Northern Hemisphere midlatitudes and subtropics, and the later chronic-phase impacts which, although milder, could occur worldwide.

Acute Phase

At the 1987 ENUWAR workshop in Bangkok, three plausible, but not limiting, smoke injection scenarios were devised; they suggest consideration of

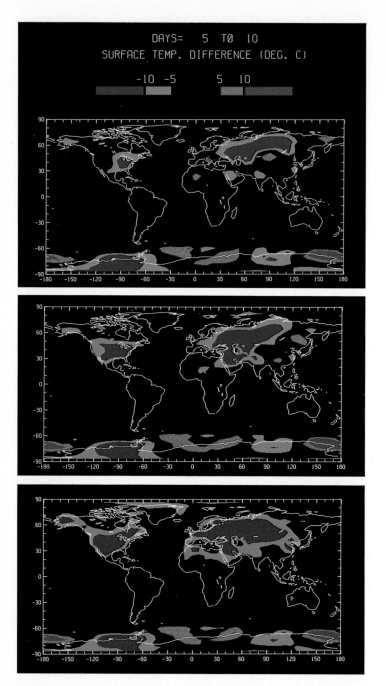

Figure 3. Simulations of the change in July surface temperature for the three Bangkok smoke injection scenarios relative to the unperturbed (control) atmosphere. (From G. Glatzmaier and R. Malone, Los Alamos National Laboratory)

smoke injections of 15, 50, and 150 Tg of dark smoke such that – if spread evenly and instantaneously over the Northern Hemisphere – the absorption optical depths would be 0.3. 1.0, and 3.0. These 'Bangkok scenarios' have now been used by several modeling groups in calibrated studies to estimate potential climatic effects, as reported at the Moscow workshop (Glatzmaier and Malone, 1988; Stenchikov and Carl, 1988; MacCracken et al., 1988). For smoke injections occurring in July, the maximum average five-day decrease in land surface temperature in the zone from 30°N to 70°N ranged from about 5° C in the case of low smoke mass, to 13° C in the intermediate case, to 22° C in the high case. Reductions are mainly in daytime temperature. Maximal short-term temperature drops within continental interiors reached 30° C or more, with widespread, but episodic, frosts and freezing events projected to occur in midlatitude regions within the first month for the intermediate and high-smoke-mass cases. Figure 3 illustrates the distribution of surface temperature changes for the three Bangkok scenarios from the GCM at the Los Alamos National Laboratory (Glatzmaier and Malone, 1988; Malone et al., 1986). For wintertime smoke injections, the estimated temperature decreases were considerably smaller. It is important to note, however, that the GCMs do not account for fog formation at the surface, which could, under some conditions, limit the initial temperature drops to the dew point (Molenkamp, 1989).

Calculated changes in precipitation also show quite dramatic changes. A 75 percent decrease in July rainfall over land at latitudes from 30°N to 70°N was projected for each of the three Bangkok scenarios (averaged over days 20 to 30 of the simulations) using the Lawrence Livermore National Laboratory GCM (Ghan et al., 1988). For latitudes of 0° to 20°N, precipitation dropped by about 25 percent in the low-smoke-mass case and by about 60 to 75 percent in the intermediate and high smoke cases. In these model simulations, the summer Asian monsoon also failed.

Reductions in light levels would also be significant. In the Livermore model, for example, the average projected decrease in solar insolation over land from 30° to 70°N was about 40 percent in the low-smoke-mass case (for days 6 to 10); about 70 percent in the intermediate case (days 6 to 10); and about 90 percent in the high smoke case (days 21 to 25). The time of the maximum solar reduction is delayed in the high smoke case because the effectiveness of smoke absorption increases as the particles become more spread out, even though it decreases as smoke amount decreases due to scavenging. At low latitudes (0° to 20°N), the corresponding light reductions were about 0 percent, 30 percent, and 50 percent, respectively. Such perturbations over such large areas would be unprecedented in human history.

These calculations, while adding greater detail, generally confirm the results described in the first edition of this volume, especially the potential for sharp reductions in temperature; to the extent that these reductions take temperatures below the dew point. However, they may be exceeding the validity range of the GCMs. The predictions of precipitation change are more uncertain, however, because of the limitations inherent in current representations of the hydrological cycle in the models (Mitchell, 1988). Significant efforts are still underway to refine estimates by improving treatments of chemistry, convection, resolution, the formation of fog, and smoke scavenging.

Smoke Lofting to High Altitudes

A key finding of recent research has been the extent to which injected masses of smoke may become lofted into the upper atmosphere after the fire plume phase (Malone et al., 1986). In the initial stages of this process, the extensive, optically dense smoke layers in the Northern Hemisphere would be heated by the absorption of solar radiation. The heated air would then rise into the upper atmosphere, displacing much of the ambient stratospheric air toward the Southern Hemisphere. Because most of the nuclear war smoke would initially be injected into the lower atmosphere (the troposphere), where the smoke lifetime associated with precipitation scavenging under normal conditions may be only a matter of days to weeks, this lofting of the smoke into a stabilized layer at stratospheric altitudes would lead to much longer smoke lifetimes of months to years.

In July simulations for the Bangkok scenarios (Glatzmaier and Malone, 1988; MacCracken et al., 1988), it was found that after 30 days, 25 to 40 percent of the initial smoke injection was stabilized in the low-smoke-mass case, 35 to 40 percent in the intermediate emissions case, and about 55 percent in the high emissions case. The atmospheric lifetime of this residual smoke was calculated to be about one year in each case. Moreover, after only a few weeks the smoke pall had spread over the entire Northern Hemisphere and was moving well into the Southern Hemisphere. These chronic phase simulations clearly indicated that the optical thickness of the stabilized smoke layer could be large enough to induce significant long-term climatic perturbations.

Chronic Phase

Effort is only beginning to focus on the longer term effects of a nuclear war. Special attention has focused on possible environmental impacts of moderately thin elevated smoke layers (of absorption optical depth of order 0.2), such as may be expected to persist in the Northern Hemisphere into the second growing season after a nuclear war, and to occur over the Southern Hemisphere (Pittock et al, 1989, 1988; Walsh and Pittock, 1988; Ackerman et al., 1988b) and the climate feedback processes (involving the oceans and ice fields) that might contribute to prolonging global climatic anomalies (Mettlach et al., 1987; Stenchikov and Carl, 1988; Stenchikov, 1987; Vogelmann et al., 1986; Covey, 1987). For the larger smoke injection scenarios, these new studies suggest that climatic cooling would extend at least into the year following the war. Ocean surface temperatures could decrease by 2° C on average, and land surfaces in the Northern and Southern Hemispheres could be several degrees cooler through their respective following summers. Significant reductions in rainfall in the tropics and monsoonal areas have also been found in calculations by Pittock et al., (1989; 1988) (see figure 4). Considerable additional research will be needed to delineate with confidence these long-term climatic effects of nuclear war.

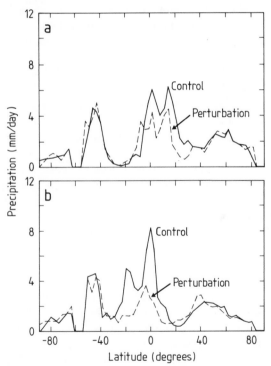

Figure 4. Zonally averaged 30-day mean precipitation rates (mm/day) over land for (a) July days 71–100, and (b) January days 31–60. Control simulations are indicated by the full lines, and perturbed simulations (globally uniform smoke AOD = 0.2) by the dashed lines (from Pittock et al., 1989)

STRATOSPHERIC OZONE DEPLETION

New simulations of perturbations to the stratospheric ozone layer caused by massive injections of smoke and nitrogen oxides (NO_x) suggest that global ozone depletions of 50 percent or more are possible, with recovery taking several years. The preliminary calculations—carried out using a GCM with interactive smoke and ozone transport, and including some of the many nitrogen oxide reactions—show that the heated nuclear smoke pall, rising into the Northern Hemisphere stratosphere, would displace the Northern Hemisphere ozone layer toward the Southern Hemisphere. The fireball-induced nitrogen oxides would intermingle with the smoke and then erode the remaining ozone by photochemical action. The fact that the smoky air would be heated by as much as 100° C would accelerate ozone destruction (Crutzen et al., 1984; Vupputuri, 1986) (also see p. 230, this volume). At the end of one 20-day simulation, the ozone layer in the Northern Hemisphere was reduced by an average of 40 percent, with half of this reduction due to ozone displacement and half due to chemical decomposition (Glatzmaier and Malone, 1988). In a similar simulation with the NCAR model, the Northern Hemisphere ozone reduction was 50 percent after 90 days (Thompson and Crutzen, 1988). The fate of the excess ozone pushed across the equator into the stratosphere over the Southern Hemisphere must also be investigated.

Large reductions in stratospheric ozone would allow high doses of harmful ultraviolet solar radiation (UV-B) to reach the Earth's surface over an extended period of time, even before the smoke has completely cleared. This is because for ultraviolet radiation the ozone absorption optical depth could be reduced more than the smoke optical depth would be increased. The ozone depletion would be expected to grow worse during the first year as the nitrogen oxides spread (Vupputuri, 1986), and recovery would likely be extended and depend on return of the natural circulation regime of the stratosphere, which could be delayed by the stabilized smoke layer.

RADIOLOGICAL EFFECTS

The Moscow workshop in March 1988 served as the most recent forum on the radiological consequences of nuclear war. It included new results on local fallout modeling, internal radiation dose, fallout from limited nuclear attacks, human health effects, the implications of nuclear arms reductions on projections of direct effects, and lessons from the Chernobyl reactor accident on the dispersal and deposition of radionuclides (Shapiro, 1988b; Shapiro, 1988c; Leaf and Ohkita, 1988).

Local Fallout Modelling

Harvey (1988) compared the graphical local fallout model used in the ENUWAR study, and its progenitor, the KDFOC2 model, with several other popular fallout models used in the United States. The calculations assumed a hypothetical 1-Mt, fission, nuclear explosion and comparisons were done for areas experiencing various dose rates at 1 hour after the explosion. These comparisons indicated no reason to adjust the SCOPE 28 (p. 229, this volume) estimates for local fallout.

Limited Nuclear Attacks

Calculations by Levi and von Hippel (1987) have focused on the consequences of 'limited' nuclear attacks on strategic nuclear targets in the U.S. and the U.S.S.R. The scenarios they studied were considerably smaller than the 6000 megaton exchange assumed in the SCOPE study. Nevertheless, projected civilian fatalities were of the order of tens of millions of people in each country (see Table 1). In their calculations local fallout accounted for a significant share of total fatalities. A similar study by Harvey et al (1987) involving a counterforce-countervalue attack on the U.S. produced similar estimates of total fatalities, but a much smaller percentage attributed to fallout. The differences are attributable mainly to differences in the manner in which the fallout codes deposit the radioactivity around the detonation site.

TABLE 1.
PREDICTED DEATHS AND INJURIES FROM RADIOACTIVE FALLOUT
AFTER AN ATTACK ON U.S. STRATEGIC NUCLEAR TARGETS WITH AIR
AND GROUND BURSTS (DAUGHERTY, ET AL., 1986)

Effects of Radioactive Fallout	Casualties (millions) Dead	Injured
Low LD_{50} (250 rads)	9–14	8–15
Medium LD_{50} (350 rads)	7–8	4–8
High LD_{50} (450 rads)	5–6	2–3

Note: The ranges for deaths and injuries predicted at each estimated value of the LD_{50} account for the effects of different wind patterns for the different seasons of the year when a nuclear attack might occur. The patterns of fallout have large uncertainties. Different computer models predict patterns differently, some estimating fallout casualties about one-half or one-third those shown here.

Internal Radiation Dose

In addition to the dose received from radionuclides external to the body, there would be an additional body and organ dose from radioactivity that could be carried inside the body by the ingestion of food and water contaminated mainly by local fallout. Such an internal dose to an individual would be sensitive to many factors that vary greatly depending on local circumstance. Peterson, et al. (1988) developed estimates for individuals in

10 locations and situations in the United States following a major hypothetical nuclear attack. Attention was focused on the potential doses from
^{137}Cs, ^{90}Sr, ^{89}Sr, and ^{131}I, as evidence from atmospheric tests indicates that
these four radioactive isotopes would contribute the largest share to the
internal dose (about 75 percent). In this study, the greatest contributor to
the committed effective dose equivalent was ^{137}Cs. The individuals were
assumed to experience differing external doses, shelter factors, availability
of foods within local distances, and numerous other variables. The calculations showed that the ratios of internal to external dose varied from less
than one percent to approximately 100 percent. Seven of the situations indicated ratios of less than ten percent. The higher ratios resulted from the
combined effects of a large shelter factor (reducing the external dose) and
consumption of contaminated foods for long periods (increasing the internal
dose). When considering the sum of internal and external radiation, which
is the important factor affecting health, the results of Peterson et al. (1988)
suggest that local fallout dose is normally dominated by the external dose.
Svirezhev (1987) calculated the dose to humans from surface deposition of
^{137}Cs and ^{90}Sr resulting from the targeting of a nuclear reactor by a nuclear
weapon. Using a different model, his results indicated that the internal dose
would dominate the external dose, with Sr^{90} being the largest contributor.
These differences suggest that further consideration of varying conditions
may be in order.

Strategic Arms Reductions

Shapiro (1988b) examined the implications of the proposed INF and START
treaties on damage projections from a major nuclear war, and concluded
that, except for quite localized situations in Europe, the INF reductions are
likely to have a negligible effect on estimates of global and local fallout.
The START reductions of approximately 50 percent in strategic weapons
could, on the other hand, result in a lowering of estimates of local fallout that range from significant to dramatic, depending upon the nature of
the reduced strategic forces and how they might be employed. An important element of this consideration is a possible shift from fixed to mobile
land-based ICBMs, which could result in a corresponding shift of targeting
strategy from ground to air bursts, which would greatly reduce local fallout.
Should a major war occur, projections of total fatalities from direct effects
of blast, thermal radiation, and fallout, and the generation of smoke leading
to the phenomenon known as 'nuclear winter' would not, however, be expected to be significantly affected by the INF and START initiatives as now
drafted.

HUMAN HEALTH EFFECTS OF IONIZING RADIATION

Lethal Dose

A recent re-examination of the Hiroshima and Nagasaki dosimetry has led to new estimates of risk coefficients and lethal dose (LD). New determinations were made of the acute radiation dose (bone marrow) that would be expected to kill 50 percent of the exposed individuals within 60 days ($LD_{50/60}$). Fujita et al. (1987) suggest a range of values from 2.2 to 2.6 Gy, varying slightly with the method of estimation used. Their estimates included deaths in the first day and those with severe injuries, burns, or trauma who survived the first day but succumbed later to their injuries. If inclusion of the latter groups biases downwards the estimate 17.5 percent or so, as one study suggests, and the range of the $LD_{50/60}$ is adjusted upwards by this amount, it would be 2.7 to 3.1 Gy. This range is described as their 'best' estimate. If the commonly accepted factor of 0.7 is applied to transform whole body external dose to bone marrow dose, a value of about 4 Gy for $LD_{50/60}$ external dose is obtained. It was also noted in the new studies that the slope of the mortality curve is shallower than would be expected from animal experimental evidence.

Rotblat (1986) has calculated $LD_{50/60}$ making the assumption that all deaths occurring within 24 hours of the explosion in Hiroshima were from burn or blast injuries and all subsequent deaths were from ionizing radiation. After some revisions due to changes in dosimetry based on results from the Radiation Effects Research Foundation in Hiroshima, he recently recalculated the bone marrow dose $LD_{50/60}$ to be 2.5 Gy (Rotblat, 1988), which translates to a body surface (external) $LD_{50/60}$ of 3.5 Gy. Table 1 illustrates quantitatively the effect of $LD_{50/60}$ on predicted casualties from radioactive fallout. The lower the value of $LD_{50/60}$, the higher are the predicted casualties.

Synergism of Effects

Appreciation of synergistic effects of the several biological effects of nuclear war has increased over the past several years. The well-studied suppression of the immune system by ionizing radiation is associated with an effect on T lymphocytes, in which helper T cells are diminished while suppressor T cells increase. In addition to ionizing radiation, other factors likely to be highly prevalent in the wake of a nuclear war such as hard ultraviolet radiation, burns, trauma, bereavement and psychological depression, and malnutrition, could suppress the immune system by having similar effects on T lymphocytes. Such suppression would be expected to greatly increase illness and mortality from associated burns, trauma, and infections among the initial casualties and to augment the later deaths from famine, epidemics, and cancer (Greer and Rifkin, 1986), which are predicted to number in the hundreds of millions.

LESSONS FROM CHERNOBYL

On April 26, 1986, a major radionuclide release occurred as a result of the accident at the Chernobyl nuclear power plant in the Soviet Union. Theoretical models simulating the transport and deposition of the radionuclides have been used in conjunction with observations to deduce how much material was released and the processes affecting its spread and removal from the atmosphere. The scales of these models can be roughly categorized as being local (up to 100 km.), European, and hemispheric.

The Evacuation Zone near Chernobyl

On the local scale, gravitational settling of the coarser material released over the first five days dominated the deposition pattern close to the site. This deposition pattern emphasizes the potential importance in nuclear accidents of the very highly radioactive ('hot') fuel particles and the need for further studies of their behavior in the environment. The calculated deposition patterns agree well with the observed radiation levels within 30 kilometers of the plant, which is the zone that was evacuated (Israel et al., 1987).

On the European scale, the plume initially traveled westward at a height of a few hundred meters in a stably stratified air mass. Convective motions then mixed the plume vertically from near the surface up to 2 to 3 kilometers. Significant amounts were also transported to even higher levels in cumulus clouds. The plume was carried northward and was incorporated in a precipitation system. This resulted in localized peaks of wet deposition in parts of central Scandinavia (Persson et al., 1987). The radioactive particles then spread from Poland southwestward across Europe. Some of this radioactivity was deposited by local precipitation in southeast Germany, additional material moved northward over the United Kingdom (ApSimon and Wilson, 1987). Other radioactivity traveled eastward across the Soviet Union and southward to Turkey and Greece. Deposition of the most important long-lived nuclide, ^{137}Cs, did not decrease smoothly with travel distance, but was enhanced in regions where rain or snow intercepted the plume (Hohenemser et al., 1986; Hohenemser and Renn, 1988). Contamination from the Chernobyl accident outside the USSR was generally low compared with normal overall exposures from natural sources of radiation, which are of the order of 10^{-7} Gy per hour.

The radionuclides lofted more than a few kilometers into the atmosphere spread rapidly around the Northern Hemisphere. Simulations indicate that one part of the radioactive material headed toward Scandinavia and central Europe, and the remainder was transported across Asia to Japan, the North Pacific, and the North American continent (see Figure 5). To be in accord with observations, these simulations indicate that this radioactivity must have been transported at elevated levels in the atmosphere—up to 4 kilometers in the Japanese estimates (Kimura and Yoshikawa, 1988), and up to 10 kilometers in the Lawrence Livermore estimates (Gudiksen et al., 1988).

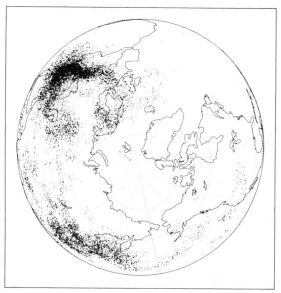

Figure 5. Calculated spatial distribution of radioactivity over the Northern Hemisphere 10 days after the Chernobyl accident as illustrated by the Lawrence Livermore National Laboratory (Lange et al., in press)

Source terms of various nuclides have been deduced by combining observations with model results. Estimates are in agreement within about a factor of 2 (Table 2). While the radionuclide emissions from the Chernobyl accident are about a thousand times larger than those released at the Windscale accident in 1957 in the U.K., and about a million times larger than the release at the Three Mile Island reactor accident in 1979 in the U.S., they are still orders of magnitude smaller (2 to 4 orders for some of the important isotopes (Gudiksen et al.,1988)) than the radionuclide debris expected from a major nuclear war. The isotopic composition and character of the Chernobyl discharge also varied from that expected from weapons fallout. However, the releases from Chernobyl affirm a conclusion from the first edition of SCOPE 28, that the direct targeting of many dozens of nuclear facilities such as reactors, spent fuel storage, fuel reprocessing plants, and high level waste facilities (a highly speculative scenario) could potentially release quantities of important fission products (e.g., Cs^{137}) comparable to that of the weapon debris from a major nuclear war.

Experimental measurements have revealed that there would be expected to be an extremely patchy deposition resulting randomly whenever precipitation happened to intercept the radioactive plume. On the local, continental and global scales, the studies of the dispersion and deposition of radionuclide releases from Chernobyl have shown that radionuclides may be transported over long distances, and that we understand this adequately for it to be simulated by mathematical models.

TABLE 2.
ESTIMATES OF EMISSIONS FROM THE CHERNOBYL REACTOR

Source of Estimate	Iodine-131 (10^{17} becquerels)	Cesium-137 (10^{16} becquerels)
USSR	2.8	4.7
MESOS model (European fallout)	1.7	3.9
Lawrence Livermore National Laboratory (Northern Hemisphere fallout)	6.0	8.9

Note: All estimates are based on values corrected for decay to 6 May 1986, as given in the Soviet report to the International Atomic Energy Agency.
SOURCE: SCOPE-ENUWAR Moscow workshop, 21–25 March 1988.

CONCLUSION

In summary, we affirm the major conclusions of SCOPE 28 concerning the serious potential threat to the global population and environment posed by nuclear war while recognizing that uncertainties that could further exacerbate or somewhat moderate our findings still remain. Our studies confirm that casualties from indirect effects (especially as a consequence of the coupled effects of infrastructure and climatic disruption) could be greater than from direct effects of nuclear war with unprecedented consequences for noncombatant and combatant countries alike. Although the predicted intensity of the global climatic cooling has been reduced from the earliest estimates and important uncertainties remain, confidence that the climatic changes will be dramatic is now much broader than was possible just following identification of this result in 1983, as also confirmed in reviews by the U.S. National Research Council (NRC, 1985), by Golitsyn and MacCracken (1987) for the World Meteorological Organization, by SCOPE-ENUWAR (1988) and others. More recent GCM simulations suggest that global ozone depletions of 50 percent or more are possible, with recovery taking several years. In addition, it now appears probable that large reductions in rainfall would occur during the acute phase in the northern midlatitudes, and that the monsoons would fail. Because the lifetime of the remaining elevated layer of smoke after the first few weeks could be of the order of a year or more, longer-term 'chronic' coolings of several degrees and significant reductions in convective rainfall are also expected, depending somewhat on smoke optical properties.

Despite gains in understanding of the atmospheric effects, there still remain gaps in many aspects of our understanding of how the projected large-scale changes in the physical environment would affect crops and ecosystems. Also little is known about the longer-term atmospheric effects, which are controlled by the interactions of the atmosphere with the oceans and land biosphere, or about changes in atmospheric chemistry.

ACKNOWLEDGMENTS

We acknowledge with appreciation the editors of *Environment* for providing the art work for the figures in this preface, most of which appeared in the June, 1988 issue (Vol. 30, No. 5).

REFERENCES

Ackerman, T.P., and Cropper, W.P. Jr. (1988a) Scaling global climate projections to local biological assessments, *Environment* **30** (5), 31–34.

Ackerman, T.P., Turco, R.P., and Toon, O.B. (1988b) Persistent effects of residual smoke layers. In: Hobbs and McCormick, (eds.), *Aerosols and Climate*, Deepak Publishing Hampton, VA., pp443–457.

Anikin, P.P., and Shukurov, A.K. (1988) Spectral attenuation of radiation by smoke aerosol. *Izvestia-Atmosphere and Ocean Physics* **25**, 247–49.

ApSimon, H.M., Gudiksen, P., Khitrov, L., Rodhe, H., and Yoshikawa, T. (1988) Modeling the dispersal and deposition of radionuclides. *Environment* **30** (5), 17–20.

ApSimon, H.M., and Wilson, J.J. (1987) Modelling atmospheric dispersal of the Chernobyl release across Europe. *Boundary-Layer Meteorology* **41**, 123–33.

Berry, M.V., and Percival, I.C. (1986) Optics of fractal clusters such as smoke, *Optica Acta* **33**, 577–91.

Bradley, M.M. (1987) Nucleation scavenging of smoke aerosol above intense fires: three dimensional simulations. *Proc. XIX General Assembly of International Union of Geodesy and Geophysics*, Vancouver, Canada, 19–22 Aug.

Bradley, M.M. (1986) Numerical simulation of nucleation scavenging within smoke plumes above large fires. *Lawrence Livermore National Laboratory Report UCRL-95306*, Livermore, CA.

Colbeck, I., Appleby, L., Handman, E., and Harrison, R.M. (1989) The optical properties and morphology of cloud processed carbonaceous smoke. *Journal of Aerosol Science* (submitted).

Covey, C. (1987) Protracted climatic effects of massive smoke injections into the atmosphere, *Nature* **325** 701–03.

Crutzen, P.J., and Birks, J.W. (1982) The atmosphere after a nuclear war : twilight at noon. *Ambio* **11**, 114–125.

Crutzen, P.J., Galbally, I.E., and Brühl, C. (1984) Atmospheric effects from post-nuclear fires. *Climatic Change* **6**, 323–364.

DASA (1966) Department of Defense Land Fallout Prediction System. *Defense Atomic Support Agency, Washington, D.C., Rept. DASA-1800-I through DASA-1800-VII* (1966-1968).

Daugherty, W.H., Levi, B.G., and Von Hippel, F.N. (1986) The consequences of 'limited' nuclear attacks on the United States, *International Security*, **10**, 4 (1986):3–45; World Health Organization, *Effects of Nuclear War on Health Services, 2d ed.* (1987) Geneva, World Health Organization.

Dod, R.L., Brown, N.J., Mowrer, F.W., Novakov, T., and Williamson, R.B. (1989) Smoke emission factors from medium scale fires: Part 2. *Aerosol Science and Technology*, **10**, 1, 20–27.

Edwards, L.L., and Penner, J.E. (1987) Potential nucleation scavenging of smoke particles over large fires: A parametric study. Lawrence Livermore National Laboratory Report UCRL-96242, Livermore, CA.

Einfeld, W., Mokler, B.V., Zak, B.D., and Morrison, D.J. (1987) A characterization of smoke particles from small, medium and large-scale hydrocarbon pool fires. In Pilat, M.J., and Davis, E.J. (eds.), *Aerosols 87*, Abstracts of the American Association for Aerosol Research Annual Meeting, Seattle, Wash., 14–17 September (Seattle).

Fujita, S., Kato, H., and Schull, W.J. (1987) The LD_{50} associated with exposure to the atomic bombing of Hiroshima and Nagasaki: A review and reassessments. *Radiation Effects Research Foundation technical report 17–87* (draft).

Ghan, S.J., MacCracken, M.C., and Walton, J.J. (1988) The climatic response to large atmospheric smoke injections: sensitivity studies with a tropospheric general circulation model. *Journal of Geophysical Research*, **93**, 8315–8337.

Giorgi, F. (1989) Two-dimensional simulations of possible mesoscale effects of nuclear war fires. 1: model description. *Journal of Geophysical Research*, **94**, 1127–1144.

Giorgi, F., and Visconti, G. (1989) Two-dimensional simulations of possible mesoscale effects of nuclear war fires. 2: model results. *Journal of Geophysical Research*, **94**, 1145–1163.

Glatzmaier, G.A., and Malone, R.C. (1988) Global climate simulations of the ENUWAR case studies. Paper presented at SCOPE-ENUWAR Moscow Workshop, 21–25 March.

Golitsyn, G.S. and Shukurov, A.K. (1987) Temperature effects of the dust aerosol as revealed by dust storms in Tadjikistan. *Doklady Acadamy of Sciences*, **297**, 1334–37.

Golitsyn, G.S. and MacCracken, M.C. (1987) Atmospheric and climatic consequences of a major nuclear war : results of recent research. WPC-142 (Geneva: World Meteorological Organization).

Golitsyn, G.S., Shukurov, A.K., Ginsburg, A.S., Sutugin, A.G., and Andronova, A.V. (1988a) Complex study of microphysical and optical properties of smoke aerosol. *Izvestia—Atmosphere and Ocean Physics*, **24**, 227–234.

Golitsyn, G.S., Gostintsev, Y.A., and Solodovnik, A.F. (1988b) Pollution of the atmosphere by mass fires. Paper presented at SCOPE-ENUWAR Moscow Workshop, 21–25 March.

Greer, D.S., and Rifkin, L.S. (1986) The immunological impact of nuclear war. In Solomon F. and Marston, R.Q. (eds.), *The Medical Implications of Nuclear War*,

Washington, DC, National Academy Press.

Gudiksen, P.H., Harvey, T.F., and Lange, R. (1988) Chernobyl source term, atmospheric dispersion, and dose estimation. Paper presented at SCOPE-ENUWAR Moscow Workshop, 21–25 March.

Hallett, J., Hudson, J.G., and Rodgers, C.F. (1989) Characterization of combustion aerosols for haze and cloud formation. *Aerosol Science and Technology*, **10**, (1), 70–83.

Hansen, A.D.A., and Novakov, T. (1988) Cloud chamber studies of the nucleation characteristics of smoke particles from liquid-fuel and wood fires. Lawrence Berkeley National Laboratory Report LBL-26016, September.

Harvey, T.F., Wittler, R., Peterson, K.R., Shapiro, C.S., and Walton, J. (1987) Local fallout risk after attack on CONUS. Lawrence Livermore National Laboratory Draft Report, Livermore, CA.

Harvey, T.F. (1988) Perspective on the local fallout model used in the SCOPE-ENUWAR Study. Paper presented at SCOPE-ENUWAR Moscow Workshop, 21–25 March 1988, Lawrence Livermore National Laboratory Report UCRL-98589, Livermore, CA.

Harvey, T.F., and Edwards, L.L. (1988) Electric field collection of charged aerosol by neutral cloud droplets. Lawrence Livermore National Laboratory Report UCRL-99887.

Hegg, D.A., Radke, T., Hobbs, P.V., Brock, C.A., and Riggan, P.J. (1987) Nitrogen and sulfur emissions from the burning of forest products near large urban areas. *Journal of Geophysical Research*, **92**, 14701–14709.

Heikes, K.E., Ransohoff, L.M., and Small, R.D. (1987) Early smoke plume and cloud formation by large area fires, *Defense Nuclear Agency Report, DNA-TR-87-176*, Washington D.C.

Hohenemser, C., Deicher, M., Ernst, A., Hofsass, J., Lindener, G., and Recknagel, E. (1986) Chernobyl: An early report. *Environment*, **28**, (5), 6–13, 30–43.

Hohenemser, C., and Renn, O. (1988) Shifting perceptions of nuclear risk: Chernobyl's other legacy. *Environment*, **30** (3), 5–11, 40–45.

Israel, Y.A., Petrov, V.N., and Severov, D.A. (1987) Radioactive fallout in the close-in area of the Chernobyl NPP. *Soviet Journal of Meteorology and Hydrology*, **7**.

Kang, S. W., (1987) A numerical simulation of the smoke plume generated by a hypothetical urban fire near San Jose, California. Paper presented at SCOPE-ENUWAR Bangkok Workshop, 9–13 February.

Kimura, F. and Yoshikawa, T., (1988) Numerical simulations of global scale dispersion of radioactive pollutants from the accident at the Chernobyl nuclear power plant. Paper presented at SCOPE-ENUWAR Moscow workshop, 21–25 March.

Lange, R., Dickerson, M.H., and Gudiksen, P.H. (1988) Dose estimates from the Chernobyl accident. *Nuclear Technology*, **82**, 311–322.

Leaf, A., and Ohkita, T. (1988) Health effects of nuclear war. *Environment* **30**, (5), 36–38.

Levi, B., and Von Hippel, F. (1987) Limited attacks on the United States and the Soviet Union. In *Effects of Nuclear War on Health and Health Services*, 2d ed., World Health Organization, Geneva.

MacCracken, M.C., Ghan, S.J., and Walton, J.J. (1988) Regional climatic effects of post-nuclear smoke. Paper presented at SCOPE-ENUWAR Moscow Workshop, 21–25 March.

Malone, R.C., Auer, L.H., Glatzmaier, G.A., Wood, M.C., and Toon, O.B. (1986) Nuclear winter: three-dimensional simulations including interactive transport, scavenging and solar heating of smoke. *Journal of Geophysical Research* **91**, 1039–53.

Marcus, S., Krueger, S., and Rosenblatt, M. (1987) Numerical simulation of near-surface environments and particulate clouds generated by large-area fires. *Defense Nuclear Agency Report DNA-TR-87-1*, Washington, DC.

Mettlach, T.R., Haney, R.L., Garwood, R.W., Jr., and Ghan, S.J. (1987) The response of the upper ocean to a large summertime injection of smoke in the atmosphere. *Journal of Geophysical Research* **92**, 1967–1974.

Mitchell, J.F.B. (1988) Presentation at SCOPE-ENUWAR Moscow Workshop, 21–25 March.

Molenkamp, C. (1989) Numerical simulation of coastal flows when solar radiation is blocked by smoke. *Journal of Applied Meteorology* **28**, (5), 361–381.

Mulholland, G.W., Baum, H., Bryner, N., and Quintiere, J. (1988) Smoke emission and optical property measurements at NBS. Progress Report, National Bureau of Standards, Gaithersburg, MD.

Patterson, M., McMahon, C.K., and Ward, D.E. (1986) Absorption properties and graphic carbon emission factors of forest fire aerosols. *Geophysical Research Letters* **13**, 129–32.

Penner, J.E., and Molenkamp, C.R. (1989) Predicting the consequences of nuclear war: precipitation scavenging of smoke. Lawrence Livermore National Laboratory Report UCRL-96916. Also *Aerosol Science and Technology* **10**, 1, 51–62.

Penner, J.E., Haselman, L.C. Jr., and Edwards, L.L. (1986a) Smoke plume distribution above large scale fires; implications for simulations of 'nuclear winter'. *Journal of Climate and Applied Meteorology* **25**, 1434–44.

Penner, J.E. (1986b) Uncertainties in the smoke source term for 'nuclear winter' studies. *Nature* **324**, 222–26.

Persson, C., Rodhe, H., and De Geer, L.E. (1987) The Chernobyl accident; a meteorological analysis of how radionuclides reached and were deposited in Sweden. *Ambio* **16**, 20–31.

Peterson, K., Shapiro, C.S., and Harvey, T. (1988) Internal dose following a large-scale nuclear war. Paper presented at SCOPE-ENUWAR Moscow Workshop, 21–25 March, Lawrence Livermore National Laboratory Report UCRL-98348. Livermore, CA.

Petryanov-Sokolov, I.V., Sutugin, A.G., and Andronova, A.V. (1988) Possible influence of mass fires on composition and optical properties of the atmosphere. Paper presented at SCOPE-ENUWAR Moscow Workshop, 21–25 March.

Pittock, A.B., Frederiksen, J.S., Garratt, J.R., and Walsh, K. (1988) Climatic effects of smoke and dust produced from nuclear conflagrations: In Hobbs and Mc-Cormick, (eds.), *Aerosols and Climate*, Deepak Publishing Hampton, VA pp395–410.

Pittock, A.B., Walsh, K., and Frederiksen, J.S. (1989) General circulation model simulation of mild 'nuclear winter' effects. *Climate Dynamics* **3**, 191–206.

Pruppacher, H.R., and Klett, J.D. (1978) *Microphysics of Clouds and Precipitation*, Dordrecht, Holland, D. Reidel, chap., 17.

Radke, L.F., Hegg, D.A., Lyons, J.H., Brock, C.A., and Hobbs, P.V. (1988) Airborne measurement on smokes from biomass burning: In Hobbs P.V. and McCormick M.P. (eds.), *Aerosols and Climate* Deepak Publishing Hampton, VA 411–422.

Robock, A. (1988a) Surface temperature effects of forest fire smoke plumes. In Hobbs P.V. and McCormick M.P., (eds.), *Aerosols and Climate*, Deepak Publishing, Hampton, VA. 435–442.

Robock, A. (1988b) China, California conflagrations cause cooling. Paper presented at SCOPE-ENUWAR Moscow Workshop; 21–25 March.

Rotblat, J. (1986) Acute radiation mortality in a nuclear war. In Solomon, F. and Marston, R.Q. (eds.), *The Medical Implications of Nuclear War*, Washington, D.C., National Academy Press.

Rotblat, J. (1988) Personal communication; 19 August 1988.

SCOPE-ENUWAR (1988) The Environmental Consequences of Nuclear War. *Environment*, **30** (5), 2–20 and 25–45.

Shapiro, C.S. (1988a) Radioactive fallout projections and arms control agreements: INF and START. Paper presented at SCOPE-ENUWAR Moscow Workshop 21–25 March, Lawrence Livermore National Laboratory Report UCRL-98234, Livermore, CA.

Shapiro, C.S. (1988b) Report on workshop on radiological effects of nuclear war, SCOPE-ENUWAR Moscow workshop, 21–25 March, Lawrence Livermore National Laboratory Report UCID-21390, Livermore, CA.

Shapiro, C.S. (1988c) Radiological effects of nuclear war. *Environment*, **30** (5), 39–41.

Shukurov, A.K., and Golitsyn G.S. (1988). Presentation at SCOPE-ENUWAR Moscow Workshop, 21–25 March.

Small, R.D., Bush, B.W., and Dore, M.A. (1989) Initial smoke distribution for nuclear winter calculations. *Aerosol Science and Technology*, **10**, 37–50.

Small, R.D., and Heikes, K.E. (1988) Early cloud formation by large area fires. *Journal of Applied Meteorology* **27** (5), 654–663.

Stenchikov, G. (1987) Climate consequences of nuclear war: the change of the land surface properties and climate variations. Paper presented at SCOPE-ENUWAR Geneva Workshop, 16–20 November.

Stenchikov, G., and Carl, P. (1988) First acute phase stress matrix calculations using the CCAS tropospheric general circulation model. Paper presented at SCOPE-ENUWAR Moscow Workshop, 21–25 March.

Stephens, S., Rossi, M.J., and Golden, D.M. (1986) The heterogeneous reaction of ozone on carbonaceous surfaces. *Int Journal of Chemical Kinetics*, **18**, 1133–49.

Svirezhev, Y.M. (1987) Ecological consequences of nuclear war. Draft presented at SCOPE-ENUWAR Bangkok Workshop, 9–13 February.

Thompson, S.L., and Crutzen, P.J. (1988) GCM Simulations of nuclear war effects on stratospheric ozone. Paper presented at Defense Nuclear Agency Global Effects Review Conference, Santa Barbara, CA., 19–21 April.

Tripoli, G.J., and Kang, S.-W. (1987) A numerical simulation of the smoke plume generated by a hypothetical urban fire near San Jose, California. Paper presented at SCOPE-ENUWAR Bangkok Workshop, 9–13 February.

Turco, R.P., Toon, O.B., Ackerman, T.P., Pollack, J.B., and Sagan, C. (1987) Climate and smoke: an appraisal of nuclear winter. Paper presented at SCOPE-ENUWAR Bangkok Workshop, 9–13 February.

U.S. National Research Council (1985) *The Effects on the Atmosphere of a Major Nuclear Exchange*, Washington, D.C., National Academy Press.

Veltishchev, N.N., Ginsburg, A.S., and Golitsyn, G.S. (1988) Climatic effects of mass fires. *Isvestia—Atmosphere and Ocean Physics* **24** , 296–304.

Vogelmann, A.M., Robock, A., and Ellingson, R.G. (1988) Effects of dirty snow in nuclear winter simulations. *Journal of Geophysical Research* **93**, 5319–5332.

Vupputuri, R.K.R. (1986) The effect of ozone photochemistry on atmospheric and surface temperature changes due to large atmospheric injections of smoke and NO_x by a large-scale nuclear war. *Atmospheric Environment* **20**, 665–80.

Walsh, K. and Pittock, A.B. (1988) The sensitivity of a coupled atmospheric-oceanic model to variations in the albedo and absorptivity of a stratospheric aerosol layer. Paper presented at SCOPE-ENUWAR Moscow Workshop, 21–25 March.

Walton, J.J., MacCracken, M.C., and Ghan, S.J. (1988) A global scale Lagrangian trace species model of transport, transformation, and removal processes. *Journal of Geophysical Research* **93**, 8339–8354.

Westphal, D.L. and Toon, O.B. (1988) Paper presented at SCOPE-ENUWAR Moscow Workshop, 21–25 March.

Westphal, D.L., Toon, O.B., and Carlson, T.N. (1988) A case study of mobilization and transport of Saharan dust, *Journal of Atmospheric Sciences* **45**, 2145–2175.

World Health Organization (1987) *Effects of Nuclear War on Health and Health Services*, 2nd Ed., Geneva, World Health Organization.

Executive Summary

This volume presents the results of an assessment of the climatic and atmospheric effects of a large nuclear war. The chapters in the volume follow a logical sequence of development, starting with discussions of nuclear weapons effects and possible characteristics of a nuclear war. The report continues with a treatment of the consequent fires, smoke emissions, and dust injections and their effects on the physical and chemical processes of the atmosphere. This is followed by a chapter dealing with long-term radiological doses. The concluding chapter contains recommendations for future research and study.

In assessments of this type, a variety of procedural options are available, including, for example, 'worst case' analyses, risk analyses, and 'most probable' analyses. All of these approaches have relevance for the subject addressed here due to the large uncertainties which surround many aspects of the problem. Some of these uncertainties are inherent in studies of nuclear war and some are simply the result of limited information about natural physical processes. In general, in making assumptions about scenarios, models, and magnitudes of injections, and in estimating their atmospheric effects, an attempt has been made to avoid 'minimum' and 'worst case' analyses in favor of a 'middle ground' that encompasses, with reasonable probability, the atmospheric and climatic consequences of a major nuclear exchange.

The principal results of this assessment, arranged roughly in the same order as the more detailed discussions contained in the body of this volume, are summarized below. The Executive Summary of Volume II (Harwell and Hutchinson, 1985), which describes the ecological and agricultural consequences of a nuclear war, is included Appendix 1 at the end of this volume. A Glossary is included as Appendix 2 and a list of participants in the study is included as Appendix 3.

1. DIRECT EFFECTS OF NUCLEAR EXPLOSIONS

The two comparatively small detonations of nuclear weapons in Japan in 1945 and the subsequent atmospheric nuclear tests preceding the atmospheric test ban treaty of 1963 have provided some information on the direct effects of nuclear explosions. Typical modern weapons carried by today's missiles and aircraft have yields of hundreds of kilotons or more. If detonated, such explosions would have the following effects:

- In each explosion, thermal (heat) radiation and blast waves would result in death and devastation over an area of up to 500 km^2 per megaton of yield, an area typical of a major city. The extent of these direct effects depends on the yield of the explosion, height of burst, and state of the local environment. The destruction of Hiroshima and Nagasaki by atomic bombs near the end of World War II provides examples of the effects of relatively *small* nuclear explosions.
- Nuclear weapons are extremely efficient incendiary devices. The thermal radiation emitted by the nuclear fireball, in combination with the accidental ignitions caused by the blast, would ignite fires in urban/industrial areas and wildlands of a size unprecedented in history. These fires would generate massive plumes of smoke and toxic chemicals. The newly recognized atmospheric effects of the smoke from a large number of such fires are the major focus of this report.
- For nuclear explosions that contact land surfaces (surface bursts), large amounts (of the order of 100,000 tonne per megaton of yield) of dust, soil, and debris are drawn up with the fireball. The larger dust particles, carrying about half of the bomb's radioactivity, fall back to the surface mostly within the first day, thereby contaminating hundreds of square kilometers near and downwind of the explosion site. This local fallout can exceed the lethal dose level.
- All of the radioactivity from nuclear explosions well above the surface (airbursts) and about half of the radioactivity from surface bursts would be lofted on very small particles into the upper troposphere or stratosphere by the rising fireballs and contribute to longer term radioactive fallout on a global scale.
- Nuclear explosions high in the atmosphere, or in space, would generate an intense electromagnetic pulse capable of inducing strong electric currents that could damage electronic equipment and communications networks over continent-size regions.

2. STRATEGIES AND SCENARIOS FOR A NUCLEAR WAR

In the forty years since the first nuclear explosion, the five nuclear powers, but primarily the U.S. and the U.S.S.R., have accumulated very large arsenals of nuclear weapons. It is impossible to forecast in detail the evolution of potential military conflicts. Nevertheless, enough of the general principles of strategic planning have been discussed that plausible scenarios for the development and immediate consequences of a large-scale nuclear war can be derived for analysis.

- NATO and Warsaw Pact nuclear arsenals include about 24,000 strategic and theatre nuclear warheads totaling about 12,000 megatons. The arsenals now contain the equivalent explosive power of about one million 'Hiroshima-size' bombs.

• A plausible scenario for a global nuclear war could involve on the order of 6000 Mt divided between more than 12,000 warheads. Because of its obvious importance, the potential environmental consequences of an exchange of roughly this size are examined. The smoke-induced atmospheric consequences discussed in this volume are, however, more dependent on the number of nuclear explosions occurring over cities and industrial centers than on any of the other assumptions of the particular exchange.

• Many targets of nuclear warheads, such as missile silos and some military bases, are isolated geographically from population centers. Nevertheless, enough important military and strategic targets are located near or within cities so that collateral damage in urban and industrial centers from a counterforce nuclear strike could be extensive. As a result, even relatively limited nuclear attacks directed at military-related targets could cause large fires and smoke production.

• Current strategic deterrence policies imply that, in an escalating nuclear conflict, many warheads might be used directly against urban and industrial centers. Such targeting would have far-reaching implications because of the potential for fires, smoke production, and climatic change.

3. THE EXTENT OF FIRES AND GENERATION OF SMOKE

During World War II, intense city fires covering areas as large as 10 to 30 square kilometers were ignited by massive incendiary bombing raids, as well as by the relatively small nuclear explosions over Hiroshima and Nagasaki. Because these fires were few in number and distributed over many months, the total atmospheric accumulation of smoke generated by these fires was small. Today, in a major nuclear conflict, thousands of very intense fires, each covering up to a few hundred square kilometers, could be ignited simultaneously in urban areas, fossil fuel processing plants and storage depots, wildlands, and other locations. Because there have never been fires as large and as intense as may be expected, no appropriate smoke emission measurements have been made. Estimates of emissions from such fires rely upon extrapolation from data on much smaller fires. This procedure may introduce considerable error in quantifying smoke emissions, especially in making estimates for intense fire situations.

• About 70% of the populations of Europe, North America and the Soviet Union live in urban and suburban areas covering a few hundred thousand square kilometers and containing more than ten thousand million tonne of combustible wood and paper. If about 25–30% of this were to be ignited, in just a few hours or days, tens of millions to more than a hundred million tonne of smoke could be generated. About a quarter to a third of the emitted smoke from the flaming combustion of this material would be amorphous elemental carbon, which is black and efficiently absorbs sunlight.

- Fossil fuels (e.g., oil, gasoline, and kerosene) and fossil fuel-derived products (including plastics, rubber, asphalt, roofing materials, and organochemicals) are heavily concentrated in cities and industrial areas; flaming combustion of a small fraction ($\sim$25–30%) of the few thousand million tonne of such materials currently available could generate 50–150 million tonne of very sooty smoke containing a large fraction (50% or greater) of amorphous elemental carbon. About 25–30% of the combustible materials of the developed world are contained in less than one hundred of the largest industrialized urban areas.

- Fires ignited in forests and other wildlands could consume tens to hundreds of thousands of square kilometers of vegetation over days to weeks, depending on the state of the vegetation, and the extent of firespread. These fires could produce tens of millions of tonne of smoke in the summer half of the year, but considerably less in the winter half of the year. Because wildland fire smoke contains only about 10% amorphous elemental carbon, it would be of secondary importance compared to the smoke created by urban and industrial fires, although its effects would not be negligible.

- The several tens of millions of tonne of sub-micron dust particles that could be lofted to stratospheric altitudes by surface bursts could reside in the atmosphere for a year or more. The potential climatic effects of the dust emissions, although substantially less than those of the smoke, also must be considered.

4. THE EVOLUTION AND RADIATIVE EFFECTS OF THE SMOKE

The sooty smoke particles rising in the hot plumes of large fires would consist of a mixture of amorphous elemental carbon, condensed hydrocarbons, debris particles, and other substances. The amount of elemental carbon in particles with effective spherical diameters on the order of 0.1 μm to perhaps 1.0 μm would be of most importance in calculating the potential effect on solar radiation. Such particles can be spread globally by the winds and remain suspended for days to months.

- Large hot fires create converging surface winds and rapidly rising fire plumes which, within minutes, can carry smoke particles, ash and other fire products, windblown debris, and water from combustion and the surrounding air to as high as 10–15 kilometers. The mass of particles deposited aloft would depend on the rate of smoke generation, the intensity of the fire, local weather conditions, and the effectiveness of scavenging processes in the convective column.

- As smoke-laden, heated air from over the fire rises, adiabatic expansion and entrainment would cause cooling and condensation of water vapor that could lead, in some cases, to the formation of a cumulonimbus cloud system. Condensation-induced latent heating of the rising air parcels would help to loft the smoke particles to higher altitudes than expected from the heat of combustion alone.

- Although much of the water vapor drawn up from the boundary layer would condense, precipitation might form for only a fraction of the fire plumes. In the rising columns of such fires, soot particles would tend to be collected inefficiently by the water in the cloud. Smoke particles however, are generally composed of a mixture of substances and might, at least partially, be incorporated in water droplets or ice particles by processes not now well understood. Smoke particles that are captured could again be released to the atmosphere as the ice or water particles evaporate in the cloud anvils or in the environment surrounding the convective clouds. Altogether, an unknown fraction of the smoke entering the cloud would be captured in droplets and promptly removed from the atmosphere by precipitation.

- Not all fires would, however, induce strong convective activity. This depends on fuel loading characteristics and meteorological conditions. It is assumed in current studies that 30–50% of the smoke injected into the atmosphere from all fires would be removed by precipitation within the first day, and not be available to affect longer-term large-scale, meteorological processes. This assumption is a major uncertainty in all current assessments. For the fire and smoke assumptions made in this study, the net input of smoke to the atmosphere after early scavenging is estimated to range from 50 to 150 million tonne, containing about 30 million tonne of amorphous elemental carbon.

- Smoke particles generated by urban and fossil fuel fires would be strong absorbers of solar radiation, but would be likely to have comparatively limited effects on terrestrial longwave radiation, except perhaps under some special circumstances. If 30 million tonne of amorphous elemental carbon were produced by urban/industrial fires and spread over Northern Hemisphere mid-latitudes, the insolation at the ground would be reduced by at least 90%. The larger quantities of smoke that are possible in a major nuclear exchange could reduce light levels under dense patches to less than 1% of normal, and, after the smoke has spread widely, to just several percent of normal on a daily average.

- Because of the large numbers of particles in the rising smoke plumes and the very dense patches of smoke lasting several days thereafter, coagulation (adhering collisions) would lead to formation of fewer, but somewhat larger, particles. Coagulation of the particles could also occur as a result of coalescence and subsequent evaporation of rain droplets or ice particles. Because optical properties of aerosols are dependent on particle size and morphology, the aggregated aerosols may have different optical properties than the initial smoke particles, but the details, and even the sign, of such changes are poorly understood. The optical properties of fluffy soot aggregates that may be formed in dense oil plumes, however, seem to be relatively insensitive to their size. This is less the case for more consolidated particle agglomerates.

• Little consideration has yet been given to the possible role of meteorological processes on domains between fire plume and continental scales. Mesoscale and synoptic-scale motions might significantly alter, mix, or remove the smoke particles during the first several days. Studies to examine quantitatively the microphysical evolution of smoke particles during this period are needed. While changes in detailed understanding are expected, a significant fraction of the injected smoke particles is likely to remain in the atmosphere and affect the large-scale weather and climate.

5. SMOKE-INDUCED ATMOSPHERIC PERTURBATIONS

In a major nuclear war, continental scale smoke clouds could be generated within a few days over North America, Europe, and much of Asia. Careful analysis and a hierarchy of numerical models (ranging from one-dimensional global-average to three-dimensional global-scale models) have been used to estimate the transport, transformation, and removal of the smoke particles and the effects of the smoke on temperature, precipitation, winds, and other important atmospheric properties. All of the simulations indicate a strong potential for large-scale weather disruptions as a result of the smoke injected by extensive post-nuclear fires. These models, however, still have important simplifications and uncertainties that may affect the fidelity and the details of their predictions. Nonetheless, these uncertainties probably do not affect the general character of the calculated atmospheric response.

• For large smoke injections reaching altitudes of several kilometers or more and occurring from spring through early fall in the Northern Hemisphere, average land surface temperatures beneath dense smoke patches could decrease by 20–40 °C below normal in continental areas within a few days, depending on the duration of the dense smoke pall and the particular meteorological state of the atmosphere. Some of these patches could be carried long distances and create episodic cooling. During this initial period of smoke dispersion, temperature anomalies could be spatially and temporally quite variable while patchy smoke clouds strongly modulate the insolation reaching the surface.

- Smoke particles would be spread throughout much of the Northern Hemisphere within a few weeks, although the smoke layer would still be far from homogeneous. For spring to early fall injections, solar heating of the particles could rapidly warm the smoke layer and lead to a net upward motion of a substantial fraction of the smoke into the upper troposphere and stratosphere. The warming of these elevated layers could stabilize the atmosphere and suppress vertical movement of the air below these layers, thereby extending the lifetime of the particles from days to perhaps several months or more.

- Average summertime land surface temperatures in the Northern Hemisphere mid-latitudes could drop to levels typical of fall or early winter for periods of weeks or more with convective precipitation being essentially eliminated, except possibly at the southern edge of the smoke pall. Cold, near-surface air layers might lead initially to fog and drizzle, especially in coastal regions, lowland areas, and river valleys. In continental interiors, periods of very cold, mid-winter-like temperatures are possible. In winter, light levels would be strongly reduced, but the initial temperature and precipitation perturbations would be much less pronounced and might be essentially indistinguishable in many areas from severe winters currently experienced from time to time. However, such conditions would occur simultaneously over a large fraction of the mid-latitude region of the Northern Hemisphere and freezing cold air outbreaks could penetrate southward into regions that rarely or never experience frost conditions.

- In Northern Hemisphere subtropical latitudes, temperatures in any season could drop well below typical cool season conditions for large smoke injections. Temperatures could be near or below freezing in regions where temperatures are not typically strongly moderated by warming influence from the oceans. The convectively driven monsoon circulation, which is of critical importance to subtropical ecosystems, agriculture, and is the main source of water in these regions, could be essentially eliminated. Smaller scale, coastal precipitation might, however, be initiated.

- Strong solar heating of smoke injected into the Northern Hemisphere between April and September would carry the smoke upwards and equatorward, strongly augmenting the normal high altitude flow to the Southern Hemisphere (where induced downward motions might tend to slightly suppress precipitation). Within one or two weeks, thin, extended smoke layers could appear in the low to mid-latitude regions of the Southern Hemisphere as a precursor to the development of a more uniform veil of smoke with a significant optical depth (although substantially smaller than in the Northern Hemisphere). The smoke could induce modest cooling of land areas not well buffered by air masses warmed over nearby ocean areas.

Since midlatitudes in the Southern Hemisphere would already be experiencing their cool season, temperature reductions would not likely be more than several degrees. In more severe, but less probable, smoke injection scenarios, climatic effects in the Southern Hemisphere could be enhanced significantly, particularly during the following austral spring and summer.

• Much less analysis has been made of the atmospheric perturbations following the several week, acute climatic phase subsequent to a nuclear war involving large smoke injections. Significant uncertainties remain concerning processes governing the longer-term removal of smoke particles by precipitation scavenging, chemical oxidation, and other physical and chemical factors. The ultimate fate of smoke particles in the perturbed atmospheric circulation is also uncertain, both for particles in the sunlit and stabilized upper troposphere and stratosphere and in the winter polar regions, where cooling could result in subsidence that could move particles downward from the stratosphere to altitudes where they could later be scavenged by precipitation.

• Present estimates suggest that smoke lofted (either directly by fire plumes or under the influence of solar heating) to levels which are, or become, stabilized, could remain in the atmosphere for a year or more and induce long-term (months to years) global-scale cooling of several degrees, especially after the oceans have cooled significantly. For such conditions, precipitation could also be reduced significantly. Reduction of the intensity of the summer monsoon over Asia and Africa could be a particular concern. Decreased ocean temperatures, climatic feedback mechanisms (e.g., ice-albedo feedback), and concurrent ecological changes could also prolong the period of meteorological disturbances.

6. ATMOSPHERIC CHEMISTRY IN A
POST-NUCLEAR-WAR ENVIRONMENT

Nuclear explosions and the resultant fires could generate large quantities of chemical compounds that might themselves be toxic. In addition, the chemicals could alter the atmospheric composition and radiative fluxes in ways that could affect human health, the biosphere, and the climate.

• Nitrogen oxides (NO_x) created in nuclear fireballs would be lofted primarily into the stratosphere for explosions of greater than several hundred kilotons. Depending on the total number of high yield weapons exploded, the NO_x would catalyze chemical reactions that, within a few months time, could reduce Northern Hemisphere stratospheric ozone concentrations by 10 to 30% in an atmosphere free of aerosols. Recovery would take several years. However, if the atmosphere were highly perturbed due to smoke heating and by injection of gaseous products from fires, the long-term ozone changes could be enhanced substantially in ways that cannot yet be predicted.

• Stratospheric ozone reductions of tens of percent could increase surface intensities of biologically-active ultraviolet (UV) radiation by percentages of up to a few times as much. The presence of smoke would initially prevent UV-radiation from reaching the surface by absorbing it. The smoke, however, might also prolong and further augment the long-term ozone reduction as a result of smoke-induced lofting of soot and reactive chemicals, consequent heating of the stratosphere, and the occurrence of additional chemical reactions.

• Large amounts of carbon monoxide, hydrocarbons, nitrogen and sulfur oxides, hydrochloric acid, pyrotoxins, heavy metals, asbestos, and other materials would be injected into the lower atmosphere near the surface by flaming and smoldering combustion of several thousand million tonne of cellulosic and fossil fuel products and wind-blown debris. Before deposition or removal, these substances, some of which are toxic, could be directly and/or indirectly harmful to many forms of life. In addition, numerous toxic chemical compounds could be released directly into the environment by blast and spillage, contaminating both soil and water. This complex and potentially very serious subject has so far received only cursory consideration.

• If the hydrocarbons and nitrogen oxides were injected into an otherwise unperturbed troposphere, they could enhance average background ozone concentrations several-fold. Such ozone increases would not significantly offset the stratospheric ozone decrease, which also would be longer lasting. It is highly questionable, however, whether such large ozone increases could indeed occur in the presence of smoke because ozone generation in the troposphere requires sunlight as well as oxides of nitrogen. It is possible that, in the smoke perturbed atmosphere, the fire-generated oxides of nitrogen could be removed before photochemical ozone production could take place.

• Precipitation scavenging of nitrogen, sulfur, and chlorine compounds dispersed by the fire plumes throughout the troposphere could increase rainfall acidity by about an order of magnitude over large regions for up to several months. This increased acidity might be neutralized to some degree by alkaline dust or other basic (as opposed to acidic) compounds.

• Rapid smoke-induced cooling of the surface under dense smoke clouds could induce the formation of shallow, stable cold layers that might trap chemical emissions from prolonged smoldering fires near the ground. In such layers, concentrations of CO, HCℓ, pyrotoxins, and acid fogs could reach dangerous levels. The potential for local and regional effects in areas such as populated lowland areas and river valleys merits close attention.

7. RADIOLOGICAL DOSE

Near the site of an explosion, the health effects of prompt ionizing radiation from strategic nuclear warheads would be overshadowed by the effects of the blast and thermal radiation. However, because nuclear explosions create highly radioactive fission products and the emitted neutrons may also induce radioactivity in initially inert material near the detonation, radiological doses would be delivered to survivors both just downwind (local fallout) and out to hemispheric and global scales (global fallout).

- Local fallout of relatively large radioactive particles lofted by the number of surface explosions in the scenario postulated in this study could lead to lethal external gamma-ray doses (assuming no protective action is taken) during the first few days over about 7 percent of the land areas of the NATO and Warsaw Pact countries. Areas downwind of missile silos and other hardened targets would suffer especially high exposures. Survivors outside of lethal fallout zones could still receive debilitating radiation doses (exposure at half the lethal level can induce severe radiation sickness). In combination with other injuries or stresses, such doses could increase mortality. If large populations could be mobilized to move from highly radioactive zones or take substantial protective measures, the human impact of fallout could be greatly reduced.
- The uncertainty in these calculations of local fallout is large. Doses and areas for single nuclear explosions could vary by factors of 2–4 depending on meteorological conditions and assumptions in the models. A detailed treatment of overlapping fallout plumes from multiple explosions could increase the areas considerably (by a factor of 3 in one sample case). Results are also sensitive to variations in the detonation scenario.
- Global fallout following the gradual deposition of the relatively small radioactive particles created by strategic air and surface bursts could lead to average Northern Hemisphere lifetime external gamma ray doses on the order of 10 to 20 rads. The peak values would lie in the northern mid-latitudes where the average doses for the scenarios considered would be about 20 to 60 rads. Such doses, in the absence of other stresses, would be expected to have relatively minor carcinogenic and mutagenic effects (i.e., increase incidence at most a few percent above current levels). Smoke-induced perturbations that tend to stabilize the atmosphere and slow deposition of radioactive particles might reduce these estimated average doses by perhaps 15%.
- Intermediate time scale and long term global fallout would be deposited unevenly, largely because of meteorological effects, leading to 'hotspots' of several hundred thousand square kilometers in which average doses could be as high as 100 rads, and, consequently, large areas where doses would be lower than the average value.

• In the Southern Hemisphere and tropical latitudes, global fallout would produce much smaller, relatively insignificant, radiological doses about one-twentieth those in the Northern Hemisphere, even if cross-equatorial transport were accelerated by the smoke clouds. Additional local fallout would be important only within a few hundred kilometers downwind of any surface burst in the Southern Hemisphere.

• Additional considerations not factored into the above estimates are possible from several sources. Doses from ingestion or inhalation of radioactive particles could be important, especially over the longer term. Beta radiation could have a significant effect on the biota coming into contact with the local fallout. Fission fractions of smaller modern weapons could be twice the assumed value of 0.5; adding these to the scenario mix could cause a 20% increase in areas of lethal fallout. General tactical and theater nuclear weapons, ignored in these calculations, could also cause a 20% increase in lethal local fallout areas in certain geographical regions, particularly in Europe. The injection into the atmosphere of radionuclides created and stored by the civilian nuclear power industry and military reactors, a possibility considered remote by some, could increase estimates of long-term local and global radiological doses to several times those estimated for weapons alone.

8. TASKS FOR THE FUTURE

Extensive research and careful assessment over the past few years have indicated that nuclear war has the potential to modify the physical environment in ways that would dramatically impair biological processes. The perturbations could impact agriculture, the proper functioning of natural ecosystems, the purity of essential air and water resources, and other important elements of the global biosphere. Because current scientific conclusions concerning the response of the atmosphere to the effects of nuclear war include uncertainties, research can and should be undertaken to reduce those uncertainties that are accessible to investigation.

• Laboratory and field experiments are needed to improve estimates of the amount and physical characteristics of the smoke particles that would be produced by large fires, particularly by the combustion of fossil fuels and fossil fuel-derived products present in urban and industrial regions. Experimental conditions should be designed to emulate as much as possible the effects of large-scale fires.

• Laboratory, field, and theoretical studies are needed to determine the potential scavenging rates of smoke particles in the convective plumes of large fires and the scavenging processes that operate on intermediate and global scales as the particles disperse.

- Further theoretical calculations of the seasonal response of the atmosphere to smoke emissions from large fires are needed, particularly of the extent of the perturbation to be expected at early times, when the smoke is freshly injected and patchy. Simulations must be made for later times from months to a year or more, when the atmosphere has been highly perturbed and a substantial fraction of the smoke may have been lofted to high altitudes. Closer attention should be paid to the possible effects in low latitudes and in the Southern Hemisphere, where the climatic effects are likely to be much more important than the direct effects of the nuclear detonations, which are expected to be confined largely to the Northern Hemisphere.
- Laboratory and theoretical studies are needed of the potential chemical alterations of the atmosphere on global and local scales, and of the extent that smoke particles could affect and might be removed by chemical reactions high in the atmosphere.
- Radiological calculations should be undertaken using models that more realistically treat the overlap of fallout plumes, complex meteorological conditions, and that consider both external and internal doses. Patterns of land use and likely targeting strategy should be used in estimating the potential significance of various scenarios. The question of the possible release of radioactivity from nuclear fuel cycle facilities in a nuclear war should be explored more thoroughly.

Environmental Consequences of Nuclear War Volume I:
Physical and Atmospheric Effects
A. B. Pittock, T. P. Ackerman, P. J. Crutzen,
M. C. MacCracken, C. S. Shapiro and R. P. Turco
© 1986 SCOPE. Published by John Wiley & Sons Ltd

CHAPTER 1
Direct Effects of Nuclear Detonations

1.1 HIROSHIMA AND NAGASAKI

1.1.1 Historical Notes

The first atomic weapon used in warfare was the bomb dropped on Hiroshima, Japan at 8:15 AM (local time) on August 6, 1945. The second, and only other, weapon so used was dropped on Nagasaki, Japan at 11:02 AM on August 9, 1945. The Hiroshima bomb (Little Boy) had an energy yield of about 15 ± 3 kilotons (kt; a one kt explosion is equivalent in energy release to the detonation of about 1000 tons of TNT; one megaton, Mt, equals 1000 kT). The Nagasaki bomb (Fat Man) had an energy yield of 21 ± 2 kilotons (Ohkita, 1985). From these unique events, much of what is known about the effects of nuclear explosions on people and cities has been learned. About 120,000 people were killed outright in both cities, and the eventual fatalities, as of 1981, were about 210,000 (Ishikawa and Swain, 1981). In Hiroshima, an urbanized area of approximately 13 square kilometers was laid to waste, while in Nagasaki, an area approximately 7 square kilometers was destroyed. Figure 1.1 starkly illustrates the extent of the devastation in central Hiroshima; only the hulks of the most resilient steel-reinforced concrete buildings were left standing.

1.1.2 Physical Effects of the Bombings

Both the Hiroshima and Nagasaki nuclear explosions were airbursts—at elevations of 580 meters and 500 meters, respectively. These heights are near optimal for thermal irradiation and blast damage, but produce relatively little radioactive fallout because the fireball does not touch the ground (see below). The most immediate consequence was the intense thermal irradiation "pulse", which caused serious skin burns and primary fire ignitions at distances of up to several kilometers from the hypocenter. The blast pressure wave that immediately followed caused severe damage to structures at distances of 2 km at Hiroshima and 3 km at Nagasaki. While many of the primary fires were suppressed by the blast winds, numerous "secondary" ignitions occurred through breaches of domestic fires, electrical short

Figure 1.1. Panoramic view of Hiroshima following the atomic bomb explosion of August 6, 1945. In the few standing structures of reinforced concrete. the interiors were totally gutted and burned. Reproduced by permission of Popperfoto

circuits, and so on. As a result of the combined thermal irradiation, blast damage and loss of water pressure, firefighting was made all but hopeless, mass fires developed in the ruins of both cities, and burned out large areas within 24 hours.

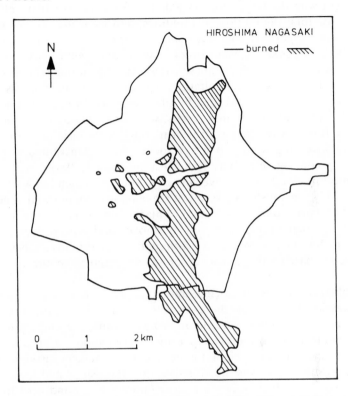

Figure 1.2. Comparison of fire-damaged areas in Hiroshima and Nagasaki (from Kiuchi, 1953, and Ishikawa and Swain, 1981). Originally published in Japanese by Iwanami Shoten, Publishers, Tokyo. Reproduced by permission of Hiroshima City and Nagasaki City

Figure 1.2 compares the burnout areas at Hiroshima and Nagasaki. These areas are often quoted as 13 km^2 and 6.7 km^2, respectively (Ishikawa and Swain, 1981). There are several important features of the fire patterns shown in Figure 1.2. At Hiroshima, with its relatively flat topography, the fire zone was roughly symmetrical about the burst point and encompassed essentially all of the area heavily damaged by the blast. Fire spread outside of this zone may have been hindered by the centrally directed winds established during the intense fire that appeared soon after the bombing. At Nagasaki, which lies at the mouth of a river valley, the fires were confined within the valley. To the east and west, hills protected the regions beyond from the most severe effects of thermal irradiation and blast. To the south, fire ignition and spread were limited by open waters. Although the fire area at Nagasaki was smaller than that at Hiroshima, the burned out area in the direction of urban development extended a greater distance from the hypocenter, which is consistent with the larger bomb yield. The fire zones reached well beyond the zones of total demolition of buildings, and even the regions of dense rubble burned vigorously. In Nagasaki, the central zone of heavy industry, with its broad open areas, was also extensively damaged by fire.

"Black rain" fell in both Japanese cities after the atomic bombings. From the recorded patterns of precipitation, it is clear that the rain was induced by convective motions established by the mass fires. In both Hiroshima and Nagasaki, typically warm humid August weather prevailed at the time of the bombings. It is believed that a "firestorm" (an intense mass fire with strongly rotating, converging winds) may have developed at Hiroshima, with attendant strong convective activity (Ishikawa and Swain, 1981). A thundering cumulonimbus cloud formed over the city. The rain which fell was, at times, black and oily, obviously as a result of scavenging of smoke and charred fire debris. The Hiroshima and Nagasaki "black rain" events provide qualitative evidence that prompt washout of smoke in city-sized fires can occur (at least in humid environments), although no quantitative measurements or estimates exist of the efficiency of smoke removal in the fire-induced precipitation.

Physiological effects of nuclear radiation were observed in Hiroshima and Nagasaki. Because of the relatively small sizes of the first atomic bombs, and the fact they were detonated as airbursts, prompt gamma rays and fast neutrons were the principal nuclear radiations to produce effects. Gamma rays and neutrons cannot penetrate long distances through air at sea level—several kilometers is the effective limit (Glasstone and Dolan, 1977). Accordingly, individuals who were close enough to ground zero to receive a lethal dose of prompt nuclear radiation were more likely to have been killed outright by the blast or thermal flash. Nevertheless, cases of radiation sickness appeared frequently among the survivors in Japan (Ishikawa and Swain, 1981).

Some deposition of radioactive fission debris ("fallout") occurred at Hiroshima and Nagasaki. The black rain in both cities apparently washed out a small fraction of the airborne radioactive aerosols (Molenkamp, 1980). The consequences of this radioactive fallout (in combination with the residual radioactivity induced by the weapon's fast neutrons) are not well defined. The maximum total whole-body gamma ray doses accumulated by survivors are estimated to have been about 13 rads in Hiroshima and 42 to 129 rads in Nagasaki (Shimazu, 1985). For such doses, physiological effects would not be readily identifiable (see the discussion of the effects of radiation on humans in Chapter 7).

1.1.3 Lessons of Hiroshima and Nagasaki

The atomic explosions in Japan in 1945 offer several clear lessons:

1. The destructive power of nuclear weapons is immense—a single bomb can destroy an entire city in a matter of seconds. This is particularly true since the yield of a typical strategic warhead is now at least ten times greater than the Hiroshima or Nagasaki bombs.
2. Nuclear weapons are efficient long-range incendiary devices—all of the area subject to blast damage is susceptible to burnout in conflagrations.
3. The atmosphere over a large region is affected by a nuclear explosion: smoke and dust are lofted, clouds form, precipitation may occur, and radioactive debris is dispersed through the environment.
4. The human impacts of nuclear explosions can be enormous—physical injuries from the blast, severe burns from the heat rays, exposure to radiation, psychological trauma—in addition to the long-term effects discussed elsewhere in this report.

A survey is made below of the basic processes of nuclear detonations that are relevant to an assessment of the potential global-scale physical effects of nuclear war. The biological implications are discussed at length in Volume II of this report.

1.2 THERMAL IRRADIATION

1.2.1 Fireball

Upon detonation, a nuclear fission weapon disassembles and vaporizes within one millionth of a second (Glasstone and Dolan, 1977). At that time, about 70 to 80% of the energy has been converted into "soft" X-rays with an effective radiation temperature of several tens of millions of degrees Celsius; most of the remaining energy comprises kinetic energy of the bomb debris. At sea level, the primary thermal X-radiation is absorbed by the air within several meters of the device, heating the air and forming an embryonic fireball. This enormously hot sphere continues to expand rapidly by radiative transfer to the surrounding ambient atmospheric gases. As the fireball grows and cools to about 300,000°C, the thermal irradiation becomes less penetrating, and the radiative fireball growth slows. At this point, a shock wave forms and propagates ahead of the fireball ("hydrodynamic separation"). The shock-heated air is opaque and luminous and shields the direct radiation of the fireball. However, as the shock wave continues to expand, the temperature of the shock-heated air decreases and it becomes less opaque. At about 3000°C, the thermal irradiation of the fireball again becomes visible through the shock front ("breakaway"). From a distance, the apparent radiation temperature then increases rapidly to the temperature of the fireball, which is about 7500°C (approximately the temperature of the Sun's surface), before decreasing again as the fireball continues to cool by radiation, expansion, and entrainment of ambient air.

Corresponding to the formation and growth of the fireball, two pulses of thermal irradiation are emitted. Together, these carry away about 35% of the total energy of the explosion, mainly as visible and near-infrared radiation (spectrally, the average emission is very similar to sunlight). The first pulse of light originates from the shock wave front (attenuated to some degree by ozone and nitrogen oxides generated ahead of the shock wave by prompt penetrating nuclear radiation). The timescale for the first emission is of the order of milliseconds, and it carries only about 1% of the total thermal energy. While this pulse has little incendiary effect, it can damage the retina of the eye. The second burst of light, the true "thermal pulse", commences as the shock wave becomes transparent and the incandescent fireball is revealed. The time scale for this emission is of the order of seconds, and its duration tends to increase with yield. Almost all of the thermal emission (about 35% of the total energy yield of the explosion) is liberated during this pulse.

For burst heights between the surface and roughly 30 km, the basic explosion phenomenology remains essentially unaltered. In this regime, the overall energy partition of a nuclear fission explosion is: thermal irradiation, 35%; blast and shock, 50%; initial or prompt nuclear radiation, 5%; residual nuclear radiation, 10%. In quoting the energy yield of a nuclear explosion, the energy of the residual nuclear radiation, i.e., that released by nuclear decay beyond the first minute, is generally omitted. For a fission weapon this is approximately 10% of the total energy yield, and for a fission/fusion weapon, approximately 5%. Typical thermonuclear devices are driven by roughly 50% fission and 50% fusion energy; most of the fission yield results from the disintegration of a heavy shield of ^{238}U used as an X-ray and neutron reflector for the fusion stage. Except for radioactive fallout, the distinction between fission and fission/fusion weapons is unimportant in the present analysis (see Chapter 7).

1.2.2 Thermal Effects

The intense thermal irradiation from a nuclear fireball, emitted at visible wavelengths, can readily ignite a fire, much as does sunlight focused by a lens. Hence, the first effect of a nuclear explosion is to ignite "primary" fires over a large area (where kindling fuels are exposed). The radiant energy from a nuclear detonation impinging on an object can be expressed as a fluence in calories per square centimeter; that is, the energy per unit area perpendicular to the surface of the object, integrated over wavelength and time for the duration of the thermal pulse. A 1-Mt airburst, which is equivalent in energy release to detonation of about a million tons of TNT, can ignite newspaper and leaves at 6 cal/cm², fabrics at 15 cal/cm², and roofing and wood at 30 cal/cm² (Glasstone and Dolan, 1977). The fluence required to ignite a material depends weakly on the explosion yield for yields ≤ 1 Mt. Generally, the lower the yield, the less fluence that is needed for ignition. This occurs because larger yield explosions have longer thermal pulses, which are somewhat less efficient at heating and igniting bulk materials at modest levels of fluence. The size, shape, color, and orientation of an object also affects its flammability by the thermal flash.

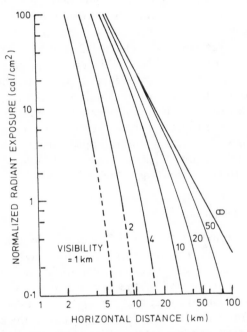

Figure 1.3. Maximum radiant exposures versus ground range for a 1 Mt airburst (detonated between about one and several kilometers altitude) as a function of the ground level visibility. The radiant exposures scale roughly with the yield in megatons for yields between 0.1 and 1 Mt (from Kerr et al., 1971)

Figure 1.3 provides an estimate of the thermal fluences associated with a 1-Mt low-altitude airburst as a function of distance from ground zero for several different atmospheric visibilities. A surface burst produces thermal fluences that are roughly 50% of those of an airburst for yields between 0.1 and 10 Mt (Glasstone and Dolan, 1977). Within several kilometers of a 1-Mt explosion, the thermal fluence can exceed 100 cal/cm². The actual energy flux on a surface may be much lower, however, because of (a) the shading of surfaces by topography, structures, and vegetation; and (b) the attenuation of the fireball radiation by steam, smoke, and dust raised from vegetation and soil by the thermal irradiation and blast. Fireball radiation can also be scattered and focused by clouds, dust and steam, accentuating thermal effects over a large area around the burst point and possibly causing isolated ignitions well beyond the perimeter of the nominal zone of thermal effects.

The thermal pulse of a 1-Mt fireball can ignite primary fires over an area of 1000 square kilometers when the atmosphere is exceptionally clear and dry flammable materials (with ignition thresholds of approximately 5–10 cal/cm^2) are present (Figure 1.3). For more typical visibilities, and for substances with greater ignition thresholds, the primary ignition zone may extend over an area of 200 to 500 km^2 for a 1-Mt airburst (equivalent to an ignition area per unit yield of 0.2 to 0.5 km^2/kt) (NRC, 1985). The fire areas at Hiroshima and Nagasaki were, correspondingly, about 0.9 km^2/kt and 0.3 km^2/kt, respectively. Generally speaking, the lower the yield of a nuclear explosion, the greater its incendiary efficiency (ignition area per kiloton). This is a result of the faster release of thermal energy and lesser impact of visibility. Hiroshima probably represents the maximum primary incendiary efficiency of nuclear weapons (approximately 1 km^2/kt), although fire spread beyond the primary ignition zone could increase the effective fire area considerably (see Chapter 3 and Appendix 3A).

1.2.3 Blast/Fire Interactions

Blast effects (described in Section 1.3) can greatly influence the course of fires initiated by a nuclear explosion (Glasstone and Dolan, 1977). The winds generated by the blast wave can extinguish flames in materials ignited by thermal irradiation. Not all of the primary fires would be blown out, however, and many ignited materials would continue to smolder and eventually rekindle flaming combustion. More importantly, the blast wave causes secondary fires, creates conditions favorable to fire spread, and hinders effective firefighting, but can also bury burnable material under non-combustible rubble. In Hiroshima and Nagasaki, secondary ignitions were apparently as important as primary ignitions in the mass fires which developed (Ishikawa and Swain, 1981). Secondary fires result from electrical short circuits, broken gas lines, breaches of open flames, and similar effects. Typically, about one secondary fire is expected for every 10,000 square meters of building floor space (Kang et al., 1985, see also Appendix 3A). In general, blast damage would facilitate fire spread and hinder efforts to suppress the fire. Fires can propagate more effectively through buildings with broken windows, doors, and firewalls, across natural firebreaks breached by flammable debris, and along flows of spilled liquid and gaseous fuels and petrochemicals. With the additional burden of large numbers of injured personnel, widespread fire ignitions, blocked streets, and loss of water pressure, meaningful firefighting efforts could not be mounted. This was precisely the situation that arose in Hiroshima and Nagasaki (Ishikawa and Swain, 1981).

In urban/industrial regions close to the explosion hypocenter, even buildings of heavy construction could be reduced to rubble. Nonflammable debris, such as concrete and steel, would cover some of the flammable materials. However, zones of thick rubble (formed in tracts that are very densely built-up) would probably account for less than 10% of the total area of destruction and fire (NRC, 1985), although they could contain a disproportionately high areal density of combustible material. However, even within this central zone, many materials would be instantly ignited by the intense thermal irradiation in "flashover" fires (an effect observed during the Encore nuclear test—27 kt, Nevada Test Site, May 8, 1953—in which an entire room was ignited simultaneously, that is, "flashed over", within seconds of irradiation). These "instantaneous" fires would continue to spread and smolder in the rubble. Because of the Encore effect and the other known incendiary effects of nuclear weapons, it is expected that all urbanized areas, from modern city centers to spacious suburban zones, from commercial tracts to industrial parks would be subject to burning by nuclear explosions (NRC, 1985).

Forests, agricultural lands, and wildlands are also susceptible to complex nuclear fire effects. Thermal irradiation not only ignites dry fuels, but also dessicates moist fuels and live vegetation (Kerr et al., 1971), making them more susceptible to fire. The blast wave extinguishes some fires, but also spreads firebrands. Blast-induced winds can knock down foliage and branches (blowdown) not usually involved in wildfires; on relatively flat terrain, a 1-Mt airburst causes such damage over an area of roughly 500 km² in foliated deciduous forests, and over about 350 km² in leafless deciduous stands and unimproved coniferous forests (Glasstone and Dolan, 1977). The simultaneous ignition of fuels over a vast area by thermal flash, the dessicating effect of the thermal pulse on vegetation, and the augmentation of ground fuel by blowdown imply that nuclear-initiated wildland fires could be more easily ignited, consume more fuel, and burn more intensely than natural wildfires (NRC, 1985). At the present time, only historical information gathered on natural and prescribed wildland fires is available to estimate the extent and effects of the wildland fires that would be ignited in a nuclear conflict. A very large seasonal variation in the susceptibility of wildlands to fire is expected, with fewer fires likely in the winter. Historically, seasonal variations in fire occurrence in urban areas have been much less significant (Chandler et al., 1963).

A number of additional technical issues related to nuclear-initiated fires and smoke production are discussed in Chapter 3.

1.3 DYNAMIC PHENOMENA OF NUCLEAR EXPLOSIONS

1.3.1 Shock Wave in Air

As explained in Section 1.2.1, the air shock wave of a nuclear explosion begins to move away from the fireball at the time of hydrodynamic separation. Thereafter, it acts as a simple pressure wave in air. For low altitude explosions, the shock wave is also reflected from the surface; the incident and reflected waves may then combine to form a "Mach stem", in which the shock pressures are roughly twice the incident values. For bursts below about 30 kilometers, approximately 50% of the total energy of a nuclear explosion is carried away by the shock waves.

One measure of the destructive power of a nuclear explosion is the peak overpressure it creates at various distances from the hypocenter. The peak overpressure in the shock wave is the maximum increase of static air pressure over ambient atmospheric pressure. The overpressure is usually measured in pounds per square inch in the American literature (psi; 1 psi = 6.9 kPa; the mean atmospheric pressure at sea level is 14.7 psi). Figure 1.4

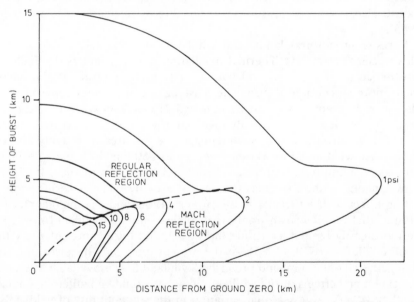

Figure 1.4.   Peak blast overpressure at the ground in pounds per square inch (psi; 1 psi = 6.9 kPa) for a 1 Mt detonation as a function of the distance from ground zero and the height-of-burst. For other explosive yields, the distance and height-of-burst scale as $Y^{1/3}$ where Y is the yield in megatons. For example, in the case of a 1 kt explosion, all lengths would be multiplied by 0.1 (from Glasstone and Dolan, 1977)

illustrates the peak overpressure produced by a 1 Mt detonation as a function of distance from ground zero and height-of-burst (HOB). For a given overpressure, there is generally an optimum HOB to maximize the range for that overpressure. However, very close to the explosion, nearly identical peak overpressures can be achieved for bursts at the surface to a moderate height above the surface.

The blast wave also produces sudden outward displacements of air and large peak dynamic (wind) pressures. A physical relationship exists between the peak overpressure, peak dynamic pressure, and maximum wind speed (Glasstone and Dolan, 1977). At a peak overpressure of 100 psi, the peak dynamic pressure is 120 psi and the maximum wind velocity is 630 m/s (2270 km/h); at 10 psi, it is 2.2 psi and 130 m/s (470 km/h), respectively; and at 2 psi, it is 0.1 psi and 30 m/s (110 km/h), respectively. (By comparison, the winds in severe tropical cyclones reach velocities of 150 to 250 km/hr). It follows that, for wind sensitive structures, the importance of dynamic pressure relative to overpressure increases with proximity to the explosion.

1.3.2 Ground Shock

Nuclear airbursts create ground shock when the blast wave impacts the surface and induces ground motions. Surface and subsurface explosions efficiently couple energy directly into the ground and create strong local ground shock. The air blast accompanying surface and near surface bursts also produces significant ground shock away from the detonation site (Glasstone and Dolan, 1977). In deep underground explosions, energy is converted directly into ground shock waves; air blast has little importance.

Underground explosions may induce aftershocks and displacements along faults originating near the detonation site (Glasstone and Dolan, 1977). However, analyses of seismic records following megaton-range deep underground nuclear explosions at the Nevada Test Site and at Amchitka Island in the Aleutians show no major anomalous earthquake activity. Another possible groundshock hazard is related to hillslope instability and landslides that might be triggered by nuclear detonations (Bennett et al., 1984). The effects of earthquakes and landslides are dependent on detailed geological conditions near the explosion sites, and must be evaluated individually.

1.3.3 Blast Damage

The atomic explosions at Hiroshima and Nagasaki starkly revealed the destructive power of nuclear blast (Figure 1.1). All structures are vulnerable (Glasstone and Dolan, 1977). Residential wood-frame houses (with wood or brick exteriors) suffer substantial damage at 2 psi peak overpressure, and are crushed at 5 psi. Glass windows are shattered at 0.5 to 1.0 psi. Concrete and steel buildings are broken apart at 10–15 psi (although the interiors and facades are destroyed at much lower overpressures). Aircraft, parked or in flight, are susceptible to significant damage at 1–3 psi. Splitting of liquid storage tanks occurs at 3 to 10 psi, depending on their size and fluid level (a tank is generally less vulnerable if it is larger and fuller).

Flying debris is a major cause of damage in a nuclear explosion. People are particularly vulnerable to flying objects. For example, while the human body can withstand substantial static overpressures (greater than 10 psi is required to produce severe injuries), serious wounds due to flying glass and rubble can occur at 1–2 psi.

Blast damage also leads to secondary fire ignition, as previously noted. From the nature of the blast damage, it follows that secondary fires can occur anywhere within the perimeter of the 2 psi zone.

1.3.4 Fireball Rise and Stabilization

The fireball of a nuclear detonation is essentially a hot buoyant bubble of air, and it begins to rise immediately after detonation. In a matter of seconds, the fireball of a 1-Mt burst attains a vertical velocity exceeding 100 m/sec. The rising sphere becomes unstable, deforming into a torus that later defines the mushroom cloud cap. The initial upward rush of the fireball creates a strong suction beneath it. At ground level in the vicinity of the explosion, the surface winds reverse direction within a second from outward, due to the blast wave, to inward, due to the fireball rise and suction. This reversal is called the negative pressure phase. The air drawn up behind the fireball forms the stem of the mushroom cloud and contains debris initially raised from the surface by the thermal pulse and blast wave.

The fireball stabilizes when its temperature and pressure become equal to those of the ambient atmosphere. Hence, fireball rise is influenced by the local atmospheric temperature structure and humidity. For explosions of less than 100 kt at mid-latitudes, the nuclear cloud stabilizes almost entirely within the troposphere (the well-mixed atmospheric layer extending from the surface to the tropopause at about 10–15 km altitude at mid-latitudes). For explosions of greater than 100–200 kt, the fireball penetrates into the stratosphere (the thermally stable atmospheric region extending from the tropopause to 50 km altitude). A 1-Mt explosion cloud would be expected to come to rest just within the stratosphere at mid-latitudes. For yields greater than 100–200 kt, the cloud stabilization height scales approximately as $Y^{0.2}$, where Y is the yield in megatons (Glasstone and Dolan, 1977). Because limited data are available from high-yield nuclear test explosions at middle latitudes, the cloud heights obtained by interpolating observations from low and high latitudes (and calibrating against limited hydrodynamic model calculations) can be uncertain by as much as several kilometers.

1.3.5 Nitrogen Oxide Production

Nitrogen oxides (NO_x) are produced when air, which consists primarily of N_2 and O_2, is heated above approximately 2000°C and then cooled rapidly. This can occur in two ways in a nuclear explosion: when air is compressed by the passing shock wave, and when air is entrained into the rising central fireball. On average, about 1×10^{32} NO_x molecules are generated for each megaton of explosive yield. The photochemical effects of this NO_x are discussed in Chapter 6.

1.3.6 Water Bursts

Nuclear explosions on water surfaces are similar phenomenologically to explosions on land surfaces. Here, water instead of soil is entrained by the fireball, and surface waves can be generated. Deep underwater explosions produce a shock wave with a greater peak overpressure and shorter duration than an equivalent shock wave in air at the same range. As in underground explosions, underwater bursts also create an airblast whenever the fireball breaks through the surface. In addition, wave trains carrying up to 5% of the original explosion energy can be generated. Waves with heights of ten meters or more can propagate away from the explosion site. These waves could cause significant destruction, particularly if they were to propagate into estuaries or harbors.

1.4 RADIOACTIVITY

1.4.1 Origins of Nuclear Radiation

In a nuclear detonation, several types of energetic ionizing radiation are produced:

1. Prompt (fast) neutrons which escape during fission and fusion reactions.
2. Prompt gamma rays created by fission/fusion processes, including neutron capture and inelastic scattering, and by early fission-product decay.
3. Delayed gamma and beta radiation from induced activity in materials bombarded by prompt neutrons.
4. Delayed gamma and beta radiation emitted through the decay of long-lived radionuclides (lifetimes greater than minutes) produced by nuclear fission and carried in the bomb residues.

At Hiroshima and Nagasaki, where the nuclear detonations were of relatively low yield, the prompt neutrons and gamma rays had important effects on survivors of the heat and blast within a few kilometers of ground zero. However, for greater yields, the prompt radiations still do not propagate beyond a few kilometers because of their strong attenuation over such pathlengths in air. Thus, with existing nuclear weapons, greater concern centers on the delayed nuclear radiation of fallout debris.

When a typical fission or fission-driven fusion weapon detonates, several hundred distinct radionuclides are generated (Glasstone and Dolan, 1977). These unstable species decay at different rates, emitting gamma rays and beta particles in the process. Inasmuch as gamma rays readily penetrate through both air and tissue, they pose a hazard even at a distance. On the other hand, beta particles are not nearly as penetrating, and thus pose a danger principally when the particle sources are close to living tissues (either externally or internally).

The fission radionuclides associated with fallout consist mainly of refractory elements that readily condense on particle surfaces as the fireball cools. Hence, any dust or debris entrained into the fireball is likely to be contaminated with radioactivity. The largest debris particles fall out quickly, while the smallest ones can remain aloft for months or years. The initial rapid deposition of the radioactive fission debris, or fallout, represents the most serious threat of delayed radiation. By contrast, gaseous radionuclides produced by fission and fusion (e.g., carbon-14 carried in carbon dioxide, and tritium carried in tritiated water vapor) and fission fuel residues (i.e., ^{235}U and ^{239}Pu) are less important, but not negligible.

The standard measure of exposure to radioactivity is the rad, equivalent to the absorption of 0.01 Joule of ionizing radiation per kilogram of material (Glasstone and Dolan, 1977). The rem is a biological dose unit equal to the absorbed energy in rads multiplied by a "relative biological effectiveness" factor for a specific type of radiation compared to gamma radiation. For gamma rays, X-rays, and beta particles, units of rads and rems are approximately equivalent. The term "whole-body" radiation is applied in cases where the entire organism is exposed to a (fairly) uniform external radiation field. For gamma rays, which are quite penetrating, all cells and organs are affected by exposure to whole-body radiation.

The impact of a radiation dose also depends on its rate of delivery. Roughly 450 rads delivered at the surface of the body within a few days time (an acute whole-body dose) would be lethal to half the exposed population of healthy adults; 200 rads would produce radiation sickness but would not by itself be lethal (Glasstone and Dolan, 1977). Such total exposures spread over a period of months or years (a chronic dose) would not cause acute effects, but would eventually contribute to a greater frequency of pathologies such as leukemia, other cancers, and birth defects. A far more comprehensive discussion of the biological effects of ionizing radiation is given in Volume II.

1.4.2 Radioactive Fallout

The Bravo nuclear test (15 Mt surface burst, Bikini Atoll, March 1, 1954) was the first to create serious fallout problems. Inhabitants of Rongelap

Atoll, which was downwind of the explosion, were inadvertently exposed to intense nuclear radiation (Figure 1.5). Even though they were evacuated soon after the event, the Marshall Islanders received substantial external and internal radiation doses—none lethal (Glasstone and Dolan, 1977). A Japanese fishing vessel, the Lucky Dragon, also found itself under the fallout plume. The fishermen, unaware that the white ash-like fallout was dangerous, took no special protective measures, As a result, one died of the exposure and a number of others received acute doses of several hundred rads.

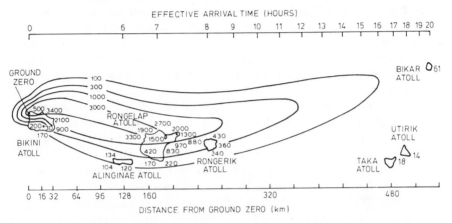

Figure 1.5. Contours of total (accumulated) fallout radiation dose in rads for continuous exposure up to 96 hours after the Bravo test explosion (15 Mt surface burst, March 1, 1954). One rad is equivalent to the absorption of 0.01 joule of radiation per kilogram of matter. The doses corresponding to longer exposure times (beyond 96 hours) would be greater than those shown (from Glasstone and Dolan, 1977)

The approximate pattern of radioactive fallout caused by the Bravo test, reconstructed from fallout measurements, is shown in Figure 1.5. The general pattern is typical of that expected for a surface detonation, although wide variations could occur in specific meteorological situations. The zone of potentially lethal radioactive fallout (for individuals continuously exposed to the fallout for up to 4 days) extended several hundred kilometers downwind of the Bravo test site, and covered an area of perhaps 5000 km². The doses illustrated in Figure 1.5 correspond to external exposure to whole-body gamma radiation. Shielding would have reduced the actual dose, although longer exposure times would have increased the dose. Additionally, there is a chronic internal radiation dose associated with the fallout due to ingestion and inhalation of radionuclides with food, water, and air. The most important of the ingested radionuclides are ^{131}I, ^{90}Sr, ^{137}Cs and ^{140}Ba (Glasstone and Dolan, 1977). Most of these elements tend to accumulate in specific internal organs (e.g., the thyroid for ^{131}I), which may thereby receive chronic doses exceeding the whole-body external dose (Lee and Strope, 1974).

Data obtained during the nuclear test series of the 1950s and 1960s (data such as that in Figure 1.5) have been used to construct standard fallout models for nuclear surface explosions (Glasstone and Dolan, 1977). These empirical radioactivity models account for the rate of fallout of contaminated debris, the decay in activity of the radionuclide mixture comprising the fallout, and the integrated exposure to the emitted gamma rays over time. For land surface bursts, approximately 40–60% of the fission products fall out the first day, constituting the early or local fallout. The approximate decay law for the radioactivity created by a typical nuclear weapon is $t^{-1.2}$. That is, the activity (and dose rate) of a fallout sample decreases by roughly a factor of 10 for every factor of 7 increase in time (e.g., between 1 hour and 7 hours, 7 hours and 49 hours, etc.). This decay law may be applied to give rough estimates over a period extending from 1 hour to 180 days after an explosion.

The settling of nuclear debris after the first day is not treated in standard fallout models. Indeed, beyond a few days, most of the residual radioactivity is deposited by precipitation. This delayed intermediate time scale and long-term radioactive fallout is borne on the smallest particles produced by a nuclear burst (carrying approximately 40–60% of the total radioactivity of a surface burst and approximately 100% of the radioactivity of an air burst). The delayed radioactivity disperses throughout the troposphere, where it may remain suspended for weeks, and the stratosphere, where it may remain for months to years (Glasstone and Dolan, 1977).

The potential radiation doses from intermediate time scale and long-term fallout are smaller than those from early local fallout because the contaminated debris is diluted over a wider area and the radioactivity decays significantly before it reaches the ground. In the long term, ^{90}Sr and ^{137}Cs are the primary sources of lingering radioactivity, with both radionuclides having a half-life of about 30 years. Because the threat of delayed contamination extends over years or decades, and the long-lived fallout may be concentrated in "hotspots" caused by precipitation and local deposition patterns, the incremental health effects of widespread fallout should not be ignored.

Detailed estimates of potential fallout areas and radiation doses in a major nuclear exchange are provided in Chapter 7. Further descriptions of the related phenomenology are also given there.

1.5 EFFECTS OF HIGH ALTITUDE NUCLEAR EXPLOSIONS

1.5.1 Electromagnetic Pulse

A nuclear explosion above an altitude of 40 km can expose a large area of the Earth to an intense pulse of electromagnetic radiation. The physical origin of the electromagnetic pulse (EMP) is illustrated in Figure 1.6.

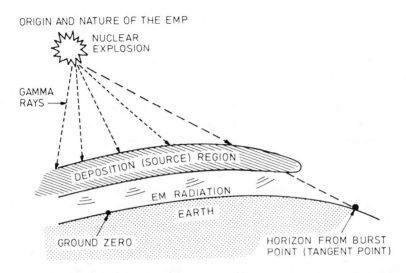

Figure 1.6. Schematic representation of the electromagnetic pulse (EMP) created by a high altitude nuclear explosion. The gamma rays emitted at the instant of detonation are absorbed in the stratosphere (the deposition region). The absorption process releases Compton electrons, whose intense currents are deflected by the geomagnetic field, creating powerful electromagnetic radiation fields (from Glasstone and Dolan, 1977)

The prompt gamma radiation from a burst above 40 km is absorbed in the Earth's atmosphere at heights of approximately 20 to 40 km. This deposition region for gamma rays is also the source region for EMP (Glasstone and Dolan, 1977). Through collisions with air molecules, the gamma rays produce high energy Compton electrons. The Compton electron currents interact with the Earth's magnetic field, thereby generating electromagnetic fields that propagate (toward the surface) as a coherent pulse of electromagnetic energy. Because the rates of gamma ray emission and deposition are so rapid, the electromagnetic pulse has an extremely short rise time (a few nanoseconds) and brief duration (a few hundred nanoseconds). The magnitude of the EMP is limited primarily by the enhanced electrical conductivity of the atmosphere caused by secondary electrons released in collisions of Compton electrons with air molecules. Nevertheless, EMP field intensities can reach several tens of kilovolts per meter over the exposed areas of the Earth. The electric field strength of the pulse can therefore be 10^9 to 10^{11} times greater than typical field strengths encountered in radio reception (Wik et al., 1985). The nuclear EMP frequency spectrum is also very broad and covers the entire radio frequency communication band.

Other forms of EMP include magnetohydrodynamic EMP (MHD-EMP), which can induce quasi-D.C. currents in very long conducting structures, and low altitude EMP, which generates very intense fields over distances of several kilometers. These are generally of lesser importance except in specific instances such as command, control and communication facilities that have been hardened against blast and thermal effects but might still be vulnerable to EMP.

Nuclear EMP induces currents in all metallic objects, which by accident or design act as antennas. Aerial and buried power and telecommunication networks in particular can collect considerable amounts of energy. Even short radio antennas and other electrical lines may experience unusual induced currents and voltages. The collected EMP energy could upset, breakdown, or burn out susceptible electrical and electronic components. Today many systems contain integrated circuits and other semiconductor devices that are subject to failure at very low energy surges (down to the order of a millionth of a joule for short pulses) (Wik et al., 1985).

In 1958 and 1962, high-altitude nuclear tests were carried out by the United States over the Pacific Ocean. During these events, some electrical and electronic systems suffered functional damage or operational upset, even hundreds of kilometers from the test sites (Glasstone and Dolan, 1977). No open reports exist on possible EMP effects during similar tests in the U.S.S.R. Lacking detailed observations, it is difficult to assess with a high degree of certainty the impacts that nuclear EMP might have on modern electronic hardware.

Apart from the difficulties inherent in designing accurate experiments of EMP effects over large spatial volumes, there are serious difficulties in applying theoretical models and calculations to real systems, which are exceedingly complex and undergo frequent modification. Most research on EMP is also classified and unavailable for analysis.

It is unlikely that EMP would incapacitate all of the exposed communication systems, power networks, and electronic equipment. However, a small number of failures distributed through a large and complex system can disrupt the entire system, or degrade its stability and performance. In this regard, power and communication networks are particularly susceptible. Moreover, the ability of nuclear power stations to withstand nuclear EMP effects safely is undetermined (Wik et al., 1985).

EMP could create confusion and isolation at precisely the time when critical decisions would have to be made regarding the use of nuclear weapons. Communications among diplomats, political leaders, and military commanders could be disrupted. EMP could also degrade sophisticated military command, control, communication and intelligence (C^3I) systems within minutes of the first detonations. Such effects could hinder a military response and/or might encourage looser control over nuclear weapons in the field.

Strategic nuclear C^3I systems are being "hardened" against EMP. At the same time, enhanced EMP weapons are being considered. Hence, the question of survivability of critical C^3I systems remains unresolved. Because telecommunications would play an important role in national and international crisis management, any major disruption of communication networks could affect the course of a nuclear conflict.

Space has a growing role in military planning for communication, navigation, and surveillance missions. Possible future deployments of space-based defensive systems against intercontinental ballistic missiles may imply an increasing potential for multiple explosions at high altitudes (tens to hundreds of kilometers) in a nuclear conflict. Hence, the importance of EMP interactions and other high-altitude effects of nuclear explosions may be increasing.

1.5.2 Radiowave Propagation and Satellite Systems

Nuclear explosions in space can disturb radiowave propagation in a number of frequency bands from tens of hertz (Hz) to tens of gigahertz (GHz). Shortwave radio signals can be degraded by power absorption, and microwave signals by phase scintillation. A high-altitude nuclear burst increases the background electron and ion densities and causes large- and small-scale structural modifications of the ionosphere (as well as longer-term chemical changes). Thus, communication, navigation, and intelligence systems may be affected intentionally or unintentionally by nuclear bursts in space. The radiowave propagation and absorption effects can lead to black-outs lasting for several hours in certain frequency bands, especially those used for long-distance high-frequency radio communications. The potential distortion of satellite signals traversing the ionized layers created by high-altitude nuclear bursts is still uncertain. Geosynchronous satellite communications near the horizon, grazing the ionosphere, at frequencies above 10 GHz might also be at risk (Wik et al., 1985). In addition, nuclear explosions would create interfering bursts of intense radio-frequency noise.

In outer space, communication satellites and other electronic systems could be exposed to direct nuclear radiation at considerable distances from a high-altitude burst. Penetrating radiations (gamma rays and X-rays) can interact with various materials to produce strong electromagnetic fields that may be incapacitating. This interaction is termed System Generated EMP (SGEMP). Space systems could also be affected by dispersed EMP (DEMP), which is associated with the propagated and reflected (dispersed) fields of the usual EMP generated by a high-altitude explosion.

Transient-radiation effects on electronics (TREE) are caused when the prompt gamma rays, neutrons, and X-radiation of a nuclear detonation interact directly with electronic parts. Transient, and sometimes permanent, changes can occur in the performance of semiconductor and optical components. For example, high-energy neutrons can displace atoms in a crystal lattice and create disabling defects. Hardening of satellites against TREE is difficult because shielding is limited by weight, and newer electronic circuits have often proven to be more vulnerable than older components (Wik et al., 1985).

The Earth's natural ionosphere and electron belts would be greatly perturbed by the widespread ionization and hydrodynamic motions associated with high-altitude nuclear explosions. Enhanced electron concentrations could be generated that might persist for months or years. Satellites operating within the enhanced ionization belts would suffer accelerated degradation due to intensified bombardment by energetic charged particles.

1.6 RESUMÉ OF NUCLEAR EFFECTS

In the previous sections, a general description was given of the most important physical effects of nuclear explosions. In this section, a quantitative summary of the spatial extent of these primary effects is provided as a function of weapon yield for air bursts and surface bursts. Estimates of the areas that would be subject to levels of thermal irradiation, blast overpressure, and radioactive fallout exceeding specific minimum values are given in Table 1.1. These estimates are approximate, and are presented only as rough indications of the potential impacts.

The sequence of physical effects that would accompany the detonation of a nuclear weapon is: thermal irradiation, blast, winds, radioactive fallout (particularly in the case of surface bursts), and fire growth and spread. In the explosion of a typical strategic nuclear warhead over a military or industrial target the effects of initial nuclear radiation (gamma rays and fast neutrons) and electromagnetic pulse, can generally be ignored, except in specific cases as already noted. The other nuclear effects occur in more-or-less distinct time intervals (over most of the area involved) (Glasstone and Dolan, 1977). The thermal pulse is delivered in the first 1–10 seconds. The blast is delayed by the travel time of the shock wave, and generally follows the thermal pulse; the positive duration of the blast wave lasts for approximately 1 second. Afterwinds then blow for several minutes. The most intense and lethal radioactive fallout occurs during the first hour after a surface detonation. Although many fires would initially be ignited in the ruins, it could take several hours for mass fires to develop. In the case of surface bursts, during the latter period, dense radioactive fallout would continue in areas downwind of the blast destruction zone.

From the data in Table 1.1, it can be seen that modern nuclear weapons (i.e., those having yields less than about 1 Mt) detonated as air bursts would create moderate to heavy blast damage over an area of approximately 500 km^2/Mt, and ignite fires over a similar area. In general, smaller weapons

TABLE 1.1.
AREAL IMPACTS OF NUCLEAR WEAPONS EFFECTS[a]

Yield (Mt)	Area (km^2) of thermal irradiation[b] 20 cal/cm^2	10 cal/cm^2	Area (km^2) of blast overpressure[c] 5 psi	2 psi	Area (km^2) of 450 rad fallout dose[d] 48 h	50 yr
0.1	35 (17)	65 (32)	34 (14)	100 (40)	(100)	(200)
0.3	105 (50)	190 (100)	70 (30)	200 (80)	(300)	(600)
0.5	160 (75)	290 (140)	100 (42)	300 (115)	(500)	(1000)
1.0	250 (120)	450 (220)	140 (65)	480 (180)	(1000)	(2000)
5.0	1150 (520)	2000 (950)	415 (190)	1410 (525)	(5000)	(10000)
10.0	2200 (1000)	3800 (1800)	660 (300)	2240 (835)	(10000)	(20000)

[a] Areas are given in square kilometers for airbursts and surface bursts (in parentheses); in the case of radioactive fallout, areas are given only for surface bursts (the early fallout from airbursts is negligible, and prompt and long-term radiation effects are ignored). Within the areas quoted, the magnitudes of the nuclear effects are greater than the limiting values shown above each column (e.g., 20 cal/cm^2); for thermal irradiation and blast overpressure, the limiting values apply at the perimeters of the circular contours centered on the explosion hypocenter which define the area of each effect. For example, the thermal irradiation (fluence) within the 20 cal/cm^2 contour is greater than 20 cal/cm^2, and can be much greater closer to the fireball. The data were obtained from Glasstone and Dolan (1977).

[b] A ground level visibility of 20 km is assumed. The thermal irradiance applies to a surface perpendicular to the fireball line-of-sight. In the outer 90 percent of the irradiated zone, more than 80% of the thermal energy is received before the arrival of the blast wave. The height of burst is chosen to maximize blast effects, as described in footnote c.

[c] For airbursts, the optimum explosion height has been chosen to maximize the area subject to the overpressure indicated.

[d] Areas are given only for surface bursts. No protection or shielding from fallout radiation is assumed. A fission yield fraction of 0.5 is adopted. A dose reduction factor of 0.7 is also applied for surface "roughness". The area in which an acute 48 hour whole-body dose of greater than 450 rad could be received is estimated from standard fallout patterns (Glasstone and Dolan, 1977). The area in which a long-term integrated total dose of more than 450 rad could result is also calculated from local fallout patterns. Cumulative global fallout is not included.

produce greater blast and thermal effects per unit energy yield than larger weapons. The area in which blast overpressures exceed a given value (e.g., 2 psi) scales approximately as $Y^{2/3}$, where Y is the yield in megatons (Glasstone and Dolan, 1977). The area affected by a specific minimum level of thermal fluence (e.g., 10 cal/cm^2) scales very roughly as $Y^{0.8}$ for yields between approximately 0.1 Mt and several megatons.

The areas of blast and thermal effects for surface bursts are about one-half the areas for airbursts of the same yield (Glasstone and Dolan, 1977). Surface bursts also create large local areas of potentially lethal radioactive fallout. Doses of up to 450 rad in 48 hours are possible over an area of approximately 1000 km^2/Mt in the fallout plumes. Lesser doses occur over much larger areas. The problem of accumulated radiation doses in overlapping fallout plumes is discussed in Chapter 7.

1.7 INTEGRATION OF EFFECTS

Previous sections of this chapter have focused on the effects of individual nuclear explosions, particularly the effects that might be relevant to an assessment of global physical and biological impacts. The effects of nuclear detonations are fairly well characterized by theoretical principles and by measurements taken during nuclear tests. The unique experiences at Hiroshima and Nagasaki have led to a general consciousness of the magnitude and power of nuclear weapons.

The damage areas summarized in Table 1.1 imply that approximately 6,000 Mt, which is less than the current world nuclear stockpile, could destroy, through direct effects alone, an area of up to about 3×10^6 km^2, assuming no overlap. This is equivalent to about 2–3 percent of the total land area of the Northern Hemisphere. Major urban zones occupy approximately 1 percent of the landmass (NRC, 1985), and thus would be directly vulnerable if attacked. Nevertheless, as will be discussed in later chapters and in Volume II, the survival of global civilization may be more dependent on the indirect effects (e.g., climate change) caused by nuclear explosions than on the direct effects (e.g., blast). The occurrence of indirect effects is obviously related to the occurrence of direct effects, which are manifested by the generation of smoke, dust, and radioactivity.

In order to integrate the individual weapons effects discussed here into a global model of the aftermath of a nuclear war, a number of additional pieces of information are needed:

1. A scenario for the nuclear exchange, including the weapon sizes, targets and heights-of-burst.
2. The physical state of the target zones, including adjacent combustible fuels, soil characteristics, and local meteorological conditions.

3. Descriptions of related physical phenomena, including fire growth and spread, smoke production and properties, microphysical evolution of smoke and dust aerosols, chemical responses, and so on.

In subsequent chapters, much of the required information is developed to the extent that is possible given current scientific knowledge, and to a depth that is consistent with the goals of this report. A detailed integration of these individual nuclear effects would require an enormous research effort and would be impractical at this time. Accordingly, the approach taken here is to consider, in each chapter, the essential global-scale consequences of specific effects of nuclear weapons; for example, fire damage in cities, climatic effects of smoke clouds, and contamination by radioactive fallout in a nuclear exchange. This approach emphasizes the plausibility of specific impacts, as well as the range of potential outcomes.

Environmental Consequences of Nuclear War Volume I:
Physical and Atmospheric Effects
A. B. Pittock, T. P. Ackerman, P. J. Crutzen,
M. C. MacCracken, C. S. Shapiro and R. P. Turco
© 1986 SCOPE. Published by John Wiley & Sons Ltd

CHAPTER 2
Scenarios for a Nuclear Exchange

2.1 INTRODUCTION

It is impossible to forecast the initiation and detailed conduct of a nuclear war. Despite this fact, all nations with nuclear weapons have elaborate plans for the deployment, targeting—and firing—of their warheads. Because of the reality of massive nuclear arsenals, and the unprecedented nature of nuclear conflict, many strategists believe that almost any use of nuclear weapons could escalate into global nuclear warfare (Ball, 1981; Bracken and Shubik, 1982). Others believe that a nuclear exchange could be controlled or limited; or if not controlled, that it might be automatically self-limiting, ending as soon as the combatants perceived their own imminent destruction, or with the fading of any rational military goals (Wohlstetter, 1983, 1985). This argument cannot be settled. Therefore, lacking solid evidence that nuclear warfare could be contained or limited in scale or magnitude, a prudent scientific approach demands that—in assessing potential long-term environmental effects—a possible and plausible nuclear exchange involving existing weapons and deployments must be considered.

2.2 WORLD ARSENALS

The actual nuclear weapons inventories of all nations are officially kept secret. Nevertheless, authoritative unclassified tabulations of existing and projected inventories are available, and these are roughly in agreement (The Military Balance, 1984; Jane's, 1984; SIPRI, 1984). Table 2.1 summarizes the principal nuclear weapons systems that have been deployed by the major nuclear alliances, or that may be deployed in the near future. Both strategic (intercontinental) and theater (intracontinental) nuclear forces are counted, but smaller tactical (battlefield) weapons and munitions—amounting to perhaps 25,000 explosives and several hundred megatons of aggregate yield—are omitted. (A typical tactical nuclear weapon has an explosive yield similar to that of the Hiroshima or Nagasaki bomb; viz, 10–20 kt.) In total, the strategic and theater nuclear arsenals hold some 24,000 warheads having nearly 12,000 Mt of explosive yield.

TABLE 2.1.
STRATEGIC/THEATER NUCLEAR WEAPONS IN
CURRENT INVENTORIES[a]

Warhead yield (Mt)	Type of system[b]	Number of warheads[c]	Aggregate yield[c] (Mt)
5.0	Bomber	280	1400
1.0	ICBM	1050	1050
1.0	SLBM	680	680
1.0	IRBM	293	293
1.0	Bomber	2520	2520
0.5	ICBM	5660	2830
0.5	SLBM	1200	600
0.5	IRBM	100	50
0.3	ICBM	1650	495
0.3	IRBM	108	32
0.2	SLBM	672	134
0.2	Cruise	1920	384
0.2	SRAM	1200	240
0.15	IRBM	1500	225
0.1	SLBM	2304	230
0.05	SLBM	3040	152
Total strategic/theater		24177	11315[d]
Tactical warheads		~25000	~300

[a] Compiled from the following reports: The Military Balance, 1984; Jane's, 1984; SIPRI, 1984.

[b] The abbreviations are: ICBM = intercontinental ballistic missile; SLBM = submarine launched ballistic missile; IRBM = intermediate range ballistic missile; Cruise = air, sea or ground launched cruise missile; SRAM = short range attack missile.

[c] These figures include the nuclear arsenals of the United States, Britain, France, and the Soviet Union.

[d] The Chinese nuclear forces are not included, as they are very uncertain at this time. Their weapons may include about 230 warheads on bombers and ICBMs with a total yield of about 500 Mt.

In studying global effects, precise information about weapons systems (warheads, launch vehicles, controls, deployments, and targets) is not really necessary as long as the general characteristics of the systems (as well as the broad strategic doctrines governing their use) are known. The weapons parameters in Table 2.1 are probably accurate to within 35 percent, and can be used as a reasonable basis for drawing implications about the use of nuclear forces.

The arsenals are constantly changing, and the present tabulation may already be outdated in some respects. Nevertheless, dramatic changes in the aggregate warhead count and yield are not expected through this decade, and likely the next, under existing development programs and treaty limitations. For example, the present Strategic Arms Limitation Treaties (SALT I and II) limit both the United States and the Soviet Union to 1200 land or sea based multiple warhead strategic missiles with 10 warheads or less per missile (except for submarine launched missiles, which can carry up to to 14 warheads). On the other hand, dramatic changes in nuclear armaments could occur if, for example, existing treaties were to lapse, if major new arms restrictions were negotiated, or if major breakthroughs in strategic defense systems triggered an offensive response.

A number of developments in nuclear warhead technology have been discussed recently (Arkin et al., 1984). The advanced concepts include "penetrators" (which can burrow into the ground before detonating, producing stronger ground shock and less fallout than surface bursts); terminal guidance and maneuvering (which allows precise targeting, and reduction in warhead yields); and enhanced radiation weapons (which generate greater neutron fluxes but require relatively less explosive energy than weapons designed specifically for blast effects). Nevertheless, despite such technological possibilities, all of the major nuclear weapons programs underway, or planned, utilize more-or-less standard nuclear fission-fusion devices exceeding 100 kt in yield. Typical examples are the MX warhead (approximately 300 kt) and Trident D-5 warhead (approximately 400 kt) of the U.S. forces, and the modern SS-18 warhead (approximately 500 kt) and the new SS-24 and SS-25 long-range missiles (probably carrying warheads in the 200–500 kt range) of the U.S.S.R. (Cochran et al., 1984; Arkin and Fieldhouse, 1985). While average strategic warhead yields had been decreasing steadily since the 1950s, that trend may now have halted.

Strategic defense systems currently under discussion have no impact on the present study. The feasibility of such systems has not yet been demonstrated, and deployment would be decades away. Moreover, it is not clear whether nuclear arsenals would decrease or increase in response to defensive deployments.

2.3 TARGETS

Nuclear weapons normally on station, and certainly those on alert in a crisis, have specific targets or missions assigned to them. Both superpowers have lists of potential targets, which probably number up to 40,000 or more (Ball, 1982). These lists are unavailable to us. Nonetheless, based on published discussions of strategic doctrine, most of the likely target categories, as well as the general targeting philosophy, can be deduced (Kemp, 1974; Katz, 1982; Ambio Advisers, 1982; Ball, 1983; Meyer, 1984; NRC, 1985; Arkin and Fieldhouse, 1985).

In most credible strategies, fixed strategic military installations garner the highest targeting priority; these include intercontinental ballistic missile silos and command centers, major airfields, nuclear submarine pens, weapons production and storage facilities, and command, control, communication and intelligence (C^3I, or C^3) centers. There are also a number of other important military targets including: mobile missiles and launchers; military formations of troops, artillery and armour; tactical weapons storage sites; support and tactical airbases; naval surface vessels and submarines at sea; other army, navy, and air force bases and logistic centers; and military satellite communication links. Targeting of warheads against military facilities—fixed and mobile—is referred to as "counterforce" targeting.

Some potential military targets might be thought of as civilian targets. For example, major airports with long runways, jet fuel supplies, and equipment that could be utilized by military forces, and industries that directly support a war effort, could be subject to nuclear attack. Among the most vulnerable industries are petroleum, oil and lubricants, electric power, steel, and chemicals (Katz, 1982). Transportation and communication nodes and principal storage sites might also be subject to destruction. These facilities represent some of the classical targets of warfare. It is also known that these facilities are included on the general nuclear targeting lists (Ball, 1982). Hence, it is possible that nations such as Japan, the Middle-Eastern countries, Australia, and South Africa might be targeted in a military campaign in order to deny their use as staging areas and forward bases, or their support through manufacturing and supply of raw materials.

Consumer-oriented production, commercial enterprises, and the infrastructure of society—concentrated in urban areas—comprise a distinct category of targets for nuclear weapons. Such "countervalue" targeting (and the implied civilian casualties) provides the basis for the deterrence doctrine of Mutually Assured Destruction (MAD). Although the publicly proclaimed strategic doctrines of the nuclear powers now place less emphasis on countervalue targeting, the direct bombing of population centers, as a final blow or as retaliation is the most fearsome potential application of nuclear weapons. Furthermore, many, if not most, major urban areas have targets within them, or nearby. Also, in the closing volleys of a major nuclear exchange, a broad range of countervalue targeting might be anticipated to cripple the ability of an enemy to recover and rebuild (and presumably, re-arm). For the same reason, targeting of noncombatants who might be perceived as a post-war threat could occur.

Some strategists believe that, in a nuclear exchange, cities would not be purposefully struck by nuclear warheads. However, cities could still suffer massive collateral damage in attacks on priority military and industrial targets. Collateral damage is the destruction caused in the area surrounding a target; with existing nuclear warheads, the zone of massive destruction (and intense radioactive fallout in the case of surface bursts) would extend far beyond the actual perimeters of most military and industrial targets (as noted in Chapter 1, typical strategic warheads are capable of devastating areas of 50 to 500 km^2). Most major cities in the U.S., U.S.S.R., and Europe have important military facilities in them or near them (bases, ports, airfields, C^{3}I facilities). Some recent assessments of potential urban collateral damage in a counterforce nuclear exchange suggest that hundreds of cities could be affected unless great restraint were exercised (e.g., NRC, 1985). Realistically, extensive urban devastation should be expected in any sizeable exchange of nuclear weapons, even perhaps in otherwise noncombatant nations such as Japan and Australia. It is well established that industrial capacity is highly correlated with population density in cities (Kemp, 1974; Katz, 1982). Hence, any attempt to cripple industrial capacity using existing nuclear weapons would cause enormous collateral physical damage and human casualties. For example, Katz (1982) estimates that approximately 300 Mt (carried by approximately 600 warheads) could destroy up to 60% of all industry in the United States and kill up to 40% of the U.S. population. Industry and population in the Soviet Union are nearly as vulnerable to nuclear attack as in the U.S. (Kemp, 1974).

Both long-range and short-range nuclear tipped missiles can suffer mechanical failure, damage, or deflection in flight. Accordingly, while the target point of a warhead can be precisely determined prior to battle, the eventual detonation point cannot. Warheads would fall at varying distances from the planned targets, some probably far off. This uncertainty in the reliability and accuracy of a strike force, together with the hardness of missile silos and the mobility of bombers and submarines, allows for the possible survival of the opposing forces. However, the uncertainty also introduces a dispersion, or randomness, into the application of nuclear force. Such randomness could increase the collateral damage in cities that are close to, but not coincident with, military targets or in forested areas adjacent to missile fields. Conversely, in the event that a missile were to go astray, the randomness might also reduce collateral damage since the warhead would be more likely to detonate over unpopulated areas than populated areas, given the much larger fractional area of the former. Factors of system reliability and accuracy (determined primarily by engineering constraints) are not explicitly defined in the scenario to be discussed. The additional destructive effects of wildly errant or deflected warheads are difficult to quantify, and will be ignored.

2.4 STRATEGIC CONCEPTS

A variety of strategies have been proposed for the use of nuclear forces. Tactical nuclear weapons (artillery shells, bombs, mines, and depth charges) could be used in the battlefield to blunt attacks, and at sea to stop ships and submarines. In space, nuclear detonations could be used to disable satellite systems with military missions. Theater nuclear weapons (on aircraft and missiles) could be used against rear echelon forces.

Strategic counterforce exchanges would involve deep missile and bomber strikes against opposing strategic forces and support facilities. Counterforce strategies also countenance strikes against key industrial elements, to blunt the capacity to sustain a war. By contrast, countervalue strategies, utilizing tactical, theater, and strategic weapons, are conceptually designed to maximize economic and civic destruction and to impede industrial and social recovery. A countervalue attack would be the ultimate cost levied in a nuclear war.

Other strategic concepts include limited nuclear warfare, flexible response, controlled escalation, launch under attack, and so on (e.g., Openshaw et al., 1983). However, since none has ever been used in actual conflict, the potential outcomes are highly uncertain. It should also be obvious that any nation suffering a massive nuclear strike might well retaliate by attacks on cities.

Much thought and concern have focused on the problem of escalation in a nuclear exchange. While some strategists argue that maintaining sufficient control of nuclear hostilities is a practical and logical goal (Wohlstetter, 1983, 1985), others question the possibility of effective nuclear battle management and argue that greater perceived control lowers the threshold for use (Ball, 1981; Carter, 1985). The official position of the Soviet Union on this matter is that controlled escalation or limited nuclear warfare is not possible (Military Encyclopedic Dictionary, 1983). We shall not pursue here the complex arguments in this debate, except to note that command and control operations in the environment of a nuclear exchange would be extraordinarily difficult and unprecedented.

A surprise nuclear attack without prior crisis or conflict is possible, but not considered very likely. Although one side might gain some military advantage in a massive first strike, the present structure of the superpower forces assures that the victim would retain a devastating retaliatory capacity. It seems more likely that a strategic nuclear exchange would follow from initial tactical or theater nuclear strikes. The doctrine of limited nuclear warfare, if adopted, might increase the possibility of initial nuclear use (or it might deter the aggression that presumably would trigger such use). Importantly, escalating nuclear conflict implies that all forces would be on alert; hence, the magnitude and speed of the eventual strategic exchange could be greatly enhanced.

While the possibility of nuclear detonations through accident or terrorism exists, it is thought that a global nuclear war caused by such events is unlikely (Wohlstetter, 1983). In normal times, and even in a crisis, there would be little reason or incentive for one side to respond immediately to isolated accidental or terrorist nuclear explosions with a nuclear counterstrike. Without strong supporting indications of a massive pre-emptive nuclear strike, massive retaliation would clearly be inappropriate. On the other hand, one cannot rule out the possibility that a series of unprecedented events and misperceptions could move the superpowers closer to the brink of a nuclear conflict, particularly during a period of confrontation or conventional warfare.

The concept of a massive pre-emptive strategic nuclear strike in a global crisis could be a real military option (Ford, 1985). Thus, if in a deepening crisis nuclear war seemed imminent, the side striking first might be expected to gain certain advantages in forces, targeting options, and C^3I operations, assuming, of course, that enough weapons of sufficient accuracy were available to destroy the key targets of the other side. (Note that this is a quite different situation from a surprise attack "out of the blue", in which an attacking nation, not under duress, risks its own destruction). The existence of pre-emptive strike options in nuclear war plans implies a fundamental potential instability in the deployment of large strategic forces; depending to some extent on the types of delivery systems, the more weapons each side has available, the greater the advantages that might accrue from a pre-emptive attack in a serious crisis. Pre-emptive nuclear strike options would seem to enhance the danger of escalation in any confrontation or conflict between the superpower alliances.

2.5 SCENARIOS

Possible scenarios for a global nuclear war are described in a number of documents (NAS, 1975; OTA, 1979; Ambio Advisors, 1982; Turco et al., 1983a; Knox, 1983; NRC, 1985). These scenarios are summarized in Table 2.2. For the most part, the scenarios are derived from analyses of nuclear weapons stockpiles, and assessments of nuclear doctrines and strategies (to the extent these are publicly available). Nevertheless, many of the scenarios have been criticized as representing extreme and unrealistic cases (Wohlstetter, 1985). The number of possibilities is obviously very large and the probability associated with any particular scenario is unknown. Hence, only the general structure of a nuclear scenario is considered here to determine if massive exchanges (amounting to thousands of megatons) could occur within the limits circumscribed by existing arsenals and deployments and the inevitable attrition of forces. Lesser exchanges would also be possible.

A hypothetical strategic nuclear exchange can be divided into four phases that might occur in an escalating conflict between NATO and Warsaw Pact forces (neglecting a possible initiating tactical phase): (1) an initial "counterforce" strike and response against key strategic military targets, with minimal

TABLE 2.2.
PUBLISHED NUCLEAR WAR SCENARIOS

Source	Description
National Academy of Sciences (1975)	10,000 Mt in 1500 detonations Warhead sizes: 1 to 10 Mt Targeting not specified
Office of Technology Assessment (1979)	7800 Mt in 8985 detonations Warhead sizes: 0.1 to 20 Mt Other parameters not described
Ambio (1982)	5742 Mt in 14,747 detonations (163 Mt on the Southern Hemisphere) Warhead sizes: 0.1 to 10 Mt 1941 Mt on cities, 701 Mt against industry
Turco et al. (1983a)	5000 Mt (baseline) in 10,400 detonations Warhead sizes: 0.1 to 10 Mt 2850 Mt in surface bursts, 1000 Mt in urban zones 3000 Mt (counterforce excursion) in 5433 detonations Warhead sizes: 0.3 to 5 Mt 1500 Mt in surface bursts, no detonations in urban zones 100 Mt (city excursion) in 1000 detonations Warhead size: 0.1 Mt 100 Mt in urban zones
Knox (1983)	5300 Mt in 6235 detonations Warhead sizes: 0.1 to 20 Mt 2500 Mt in surface bursts
National Research Council (1985)	6500 Mt in 25,000 detonations Warhead sizes: 0.05 to 1.5 Mt, plus tactical 1500 Mt in surface bursts, 1500 Mt in urban zones, 500 Mt tactical

direct destruction of population centers; (2) extended counterforce attacks against secondary military bases to disable support and logistics missions, which would necessarily involve some collateral damage to urbanized areas (Kemp, 1974); (3) massive strikes against the industrial base which supports military operations; and finally, (4) direct attacks against economic infrastructures to retaliate or retard postwar recovery. A strategic conflict could escalate within a matter of days from one phase to the next, although termination is possible at each phase, at least in theory.

The important characteristics of such a hypothetical escalating nuclear exchange are summarized in Table 2.3. Two general target categories are identified: military, and industrial/urban. Collateral damage to urban areas caused by strikes against military targets near or in cities has been counted. Detonation heights are divided into two regimes: airbursts (in which the fireball does not touch the ground, although the explosion occurs within several kilometers of the surface), and surface bursts (in which the fireball is in contact with land or water). Airbursts maximize the area of damage from blast and thermal radiation, but minimize contamination from early local radioactive fallout (although delayed global fallout may be enhanced). Surface bursts maximize the damage to nearby "hard" targets and reduce the overall area of thermal (fire) effects, but also contaminate large areas with lethal doses of radioactive fallout.

TABLE 2.3.
NUCLEAR EXCHANGE SCENARIO[a]

Phase of the exchange	Aggregate weapon yield (Mt)	Number of warheads	Military yield (Mt)		Industrial and/urban yield[b] (Mt)	
			Air	Surface[c]	Air	Surface[c]
Initial counterforce and response	2000	5000	1000	1000	0	0
Extended counterforce	2000	3800	750	750	250	250
Industrial	1000	1200	250	250	500	0
Final phase	1000	2600	250	250	500	0
Total[d]	6000	12600	2250	2250	1250	250

[a] Tactical weapons are not included. These could add 100–500 Mt in the less than 50 kt yield range. The warhead yields and numbers are taken from Table 2.1. It is assumed that the weapons have a fission yield fraction of 0.5.

[b] Includes weapons directed at industrial and economic targets as well as weapons directed at military targets that would generate significant urban collateral damage.

[c] Land surface.

[d] Cumulative targets include:

  2500 missile silos and command centers (2 warheads per silo)
  1100 military facilities and airfields throughout NATO and the Warsaw
          Pact (2 warheads per target)
   100 naval targets
   500 mobile missiles (barraged by 1200 warheads)
  1100 miscellaneous military detonations
  3000 military/industrial and energy resource sites worldwide.

The illustrative scenario in Table 2.3 was constructed using the following general guidelines (details will not be given here):

1. The weapons employed reflect the data in Table 2.1.
2. The targets within the NATO and Warsaw Pact countries include (at different phases of conflict):
   - Fixed and mobile strategic and theater missiles
   - Strategic airfields and submarine bases
   - Other military (air force, army, navy) bases
   - Military units in the field and vessels at sea
   - Logistics and communications centers
   - Military satellites
   - Nuclear weapons production and storage sites
   - Civil airfields having potential military utility
   - Fossil fuel and nuclear energy facilities
   - Cities with key industrial and/or economic functions.
3. The conflict develops over time (approximately days to weeks) from confrontation to crisis to conventional hostilities to tactical nuclear strikes, so that all major military forces are on alert and can respond in short order; a precipitous strike without warning is not considered.
4. Forces are assumed to be destroyed during the early phases of conflict (and thus do not deliver their weapons) as follows (roughly half of the nuclear weapons are depleted in this manner):
   - 90% of unfired ICBMs
   - 1/3 of strategic bombers
   - 1/3 of nuclear submarines with unfired missiles
   - 1/3 of reserve mobile missiles
   - 1/3 of reserve tactical bombers
5. Damage to industrial/urban areas, either through direct or collateral effects (neglecting radioactive fallout) is caused by a fraction of the explosive power during each phase of conflict as follows:

Initial counterforce:	0%
Extended counterforce:	25%
Industrial phase:	50%
Countervalue phase:	50%
Average over all phases:	25%

   The average over all phases is determined by multiplying the fraction for each phase by the number of megatons detonated in that phase, adding, and then dividing by the total megatonnage.
6. Each side would retain only a relatively small reserve force, consisting mainly of spare missiles and warheads.

Although each of these assumptions can be argued, the overall scenario appears to be consistent with the technical facts and strategic concepts reviewed earlier. For example, a cursory analysis of the present balance of nuclear forces indicates that massive nuclear exchanges are clearly possible (even if, as some believe, not likely). Moreover, it is plausible that many urban and industrial zones would be destroyed by volleys of nuclear weapons, given the collocation of military and industrial targets with population centers. Although each of the conflict levels in Table 2.3 is enormous with respect to the destructive power of previous wars, it is not so with respect to the actual destructive potential of the current and projected nuclear arsenals (Arkin et al., 1984). While one could propose smaller nuclear exchanges, or perhaps isolated tactical phases, it is appropriate to remain skeptical of controlled or limited nuclear warfare. It should also be noted that the scenario in Table 2.3 is not the worst possible case since less than one-half of the existing arsenals are assumed to be detonated; scenarios may be envisioned in which larger fractions of the arsenals could be detonated.

2.6 IMPLICATIONS

The full-exchange scenario in Table 2.3, which was assembled through an analysis of nuclear forces, target categories, and stated strategies, is similar to the scenario developed by the U.S. National Research Council panel (NRC, 1985) through different lines of reasoning. While this agreement does not validate either scenario, it reinforces the credibility (if not the probability) that such an outcome is possible in a nuclear conflict between the superpowers.

The critical parameters in Table 2.3 (from the perspective of an assessment of global effects) are:

1. The total yield in surface bursts with yields greater than 100 kt; these explosions lift dust into the upper troposphere and lower stratosphere, and produce large plumes of radioactive fallout. The cumulative yield of these dust-raising and fallout-generating bursts ranges from 1000 to 2500 Mt, depending on the phase of the exchange (virtually all ground-burst strategic warheads are included in this category). In the case of local fallout, it is also important to know the fission yield fractions of the weapons, and the proximity of population to the fallout plumes.
2. The total yield detonated in industrial/urban zones; these explosions ignite fires in the highly combustible and soot-generating materials accumulated in urbanized areas, including fuel storage sites. The fire-ignition yield varies from 0 to 1500 Mt, and is associated primarily with air bursts.

The tactical component of an exchange is not included in these figures, but is important in its own right, particularly in densely populated and industrialized areas such as Europe. Tactical explosions, perhaps numbering in the thousands, could produce extensive fires and radioactive fallout, and might represent the trigger for a strategic exchange. Hiroshima and Nagasaki provide examples of the potential destructiveness of modern tactical weapons (see Chapter 1).

Of all the explosions in a nuclear war, only relatively few might be detonated in the upper atmosphere (to create electromagnetic pulse, EMP) or on ocean surfaces (to destroy ships). The high altitude explosions could have an importance exceeding their relative number if they were to encourage nuclear escalation through the disruption of communications and control networks. Future ballistic missile (and other) defensive systems might one day lead to a military posture in which bursts above the atmosphere were predominant. For the present study, however, it is reasonable to assume that the fraction of such bursts is fairly small.

Turco et al. (1983a) have suggested that the number of nuclear explosions required to create severe climatic disturbances (their "nuclear winter") may be relatively small. They suggest that on the order of one thousand 100-kt detonations over major industrial and urban centers might be sufficient, because the greatest fuel densities are contained in a rather small number of urban industrial complexes and fossil fuel storage sites. According to Chapter 3 and the NRC (1985) report, about 10 percent of the total urban area may hold 50 percent of the total urban combustible material. These areas also might be subject to collateral damage in attacks against critical strategic targets, even if the cities are not bombed purposefully. Because of this concentration of fuel in a relatively few target regions, it is possible that more restricted scenarios (phases) than those described in Table 2.3 could lead to major environmental impacts, depending much more, however, on the details of the targeting and the uncertainties in the physical outcomes. Likewise, the significance of tactical explosions could be amplified to the extent that cities and industries, and fire-susceptible natural environments, were subject to collateral effects.

This general discussion of scenarios for nuclear warfare is only meant to provide information and guidelines, and to establish plausibility. The scenario described in Table 2.3 is not explicitly used, except in its most general aspects, in the following analyses of fires, smoke, dust, and climatic responses. Indeed, most of the climatic impact studies surveyed in the following chapters are not predicated on any particular targeting scenario, but rather on a particular initial amount of smoke and/or dust injected into the atmosphere. In the case of radioactive fallout, the scenario just described is used in Chapter 7 to provide an example of potential nuclear radiation effects.

There is no objective way to attach a probability to any particular scenario describing a nuclear war. For the present purposes, it is sufficient to determine whether a massive nuclear exchange is credible or not credible. Although the concept of nuclear warfare involving the use of many nuclear weapons seems incredible and even irrational, the weapons for conducting such a war have been deployed and elaborate plans of action exist. It is unacceptable simply to dismiss the potential for global nuclear conflict on philosophical grounds. The deployment of nuclear warheads implies, in a very real sense, the possibility of their use.

Environmental Consequences of Nuclear War Volume I:
Physical and Atmospheric Effects
A. B. Pittock, T. P. Ackerman, P. J. Crutzen,
M. C. MacCracken, C. S. Shapiro and R. P. Turco
© 1986 SCOPE. Published by John Wiley & Sons Ltd

CHAPTER 3

Sources and Properties of Smoke and Dust

3.1 INTRODUCTION

Recent studies have emphasized the large changes in atmospheric optical, meteorological, and chemical conditions that can result from fires started during a nuclear war (Crutzen and Birks, 1982; Turco et al., 1983a,b; Crutzen, Galbally and Brühl, hereafter CGB, 1984; NRC, 1985). In particular, attention has been focused on the effects of smoke formed in flaming combustion. This smoke has high concentrations of amorphous elemental carbon which strongly absorbs solar radiation. In this chapter, estimates are made of the quantities of smoke that might be produced by both urban/industrial and wildland fires started by a nuclear war such as described in Chapter 2. The microphysical and optical properties of the particles which comprise the smoke are also discussed and estimates of the attenuation of sunlight by the smoke are made. A brief discussion of dust raised by surface bursts is also included. Finally, the related, but special, issue of urban fire spread and urban fire modelling is included in an Appendix.

3.2 THE ORIGIN OF SMOKE IN COMBUSTION

Smoke is formed from the burning of organic materials. The burning process can be conveniently differentiated into two phases: high-temperature, flaming combustion during which sooty smoke is formed, and low-temperature, smoldering combustion during which primarily hydrocarbons are formed. The sooty smoke is of greater significance for the atmospheric radiation balance because such smoke absorbs sunlight very efficiently. Smoldering combustion, on the other hand, produces an aerosol that predominantly scatters rather than absorbs sunlight.

In flames, temperatures are sufficiently high that organic molecules can lose a large fraction of their hydrogen atoms by pyrolysis, leading to radical and ionic molecules with high C/H ratios. The free hydrogen which is created may be oxidized or may escape to the atmosphere. These soot precursors condense into hexagonal crystallites with dimensions on the order of 1 to 5 nm, which then grow further into embryonic amorphous elemental carbon spheroids with radii of the order of a few tens to a hundred nanometers and an average C-to-H atomic ratio of about 10 to 1. If these soot spheroids escape the flame, they cannot be oxidized further since oxidation requires temperatures above about 1500 K (Gaydon and Wolfhard, 1970; Wagner, 1980). Outside the flames, these carbon spheroids can aggregate into chain-like structures, sometimes consisting only of a few spheroids, sometimes forming fluffy agglomerates up to 100 μm in size (Russell, 1979; Day et al., 1979; Bigg, 1985). The surface to mass ratio of the agglomerates, however, remains relatively constant, as they grow in size.

Since in actual fire situations combustion is often not complete, the soot and smoke particles that are emitted into the atmosphere contain a substantial fraction of unburned organic matter. Soot is, therefore, a complex mixture mainly consisting of amorphous elemental carbon and oily material. Generally, smokes with higher elemental carbon contents appear blacker.

The smoke yield (defined as the mass of smoke produced per mass of material burned) and the elemental carbon fraction of the smoke are strongly dependent on the nature of the fuel and the mode of burning. Partly oxidized wood (e.g., forest materials and construction wood) produces much less elemental carbon than fossil fuel and fossil fuel-derived products. Smoldering combustion produces virtually no elemental carbon because temperatures are too low for hydrocarbon dehydrogenation to occur. Estimates of smoke yields and elemental carbon fractions for various combustible materials are given in the following sections.

3.3 SMOKE EMISSIONS FROM URBAN/INDUSTRIAL FIRES

The difficulties and uncertainties in estimating the smoke emissions from a single, large urban fire are very great. The first problem is to estimate the quantity of combustible material available and the fraction of that material that will burn. The latter factor is a complicated function of the fuel loading and type, the behavior of mass fires in urban areas, and the meteorological conditions which exist at the time of burning. Furthermore, quantitative information on smoke and amorphous elemental carbon production from fires in various materials is still limited and mainly available from test fires that were conducted under controlled laboratory conditions with small quantities of combustible material. Consequently, there is a question about the applicability of these data to mass fire conditions. The smoke emission problem is compounded when the effects of a hypothetical nuclear war are considered, due to the large number of possible urban and industrial targets. Some of the problems associated with selecting a scenario for a nuclear war, including the extent to which urban and industrial complexes and fossil fuel storage facilities are targeted, were discussed in Chapter 2. Estimating the smoke emission from fires ignited in the vicinity of these various targets is made more difficult by the variability in fuel loadings and types and in the prevailing meteorological conditions.

Quantitative information from mass fires is largely lacking despite several examples of large city fires during the Second World War, including those following the nuclear attacks on Hiroshima and Nagasaki. Studies of the destruction of Nagasaki and Hiroshima show that essentially complete burnout occurred wherever the thermal energy fluence from the nuclear explosions exceeded 20 and 7 cal/cm^2 respectively. No quantitative information is available on the smoke production caused by these fires.

Despite the substantial difficulties and uncertainties involved, three studies have attempted to make some estimates of smoke production from fires in a nuclear war (Turco et al., 1983a,b; NRC, 1985; CGB, 1984). The results from the studies by Turco et al. (1983a) and the NRC (1985) are very similar. Therefore, the following discussion will concentrate on a comparison of the NRC and CGB studies.

3.3.1 Urban Areas and Combustible Burdens

In both the NRC and CGB studies, it was assumed that a nuclear heat pulse of at least 20 cal/cm^2 is required for mass fires to occur in cities. This may be a conservative assumption, as only 7 cal/cm^2 reached the perimeter of the burnout area in Hiroshima (Glasstone and Dolan, 1977). In fact, the NRC report mentions that, even in Nagasaki, in directions unobscured by hills, total burnout occurred at all sites where the heat pulse exceeded about 10 cal/cm^2 (see Chapter 1 of this Volume for further discussion). The potential fire area per megaton of explosive yield, roughly corresponding to the 20 cal/cm^2 irradiation zone, were taken to be 250 km^2/Mt by Turco et al. (1983a) and by the NRC and 375 km^2/Mt by CGB. These area per Mt estimates were used both for urban and wildland fires.The values differ somewhat because of differences in assumptions about atmospheric visibility, overlap of fire zones, and fire spread. In many urban fire situations, firespread would be expected to contribute significantly to the total fire area (see Appendix 3A), thus increasing the area burned per Mt. Both the area estimates given above appear to be conservative for isolated bursts if a burnout criterion of 10 cal/cm^2 is adopted. This latter value of the fluence corresponds roughly to that experienced at Hiroshima and Nagasaki, for which equivalent fire areas of 300 to 1200 km^2/Mt occurred (see Chapter 1).

A weapon yield of 0.4 Mt corresponds approximately to the average yield of nuclear weapons that might be used in attacks on targets located near or in cities (see Table 2.1, Chapter 2). In the NRC study it was assumed that a total yield of 1500 Mt of nuclear weapons, out of a total of 6500 Mt, might be used against military command and industrial targets that are co-located near or in large cities in the Warsaw Pact and NATO countries. Taking into account a factor of 1.5 overlap of potential fire areas, a total urban area of 0.25×10^6 km^2 was assumed to burn, which corresponds to about half of the area of the 1000 cities with more than 100,000 inhabitants in both combatant blocks (Turco et al., 1983b; NRC, 1985). In the CGB scenario, 300 of the most important urban industrial centers of the NATO and Warsaw Pact nations were assumed to be targeted with about 800 Mt of nuclear weapons, each having an average yield of 0.4 Mt. Allowing for a factor of three overlap between potential fire areas, the corresponding total urban area in the CGB study is likewise about 0.25×10^6 km^2. It is estimated that the number of potential human casualties in the destroyed cities would be about 250 million, which is 30% of the total urban population in the combatant nations (UN, 1980b).

TABLE 3.1.
POPULATION AND NUMBER OF CITIES IN THE DEVELOPED
WORLD IN GIVEN SIZE CLASSES (UN, 1980B)

Size class (millions)	Number of cities	Total population (millions)
>4	16	142
2–3.9	27	73
1–1.9	74	99
Sum	117	314
Total urban		834

The urban population of the industrialized nations is between 65 and 75% of the total population of these nations (UN, 1980b). Statistics on the populations of the largest cities in the developed world are given in Table 3.1. These statistics indicate that 40% of the total urban population live in the largest 120 industrial and commercial centers of the combatant nations. The total destruction of all of these urban areas, therefore, implies the potential burning of at least 40% of all processed combustible materials in the NATO and Warsaw Pact nations. Katz (1982) estimated that attacks with 600 warheads carrying a total of 300 Mt of nuclear weapons could destroy up to 60% of all U.S. industry and 40% of its population. An analysis of the U.S.S.R. population distribution and industrial capacity (Kemp, 1974) showed that the largest 50 cities contained 33% of the urban population and 40% of the industrial capacity.

According to the 1980 U.S. Census, about half of the total U.S. population lives in the 62 largest metropolitan centers. Of these urban dwellers, about

50% live in about 10% of the total developed areas. This implies that urban combustibles are concentrated in relatively small areas. Hence, the total built up area that might be subject to direct or collateral damage in a nuclear war is of great importance with regard to the quantification of the combustible material which might burn. In some cases, the cores of the central cities provide sufficient fuel loading to support "firestorm" conditions that might lift smoke to the upper troposphere and lower stratosphere (see Appendix 3A and Chapter 4). As pointed out by Turco et al. (1983a), the complete burning of a hundred large population and commercial centers could, within the range of current uncertainties, produce about as much smoke and soot as that estimated in the baseline cases which were adopted by Turco et al. (1983a), NRC (1985) and CGB (1984).

An important element of these studies is the need to estimate explicitly the amounts of combustible material in the urban and industrial centers. Here the NRC and CGB studies followed different approaches. Using the limited information available from surveys of U.S. cities by FEMA (1982) and statistics on the world production of combustible materials, the NRC study adopted an average combustible material loading of 40 kg/m^2 in urban areas, representing a weighted mean between heavy loading of the order of several hundred kg/m^2 in the cores of cities and a much lighter loading down to 5 kg/m^2 in the suburbs. This leads to an estimated total of ten thousand million tonne (10^{16} g) of material that could be consumed in fires. Three quarters of this, or 7.5 thousand million tonne (7.5×10^{15} g), was assumed to burn (NRC, 1985). This quantity of combustibles was assumed to consist of 5 thousand million tonne (5×10^{15} g) wood, 1.5 thousand million tonne (1.5×10^{15} g) liquid fossil fuels, and one thousand million tonne (10^{15} g) of industrial organochemicals, plastics, polymers, rubber, resins, etc.

The CGB estimates were made differently. Material production statistics in the developed world were assembled from the United Nations and other sources, as reproduced in Table 3.2. Based on these data and the assumed average lifetimes of wood in constructions and furnishings, CGB estimated that the urban centers that would be targeted could contain about 4000 million tonne (4×10^{15} g) of cellulosic materials in constructions, furnishings, plywood, books, etc. This amount is about equal to 25% of all available cellulosic materials in the developed world, and about 35% of all such materials in the urban centers of these countries. It was assumed that half of this, i.e., 2000 million tonne (2×10^{15} g), would burn in flaming combustion (see Appendix 3A). The remaining material would not burn, or would smolder over days to weeks (see also Chandler et al., 1963). Since our primary concern here is the sooty smoke that is produced by flaming combustion, only the 2×10^{15} g of cellulosic materials that would burn quickly will be considered. In Chapter 6, the potential chemical effects of the smoke produced by smoldering fires will be discussed.

TABLE 3.2
ANNUAL PRODUCTION OF VARIOUS COMBUSTIBLE MATERIALS
AND ESTIMATED ACCUMULATED QUANTITIES IN DEVELOPED
WORLD (SEE TEXT, AND UN, 1980B; 1981; FAO, 1976; WORLD BANK,
1978). FOR FOODSTUFFS, ONLY STORAGE IN URBAN AREAS WAS
CONSIDERED (SEE VOLUME II)

Material	Production (g/y)	Accumulation (g)
Liquid fuels	3.1×10^{15}	$1.1–1.5 \times 10^{15}$
Coal, lignite	3.5×10^{15}	$\sim 10^{15}$
Natural gas and liquids	8.9×10^{14}	1.5×10^{14}
Sawnwood, panels, etc.	3.4×10^{14}	1.4×10^{16}
Pulp, paper, paperboard	9×10^{14}	$\sim 10^{15}$
Bitumen, total	(7×10^{13})	$(1–1.5 \times 10^{15}$ g$)$
roof protection	10^{13}	$\sim 2 \times 10^{14}$
city roads	3×10^{13}	6×10^{14}
Organic polymers	(7×10^{13})	(4.6×10^{14})
plastics	4×10^{13}	2×10^{14}
resins and paint	1.2×10^{13}	1.2×10^{14}
fibers	1.4×10^{13}	1.4×10^{14}
Cotton	10^{13}	10^{14}
Fats and oils	7×10^{13}	2×10^{13}
Cereals	3×10^{14}	$0.5–2 \times 10^{14}$
Sugar	5×10^{13}	2×10^{13}

CGB included fossil fuel and fossil fuel-derived products as a separate category of combustibles and showed that combustion of these materials most likely would have the gravest potential optical effects. According to information supplied by the Oil Market Division of the OECD in Paris, the total oil stocks at the primary level in ports and refineries in OECD countries in 1984 were equal to 420 million tonne (4.2×10^{14} g), including 70 million tonne (7×10^{13} g) of strategic stocks in the U.S., West Germany, and Japan. Underground storage of oil amounts only to 60 million tonne (6×10^{13} g), mainly in the U.S. and West Germany. At the distributor and user level, additional stocks of petroleum and petroleum-derived fuels vary between 40% and 100% of those at the primary level. The quantity of oil at sea is about 100 million tonne (10^{14} g). The OECD countries represent about 60% of the total world oil trade. From this information, CGB estimated, therefore, that the total amount of oil that is currently stored in the developed world is in the range of 1.1–1.5 thousand million tonne ($1.1–1.5 \times 10^{15}$ g). About one thousand million tonne (10^{15} g) might be readily available for burning following attacks on targets co-located with urban areas. To this must be added similarly large quantities of coal stockpiled at mines, power stations, and elsewhere, although the ignition and free burning of coal is probably less significant.

About 70 million tonne (7×10^{13} g) of the annual crude oil production goes into asphalt production (UN, 1981). An analysis of the bitumen production (UN, 1981) indicates that 1.0–1.5 thousand million tonne (1.0–1.5×10^{15} g) bitumen has accumulated in the developed world. The flaming point of bitumen is between 210 and 290°C (Güsfeldt, 1974), so that this material can readily burn in intense fires. In West Germany almost 30% of the bitumen is now used in the building construction industry, including 15% for roof insulation purposes (Arbeitsgemeinschaft der Bitumen-Industrie, personal communication). In the U.S., the proportion is about 19% (B. Williamson, personal communication). The rest of the bitumen has been applied on roads and highways. The urban fraction of this in West Germany is about 60% (Arbeitsgemeinschaft der Bitumen-Industrie, personal communication). This implies that together about 700 million tonne (7×10^{14} g) of bitumen might be available for burning in the urban areas of the developed world. Of this quantity, the nearly 200 million tonne (2×10^{14} g) of bitumen on roofs would be particularly easy to burn.

The world production of petroleum-derived organic polymers in 1980 amounted to about 60 million tonne (6×10^{13} g) (UN, 1981; Weissermel and Müller, 1981). Extrapolating from West German conditions, the synthesis of polymers from natural gas as the feedstock adds another 10 million tonne (10^{13} g) of organic polymers to the 1980 production figure (Hofmann and Krauch, 1982). The total of 70 million tonne (7×10^{13} g) went mainly into the production of 40 million tonne (4×10^{13} g) of plastics, 12 million tonne (1.2×10^{13} g) of synthetic resins and paints, and 14 million tonne (1.4×10^{13} g) of synthetic fibers (UN, 1981; Hoechst A.G., private communication). Assuming average lifetimes of these products of 5, 10, and 10 years, respectively, the amount of stored organic polymers in the developed world could be equal to about 400 million tonne (4×10^{14} g), indicating a ratio of about 1 to 10 between synthetic organic polymers and wood products. According to information from Verband der Sachversicherer (personal communication) the ratio of these products in West Germany is in the range of 1 to 10 or 20 and growing. There are also many additional, individually smaller quantities of materials, such as foodstuffs (W. Cropper, personal communication; see also Volume II of this report) and rubber, that are stored in the developed world. These are summarized in Table 3.2. The total may add up to a hundred million tonne (10^{14} g).

According to the above analysis, in the developed world a total of about 140 million tonne (1.4×10^{14} g) mainly bitumen and synthetic organic polymers, go each year into long-lived products. This compares well with the estimated annual production of 170 million tonne (1.7×10^{14} g) of slowly oxidizing fossil fuel products of Marland and Rotty (1983).

CGB assumed in their calculations that about 700 million tonne (7×10^{14} g) of fossil fuel and fossil fuel derived products might burn in a nuclear war. This is about 25–30% of the materials readily available for combustion in this category of fuels, excluding coal. The actual amount would, of course, depend on the adopted nuclear war scenario. Nevertheless, because fossil fuel processing facilities and storage depots themselves are likely targets in the case of a major nuclear confrontation (OTA, 1979; see also Chapter 2), the assumption that 700 million tonne (7×10^{14} g) of fossil fuel and fossil fuel derived materials might burn in a major nuclear war seems entirely plausible and serves as a reasonable working hypothesis. Because about half of the available fuel in this category has accumulated in densely populated Europe, a nuclear war limited to this region alone could lead to the hypothesized fuel combustion if 50% of the available fuel were to burn.

The information contained in this section may be summarized as follows. NRC (1985) estimated that altogether 7.5 thousand million tonne of combustibles could burn as a result of a nuclear war. Of this amount, two-thirds were assumed to be wood and wood products, and one-third to be fossil fuel and fossil fuel products. CGB (1984) assumed that 25–30% of all available combustible materials in the developed world would burn, leading to the flaming combustion of two thousand million tonne of wood and wood products, and 700 million tonne of fossil fuel and fossil fuel derived products. From the information given in this section, the burning of these quantities of material could be achieved by nuclear attacks that burned essentially all of less than one hundred of the most important industrial and commercial centers of the developed world.

3.3.2 Smoke Emission Factors

In the NRC study, the average smoke emission factor for all fires was set at about 4% (0.04 g smoke emitted per g fuel), which is the weighted mean of two-thirds cellulosic materials with an emission factor of 3% and one-third liquid fuels and synthetic organics with an emission factor of 6%. The average elemental carbon content of the smoke was taken to be about 20%, which was considered to be a conservative assumption. As a baseline, the NRC study derived total smoke and elemental carbon emissions of about 300 million tonne (3×10^{14} g) and 60 million tonne (6×10^{13} g), respectively. It was assumed that 50% of the smoke would be promptly removed by precipitation in the fire plume, implying net emission values of 150 million tonne (1.5×10^{14} g) for smoke and 30 million tonne (3×10^{13} g) for elemental carbon.

In the study of CGB, smoke emission factors based on the data in Tables 3.3–3.5, were taken to be 1.5% for construction wood, 7% for fossil fuel and asphalt, and 5% for plastics. The corresponding elemental carbon contents were 33, 70, and 80%, respectively. The total estimated smoke emission from urban fires was then 80 million tonne (8×10^{13} g), of which 45 million tonne (4.5×10^{13} g) was elemental carbon. The largest single contribution to the elemental carbon total is from fossil fuel burning. CGB assumed that one-third of the smoke would immediately rain out in the convective fire columns, leading to the net injection into the background atmosphere of 53 million tonne (5.3×10^{13} g) of smoke, containing 30 million tonne (3×10^{13} g) of elemental carbon.

TABLE 3.3.
LITERATURE SURVEY ON CHARACTERISTICS OF AEROSOL
PRODUCED BY BURNING OF WOOD (FROM CRUTZEN, ET AL. 1984).
REPRODUCED BY PERMISSION OF D. REIDEL PUBL. COMPANY

Ref. No.	Type	Aerosol yield	Elemental C	Extinction m^2/g fuel
1	Fireplace, softwood	9 g/kg	33% of aerosol	
	Fireplace, hardwood	10 g/kg	8% of aerosol	
2	Residential wood		13% of aerosol	
3	Test fires		50% soot	
	free burning			0.023
	ventilation			0.15
	controlled			
4	Test fires			
	hardwood	0.085–0.16%		
	fiberboard	0.75%		
5	Test fires	1.0–2.5% (flaming) 3.1–16.5% (nonflaming)		
6	Test fires	1.5%	40% of aerosol	0.11
7	Test fires	0.2–0.6%		
"Average"		1.5%	33%	(0.10)

References:

1. Muhlbaier-Dasch, 1982;
2. DeCesar and Cooper, 1983;
3. Rasbash and Pratt, 1979, and private communication D.J. Rasbash;
4. Hilado and Machado, 1978;
5. Bankston et al., 1981;
6. Tewarson, private communication;
7. Seader and Einhorn, 1976.

The estimated emissions of amorphous elemental carbon in the baseline NRC and CGB studies are, therefore, equal, although more smoke is emitted in the NRC scenario. As will be discussed in section 3.6, the climatic effects of nuclear war are mainly determined by the emissions of the strongly

48 *Physical and Atmospheric Effects*

TABLE 3.4.
LITERATURE SURVEY ON CHARACTERISTICS OF AEROSOL
PRODUCED IN OIL AND GAS BURNING. NOTE THAT REFERENCES
1-4 ALL REFER TO CLEAN BURNING IN HOUSEHOLD EQUIPMENT
AND ARE NOT REPRESENTATIVE FOR FREE BURNING (FROM CRUTZEN
ET AL.. 1984). REPRODUCED BY PERMISSION OF D. REIDEL PUBL.
COMPANY

Ref. No.	Type	Elemental C	Extinction m^2/g fuel
1	Residual oil in burner	31% of aerosol carbon	
2	Diesel engine	80% of aerosol	
	Gas furnace	90% of aerosol carbon	
3	Light oil in burner	40–70% of aerosol	
	Natural gas furnace	40–70% of aerosol	
4	Light oil in burner	40% of aerosol carbon	
5	Oils, rubber	100% soot	0.7–1.2
6	Oil slick	2–6% of fuel burned	
7	Natural gas diffusion flames	all emissions as soot	3
	Heavy fuel oil diffusion flames	all emissions as soot	2
8	Aliphatic oils	3–10% of fuel burned	
9	Benzene, styrene		0.8
"Average"		5% of fuel burned	(0.7)

References:
1. Cooper and Watson, 1979;
2. Muhlbaier-Dasch and Williams, 1982;
3. Nolan, 1979;
4. Wolff et al., 1981;
5. Rasbash and Pratt, 1979; Rasbash, private communication;
6. Day et al., 1979;
7. Maraval, 1972;
8. Rubber and Plastics Research Association of Great Britain, letter to authors;
9. Tewarson, private communication.

light-absorbing elemental carbon. Thus, the climatic consequences of the estimated emissions, which are discussed in the following two chapters, will be similar regardless of whether the NRC or CGB scenario is used.

It is clear from the information gathered in Tables 3.3–3.5 that, even for simple test fires, there is at least a factor of two uncertainty in the smoke and elemental carbon emission yields for each category of combustibles. Further uncertainties are connected with the applicability of this smoke emission data to the mass fires which could develop in large cities in a major nuclear war situation. Various factors, such as the greater intensity of the large fires and the generation of strong convective motions that could loft

TABLE 3.5.
LITERATURE SURVEY ON CHARACTERISTICS OF AEROSOL
PRODUCED BY BURNING OF PLASTICS (FROM CRUTZEN ET AL.,
1984). REPRODUCED BY PERMISSION OF D. REIDEL PUBL. COMPANY

Ref. No.	Type	Aerosol	Elemental C	Extinction m^2/g fuel
1	Plastics		100% soot	0.2–1.6
	Rubber			1.0
2	Various plastics	5–50% soot	2–40% of fuel	
3	Polyethylene, styrene			
	P.V.C. (flaming)	1.2–3.2%		
	Polyurethane (flaming)	9%		
4	Plastics	6–20%		
5	Plastics	3–5%	100% soot	
6	Plastics	11–20%	75% soot	
7	Plastics			0.3–1.2
8	Automobile components	5%		
9	Plastics		60–100%	
10	Polystyrene	3–10% soot		
	Polyethylene	5–8.3%		
	Polyisoprene	19.4%		
	Polystyrene	21.0%		
11	Various plastics	6.4–9%		0.2–1.7
"Average"		5%	80%	(0.6)

References:
1. Rasbash and Pratt, 1979, and D.J. Rasbash, private communication;
2. Morikawa, 1980;
3. Bankston et al.,1981;
4. Hilado and Machado, 1978;
5. J.E. Snell, private communication;
6. Tewarson et al., 1981;
7. Tewarson, 1982;
8. EPA,1978;
9. Seader and Ou, 1977;
10. Rubber Plastics Research Association of Great Britain, letter to authors;
11. Tewarson, private communication.

debris, argue for larger emission factors in the case of the large fires. There are strong indications that ventilation-controlled fires produce much more smoke than free burning test fires (e.g. Rasbash and Pratt, 1979). Also, in a nuclear war, much of the burning of liquid fuels would occur from ruptured fuel containers which would create pool fires. These fires and the burning of street asphalt may produce substantially more than 7% smoke (Rasbash, private communication). High coagulation rates in the dense smoke plumes of oil fires might lead to fluffy soot particles in the supermicron range. (The microphysical and optical properties of these large particles are discussed below.) It is also possible that in large pool fires an appreciable fraction of the oil would be volatilized without burning and then condense as oil droplets a few millimeters in size; such large drops would be removed rapidly from the atmosphere by gravitational settling. Observations of the size and optical properties of the aerosol generated in large oil fires are, unfortunately, not available, leaving this as one of the most critical sources of uncertainty in the estimates of smoke emission. Further research on the emissions from large fires is clearly needed (see Chapter 8).

3.4 PARTICULATE EMISSIONS FROM FOREST AND WILDLAND FIRES

Both the NRC (1985) and the CGB (1984) studies considered as baseline cases that 0.25×10^6 km^2 of forest could burn in a nuclear war. The NRC study assumed that on a purely random basis about 40% of the attacks on military targets could occur above forests and 40–50% over brushlands and grasslands. According to a meteorological analysis by Huschke (1966), about 50% of such areas are medium to highly flammable during the summer. Because effective fire fighting would not be possible during and after a nuclear war, such forest areas, once ignited, could spread over larger areas than what is now normally the case. Assuming that fires would burn over all areas receiving thermal pulses larger than 20 cal/cm^2, corresponding to a fire area of 250 km^2/Mt, the total forest area that could burn during summer and fall would be equal to 250,000 km^2, if neither overlap of fire areas nor fire spread are considered. In an alternative calculation the NRC panel considered the effects of attacks on missile silo fields (2000 Mt) and other military targets (3000 Mt). The missile silo fields occupy an area of 250,000 km^2, of which it was estimated that 50,000 km^2 were located in forested areas. It was assumed that this entire forest area would be totally incinerated. The remaining 3000 Mt on military targets could lead to fires in 150,000 km^2 of forest. According to NRC, if some fire spread is considered, a total of 250,000 km^2 of forest could burn in a nuclear war.

In the CGB study, which was based on the Ambio scenario, it was assumed that a statistical average of 22% of the total megatonnage used in the war (i.e., 1000 Mt) would explode on forest lands and 43% in brushlands and grasslands (Galbally et al., 1983). Prominent among the targets are ICBM silos that receive two 0.5 Mt weapons per site and, altogether, 70% of the total megatonnage (about 2000 Mt). The average spacing between Minuteman silos in the U.S. corresponds to an average area around each silo of about 100–150 km^2. If such close spacing also applies in the U.S.S.R., the area of forest burning near ICBM sites could be no more than 50,000 km^2.

The remaining 30% of the megatonnage (about 900 Mt) would consist of 0.2–0.3 Mt weapons, used mainly against single military targets, such as army and air bases and command posts. Extrapolating from data quoted by Hill (1961) which include estimates of firespread, the minimum fire spread area for a 0.25 Mt weapon would be 200 km^2. With about 1400 explosions, it was, therefore, concluded that 200,000 km^2 of forest land could burn near military targets other than ICBM silos. Considering that additional forest fires could start near cities and industries, a forest fire area of 500,000 km^2, according to CGB, would be possible. The total forest fire area estimated by CGB would have been appreciably reduced to 100,000 km^2 if a more conservative burnout criterion of 250 km^2/Mt, which neglects firespread, had been adopted.

The average load of combustibles in temperate forests is about equal to 20 kg/m^2, of which it was assumed that about 20% could be consumed by the fires. This assumption is somewhat larger than in normal forest fires (Safronov and Vakurov, 1981), but takes into account the effects of simultaneous ignitions, which could lead to mass fires, and debris formation by nuclear blasts. These figures lead to a total forest fuel consumption of 10^{15} g, as calculated by both NRC and CGB, for a total forest fire area of 250,000 km^2. On the average, half of the fuel would be consumed in flaming and half in smoldering combustion (Chandler et al., 1963; Wade, 1980). The smoldering combustion produces about 5 times more particulate matter than flaming combustion (Wade, 1980). Based mostly on the compilations by Ward et al. (1976), the smoke emission factor (mass of smoke produced per mass of fuel) was assumed to be 3% in the NRC study and 6% in the CGB study. Of this, 10% was assumed to be amorphous elemental carbon. This leads to total smoke and elemental carbon emissions of 3×10^{13} g and 3×10^{12} g (NRC, 1985) and 6×10^{13} g and 6×10^{12} g (CGB, 1984), respectively. These quantities are appreciably smaller than the potential smoke emissions from urban and industrial targets.

Some recent studies indicate that the effects of forest fires may have been overestimated in the NRC and CGB studies. Patterson and McMahon (1984) inferred elemental carbon fractions ranging from 0.5 to 20% from light absorption measurements in smoke produced from forest fuels in laboratory experiments. As expected, the smaller values were associated with smoldering combustion and the larger with flaming combustion. Emission factors for the flaming fires were found to be between 0.8 and 2%, and for the smoldering fires were about 5 to 6%. In combination, these factors gave estimated elemental carbon emission factors from laboratory pine needle fires ranging from 0.07 to 0.25%. In considering data from low-intensity field burns with forest fuels, Patterson and McMahon (1985, 1986) inferred an elemental carbon fraction in the smoke of at most 8%. Field burn experiments of logging residues gave values of the emission factor of elemental carbon in the range of 0.08 to 0.12% Considering both laboratory and prescribed field burns, Patterson and McMahon (1985, 1986) proposed an elemental carbon emission factor of 0.14%. In comparison, based on earlier compilations, the NRC and CGB studies adopted emission factors for elemental carbon of about 0.3 and 0.6%, respectively. In addition, the data of Patterson and McMahon (1985, 1986) suggest that total smoke emission factors may also have to be reduced. Similarly, from a series of 35 measurements in 6 prescribed burns of forest products in the states of Washington and Oregon in the U.S., Hobbs et al. (1984) deduced an average smoke emission factor of only 0.4% for particles with diameters less than 2 μm.

Unfortunately, the results quoted above are all for small laboratory fires or for prescribed burns of logging residues, which are relatively low-intensity fires. Major forest fires may produce more smoke per mass of fuel and more intense fires are almost certain to produce more elemental carbon. Patterson and McMahon (1985, 1986), in fact, show that the light absorption increases with fire intensity, indicating more elemental carbon emission. Clearly, field observations of these larger fires are urgently required. Also, Patterson and McMahon (1985, 1986) mention that the burning of organic gases that are driven out of live vegetation might produce considerable soot. This could be an important source of elemental carbon in a nuclear war in which large amounts of live vegetation in forests and croplands are burned.

A large reduction in the estimates of the potential atmospheric optical impact from forest fires would follow from a study by Small and Bush (1985). Rather than assuming a wildland fire area proportional to total yield and global statistical coverage of forest, brush, grass, and agricultural lands, Small and Bush (1985) attempted to identify the exact locations of potential military targets (missile silos, air bases, radar sites, weapon storage depots, communication centers, etc.) and to calculate the ignition area, type of wildland and its fuel loading, firespread, and smoke production. The results of the calculation vary seasonally, but in all seasons Small and Bush (1985) found at least an order of magnitude less smoke production than estimated by Crutzen and Birks (1982), Turco et al. (1983a), CGB (1984), and NRC (1985).

Small and Bush (1985) point out that most military targets are not distributed randomly over the various ecosystems of the U.S. and U.S.S.R. but are either concentrated in a few missile fields, or located along major transportation arteries. According to their analysis, the greatest number of targets are located in agricultural and grasslands, while 14% are in forest lands, mostly in the Soviet Union. Burnable fuel loadings in each of the categories of wildlands and agricultural lands were taken from analyses of the U.S. Forest Service (Deeming et al., 1977). Small and Bush also assumed that croplands would not be in a condition to burn for most of the year, but only when grains have ripened and have not yet been harvested, a period of about two weeks. Consequently, although only 10% of the military targets are located in forested lands, forest fires would still account for 40% of the total area burned.

Weather conditions could have a large influence on firespread, leading to substantial seasonal variations, with maxima during summer. In the study of Small and Bush (1985), average climatic conditions necessary for calculations of ignition radius (visibility and fuel moisture) and probability of firespread (temperature, relative humidity, fuel moisture, and winds) were obtained for weather stations closest to the potential targets. Altogether, Small and Bush (1985) derived a maximum forest fire area of 70,000 km^2 in summer time, about 30% of that adopted by the NRC and CGB studies, and total smoke emissions of at most 3×10^{12} g, an order of magnitude less than that derived in the earlier studies.

The study by Small and Bush (1985) currently is the most complete analysis of the possible wildland fire areas following a nuclear war. However, very little detail about their analysis procedures has been provided. Furthermore, there are critical factors and assumptions entering into their analysis that indicate that their estimates represent lower bounds to the smoke that would be produced from wildlands in a nuclear war.

First of all, the applicability of the adopted fire ignition and spread model, in which average meteorological conditions are assumed, is open to question. For instance, it is conceivable that large tracts of the Soviet Union are in a condition of drought, while those in the U.S. are not (and vice versa). The probability that major forest fires may spread over large areas of any one of the combatant nations due to regional drought conditions should be considered. It is exactly this factor that leads to large-scale forest fires during unfavorable years. For instance, there are reports of forest fires lasting for months and burning for some ten million hectare in Siberia (Shostakovitch, 1925). It is clear that firespread must have played a large role in this. The potential effects of fires started by tactical nuclear weapons (such as the 30,000 such weapons in the European theater alone) should also be taken into account.

In the study by Small and Bush (1985), firespread accounted for less than 7% of the total fire area. This is an extremely low number in light of the tabulated frequency distribution of fire danger indexes by Schroeder and Chandler (1966), partly reproduced in Table 3.6. According to this tabulation, in the period April–October, the probability of "critical fire conditions" in the Northern Plains regions would be between 3–23% and for "actionable conditions" near 50%. Critical conditions imply that any fire would be uncontrollable and could spread, until the weather changes. Actionable conditions imply that the fires can be controlled with fire fighting efforts. However, effective fire fighting would be very unlikely in a nuclear war situation.

TABLE 3.6.
EXPECTED FIRE BEHAVIOR (SCHROEDER AND CHANDLER 1966)

Fire condition[a]	January	April	July	October
Northeastern Plains Region				
FO	.93	.35	.23	.29
NS	.06	.22	.32	.25
Act	.01	.40	.43	.42
C	.00	.03	.02	.04
Northwestern Plains Region				
FO	.73	.34	.12	.22
NS	.15	.19	.13	.14
Act	.12	.41	.57	.50
C	.00	.06	.18	.14
Northern Rockies & Northern Intermountain Region				
FO	.90	.23	.03	.30
NS	.06	.31	.07	.20
Act	.04	.45	.72	.46
C	.00	.01	.18	.04
Central Intermountain Region				
FO	.72	.16	.02	.10
NS	.17	.15	.04	.11
Act	.11	.55	.43	.64
C	.00	.14	.51	.15

[a] The fire conditions considered are:

FO = Fire Out (fire won't start)
NS = No Spread (fire will start but won't spread and will go out if weather stays the same)
Act = Actionable (fire needs action and is controllable)
C = Critical (fire uncontrollable until weather changes)

Secondly, the fuel contained in the U.S. Forest Service Fire Danger Model, which was used by Small and Bush, accounts only for a small fraction of the potentially available fuels, i.e., only the "fast burning" fuels that mainly contribute to fire intensity. They suggest that the fires would consume 0.5–1.6 kg/m^2 in forests, which accounts for only 3–10% of the available biomass density. A range of 2–4 kg/m^2 seems much more realistic, especially since trees would be shattered by blast waves over wide areas.

Furthermore, it is a distinct possibility that the simultaneous ignition of forest and other wildland materials following nuclear attacks would become more efficient in a multiple burst scenario than under normal conditions because the overlapping blast wave and thermal radiation zones can shatter and dry out live and moist fuels, making them more susceptible to burning.

Finally, the analysis of fire ignition area and land use characterization derived by Small and Bush (1985) is not uncontested. A recent evaluation of land use in and around U.S. ICBM silo fields by Ackerman et al. (1985b) indicates that as much as 150,000 km^2 of vegetation could be affected in the summer half-year by attacks on U.S. silos alone, which should be contrasted with the 190,000 km^2 derived by Small and Bush (1985) for *all* military targeting. The difference in these estimates is partly related to the assumed incendiary efficiency of multiple nuclear bursts over missile fields.

The land use data derived by Ackerman et al. (1985b) from Landsat imagery and U.S. Geological Survey land use maps are reproduced in Table 3.7. Making use of a new survey of biomass loadings (see also Volume II of this report), these authors calculated the smoke emissions from forest and crop fires to be from 5 to 10 times greater than those derived by Small and Bush (1985). This difference is related to the larger area assumed affected as well as the larger estimates of readily combustible biomass. As a further complication, there are indications that more Soviet silo fields and military bases are located in forested areas than are U.S. bases.

TABLE 3.7.
LAND USE IN U.S. MISSILE BASES AS FRACTIONS OF THE TOTAL
AREA (FROM ACKERMAN ET AL., 1985B)

Missile Base	Grass-lands	Deciduous Forest	Coniferous Forest	Agriculture	Other[a]
Ellsworth, SD	.74	—	.03	.22	.01
Grand Forks, ND	.03	.03	—	.94	—
Malmstrom, MT	.48	—	.12	.40	—
Minot, ND	.22	—	—	.73	.05
Warren, WY	.54	—	—	.45	.01
Whiteman, MO	—	.11	—	.87	.02

[a] Mainly water or barren land.

The estimations of the possible contributions of wildland fires to smoke production in a nuclear war remain, therefore, still uncertain. Following the studies of Hobbs et al. (1984), Patterson and McMahon (1985, 1986), and Small and Bush (1985), it seems likely that their importance was somewhat overestimated in the previous studies. However, this does not significantly alter the total estimated emissions, which are dominated by the emissions from urban and industrial fires.

Finally, it should be noted that forest fires in the postwar environment might lead to important effects. It is conceivable that large quantities of unburned and dead forest material could accumulate due to the combined effects of climate changes and the release of intense radioactivity (Woodwell,

1982) and hazardous chemicals from industrial and urban targeting. If ignited some months after the nuclear war, these areas could contribute large amounts of smoke, although by this time it may not be a significant contributor to climatic effects. Peatbog fires in north temperate latitudes, which can last for months (Shostakovitch, 1925; Safronov and Vakurov, 1981) are another potential source of large amounts of smoke.

3.5 MICROPHYSICAL PROCESSES

In the preceding sections, estimates of smoke emission from fires were given. The particles or aerosols, comprising this smoke (and the dust raised by surface bursts—see section 3.8) interact with each other and with background particles, clouds, and precipitation as they evolve in the atmosphere. Accordingly, a variety of physical processes involving heterogeneous mixtures of airborne particles are of interest here. The properties that define a particle include its size, shape, structure (morphology), composition, density or mass, index of refraction, and response to humidity. These properties can also change in time through the physical and chemical transformations outlined below.

3.5.1 Interactions With the Environment

Smoke and other aerosols in the atmosphere are influenced by gravity and interact with ions, gases, solar radiation, and, near the ground, physical surfaces. Gravitational sedimentation occurs when a particle falls relative to the surrounding air because of its much higher density (compactness). Sedimentation may generally be neglected for submicron particles in the troposphere and lower stratosphere (Twomey, 1977) because vertical atmospheric motions transport such particles more rapidly than sedimentation. For larger particles, however, sedimentation enhances the rate of removal by bringing particles to lower altitudes or to the surface. An exception would be loosely aggregated clusters (e.g., soot-like chains), with very low effective densities and large effective aerodynamic cross-sections. Such particles would readily be carried by winds, and would settle out of the atmosphere more slowly than compact particles of the same mass.

Near the ground, particles can diffuse (and adhere) to a variety of surfaces, including soil, water, and vegetation (the latter in particular may provide an enormous collection area in heavy overgrowth). Particles of different sizes attach to surfaces with different efficiencies. Notably, aerosols with radii in the range of 0.1 to 1.0 microns have very small "deposition" velocities (less than 0.001 m/sec; Slinn, 1977). The deposition velocity is defined in terms of the net flux of a substance carried to a surface by all active microscale processes. Nevertheless, for smoke particles in the submicron size range, dry deposition can be an important secondary removal process.

Aerosols exist in close thermal equilibrium with the atmosphere. The temperature deviation is usually negligible because the energy absorbed by small particles (e.g., solar radiation) is quickly transferred to the surrounding air by thermal diffusion. Hence, nonequilibrium heating or cooling of smoke particles or fine dust particles may be ignored in the lower atmosphere. Above the middle stratosphere (roughly 35 km), particles may heat up by several degrees above the surrounding air temperature.

Smoke interacts with gaseous chemical constituents in the atmosphere. Some of the reactions alter the surface composition of the particles, while others may actually consume the smoke. Adsorption of vapors from the environment can also change the composition and size of particles as well as their hygroscopic properties and index of refraction. The reactions of oxidants such as ozone and hydroxyl radicals with carbonaceous soot particles may deplete the soot mass; such reactions could be particularly important in the stratosphere where a long residence time for soot would otherwise be expected. However, reactions that are sufficiently rapid to merit attention in the atmosphere have not been identified (see Chapter 6). Therefore, significant soot consumption by photochemical processes must be considered speculative at this time. Moreover, particles would be coated by oily material or, after a day or so in the atmosphere, with a number of inorganic compounds such as sulfates and water, thereby isolating the carbon surfaces from direct chemical attack.

Ions are present at all levels in the atmosphere. In the troposphere, ion concentrations of $10^9/m^3$ are typical. Because the mobilities of positive and negative ionic species are generally unequal, aerosols immersed in an ion plasma accumulate a net charge. The charge is small enough that its effects on aerosol microphysical processes can usually be ignored (Twomey, 1977). This situation is quite different from that which applies in powerful convective storm systems (and possibly large fire plumes), where ice processes lead to strong electrification of cloud droplets and aerosols (Pruppacher and Klett, 1980). Because air is a weakly ionized plasma with a small but finite conductivity, any highly charged objects immersed in it will tend to discharge over time.

Photophoresis describes the force exerted on a particle as a result of nonuniform heating by solar (or other) radiation. If, in absorbing an incident beam of radiation, a particle is heated preferentially on one side, the diffusion of heat away from the particle creates a thermal gradient in the surrounding air which exerts pressure in the opposite direction. By its nature, the photophoretic force is very weak. However, for small particles illuminated by sunlight, it can exceed the force of gravity. Recently, Sitarski and Kerker (1984) proposed that photophoresis may cause soot particles to levitate in daylight, and might explain the long lifetime of the Arctic haze aerosol. Depending on the size and composition of the particles, photophoresis could either increase or decrease the vertical (settling) velocities of soot aerosols. However, there are a number of reasons to believe that photophoresis is unlikely to be important for soot under atmospheric conditions: Brownian rotation of small particles reduces their nonuniform heating by sunlight; the diurnal variation of solar insolation reduces the average photophoretic force by a factor of about four relative to the gravitational force; the irregular shape of typical soot particles disturbs the required pattern of heating; upwelling shortwave radiation (scattered and reflected) heats the particles in the opposite sense from the direct solar radiation; and the solar beam could be significantly attenuated in certain circumstances following a nuclear war.

3.5.2 Agglomeration

A homogeneous or heterogeneous aerosol mixture will coagulate through various mechanisms to form aggregated particles. Collisions between particles are induced by thermal Brownian motions, winds and turbulence, and gravitational settling. Coagulation due to Brownian motion is most effective for submicron particles and is generally less significant for supermicron particles (Pruppacher and Klett, 1980). Collisions between particles traveling at different relative speeds in laminar flows and in turbulent flows are most important for larger particles, which can experience differential accelerations due to wind shears, and can cross streamlines during curvilinear acceleration. Gravitational coalescence involves differences in particle fallspeeds, in which a larger particle overtakes and intercepts a smaller particle. To be effective, the fallspeeds must be substantial; accordingly, at least one large particle must be involved in the collision process. The relative importance of the various aerosol agglomeration mechanisms is illustrated in Figure 3.1.

If an encounter between two particles is to result in coagulation, the particles must touch and adhere. Submicron particles with small Stokes numbers tend to flow around obstructions in the airstream. To impact a surface, such particles must diffuse through a laminar boundary layer separating the surface from the deflected airflow. However, because aerosols have (relatively) small diffusion coefficients, they are hindered in reaching the surface during the brief duration of an encounter. Larger particles with greater inertia can cross streamlines in the flow and impact the surface directly.

A variety of forces can act to hold the aerosols together following a collision. Van der Waal's surface forces can hold dry, submicron aerosols together. Droplets can coalesce into larger droplets under the influence of surface tension, and dry particles can be wetted in this manner. For large dry aerosols, electrical coulombic forces can effectively bond particle clusters, given a sufficient charge. Chemical substances condensed on surfaces and in crevices can act to cement and strengthen particle aggregates.

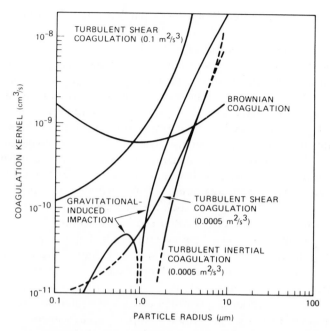

Figure 3.1.   Comparison of coagulation kernels for various aerosol collision mechanisms for a spherical particle having a radius of 1 μm and a density of 1 g/cm^3 interacting with spherical particles having radii between 0.1 and 10 μm. The dotted lines indicate regions where complex hydrodynamic interactions between the particles invalidate the theoretical treatment used. The assumed turbulent energy dissipation rates are given in brackets. (Hidy, 1973, adopted from Pruppacher and Klett, 1980)

However, when particles impact with great force, they may not stick together. The particles can rebound elastically or, if there is enough force, can break off or knock loose smaller particles (Rosinski and Langer, 1974).

As already noted, large soot aggregates have been observed in the burning of oil and plastics (e.g., Day et al., 1979). In regions of flames that are hot, rich in organic molecules, and lacking oxygen, soot is initially generated as a concentrated aerosol of very small amorphous carbon spheroids roughly 50 nm in diameter (Wagner, 1980). The spheroids, which appear to be charged, coagulate to form chain structures, and the chains may later coagulate to form fluffy aggregates. If the sooty smoke is rapidly diluted, the chains are "frozen out" at fairly small sizes (less than 1 μm). If the smoke is very dense, the chains can aggregate to much larger sizes (greater than 10 μm). The rate of dilution of the smoke with clear air is an important factor in controlling the aggregate sizes (NRC, 1985). Typical dilution rates normally limit the extent of aggregation to relatively small sizes. Nevertheless, in large oil fires, where soot yields might exceed 10%, many supermicron agglomerated particles would be expected. Such large particles may be efficiently scavenged by falling raindrops. These problems have not yet been quantified either experimentally or theoretically.

It is important to note that soot agglomerates have very different physical and optical properties than compact spheres of equivalent mass. Physically, soot agglomerates have abnormally large aerodynamic cross sections for their mass. Hence, they have lower settling velocities and smaller impact efficiencies than is suggested by their physical dimensions. The exaggerated cross sections of soot particles can also lead to accelerated coagulation rates (Baum and Mulholland, 1984). Instruments designed to measure aerosol sizes by optical means or by mobility analysis have, to our knowledge, never been calibrated against soot agglomerates. Accordingly, in situations where soot clusters are expected to form, such measurements must remain suspect. The optical properties of soot are discussed in the section 3.6.

In the plumes of large fires, the powerful winds that are induced loft ash, dust, and fire debris along with smoke and soot. Sub-micron smoke particles may be captured and removed by large supermicron particles that may also be lofted. In their global scale calculations, Turco et al (1983a, b) included the effects of modest quantities of fire-generated ash particles and large quantities of explosion-generated dust particles as scavengers of smoke. The collection processes that they treated included Brownian coagulation, gravitationally-induced impaction, and turbulent shear, and inertial coagulation (Pruppacher and Klett, 1980). For assumptions of both instantaneous and delayed dispersal of the mixed smoke and dust clouds, Turco et al. found these processes to be of minor importance.

Porch et al. (1986) have further examined the potential for scavenging by large particles, especially in firestorm environments. They considered turbulent energy dissipation rates ranging from $0.1 \text{ m}^2/\text{sec}^3$ typical of thunderstorms to $0.8 \text{ m}^2/\text{sec}^3$ scaled from the values for an intense fire plume modeled by Cotton (1985). The Porch et al model included a simplified turbulent coagulation theory, but did not account for the hydrodynamic interactions between large and small particles which generally reduce particle collection efficiencies (Pruppacher and Klett, 1980). They also did not include, however, the extra surface area of chained aggregates and the effects of particle charge, which may increase collection efficiencies. For levels of large particle concentrations ($\sim 0.1 \text{ g/m}^3$) found in a modest-sized fire by Radke et al (1983), there is very little reduction of the optical depth over a thirty minute period, which may actually be longer than typical smoke parcels would remain in the highly turbulent regions of the fire plume. In their model, large particle concentrations must be increased by about a factor of 50 (i.e., to levels observed near the ground in modest-sized dust storms) for there to be a reduction by a factor of 2 in optical depth. Maintenance of such high concentrations of large particles would require relatively high windspeeds to scour the surface and loft dust, char, and other materials. Such conditions might occur as a result of the high velocity rotating winds induced in an organized urban firestorm (e.g., as may have occurred in Hamburg). If such events are as relatively rare as was the case during World War II, the overall effect on the total smoke burden as a result of scavenging by large particles would not be substantial; if such events are frequent, further consideration of this process may be warranted.

3.5.3 Precipitation Scavenging

The primary means by which submicron aerosols are removed from the atmosphere is through incorporation of aerosols into cloud water by nucleation and phoretic scavenging, followed by cloud water coalescence and precipitation to the ground. These same processes would be the primary removal mechanisms following a nuclear war. However, in discussing scavenging and aerosol removal, the prompt scavenging of smoke and dust by precipitation that may be induced in the convective fire plumes must be distinguished from the synoptic scale scavenging processes, which would occur after the smoke plumes had dispersed into the background atmosphere.

3.5.3.1 Observations

One of the most critical problems in the estimation of the long term climatic effects of large urban fires is the extent to which smoke particles could be removed by precipitation scavenging in the convective plumes that accompany the fires. The "black rains" that followed the nuclear explosions in Hiroshima and Nagasaki amply demonstrated that smoke can be removed from plumes by fire-induced convective clouds. A graphic illustration of this process can be seen in the Peace Memorial Museum at Hiroshima, where a section of the white wall of a house covered with streamers of ink-like smoke residues is displayed (Ishikawa and Swain, 1981).

Nevertheless, quantitative information on the efficiency of smoke removal by fire-induced precipitation is lacking. The efficiency is unlikely to be close to 100%. For most convective storm systems, the precipitation efficiency, which is roughly defined as the ratio of the precipitation rate at the ground to the water condensation rate in the cloud, is typically between 15% and 65% (Foote and Fankhauser, 1973; Marwitz, 1974; Hobbs and Matejka, 1980). Strong updrafts carry some of the condensed water to high altitudes, where it detrains from the clouds and reevaporates. Much of the rainfall below the cloud-base is known to evaporate as well (amounting to about 40% of the vapor flux into the cloud-base); however, this would have a lesser effect on the re-injecton of aerosols since few droplets evaporate completely. Overall, continental cumulonimbus systems typically convert less than 50%

of the moisture entrained into precipitation at the ground. In the case of very powerful cumulonimbus systems in highly sheared environments, the precipitation efficiency can be less than 15% (Fritsch and Chappell 1980). For one thing, the strong updrafts restrict the formation of large ice particles. Because precipitation is less efficient in this case, larger quantities of ice are injected into the anvil.

The ratio of condensible water mass to smoke mass in a large fire plume in a moist ambient environment could be 1000 or more (NRC, 1985). Accordingly, there is sufficient water available to remove most of the smoke—if the cloud were to rain and if the rain were efficient at scavenging the smoke. As just noted, intense cumulonimbus systems are generally inefficient generators of precipitation. Even so, enough rain/hail could form (in humid environments) to remove a significant fraction of the smoke.

Fires do not always produce intense convective plumes, rainfall, and efficient smoke scavenging. This is clearly illustrated by observations of major forest fires. In September 1950, the smoke plumes from more than 100 forest fires in Alberta, Canada resulted in the "Great Smoke Pall" over North America. Sunlight was attenuated over much of Canada and the eastern one-third of the U.S. (Wexler, 1950). The reported altitude of the smoke cloud was between 2.5 and 4.5 km. One week later, the smoke clouds were visible over several countries in Western Europe, where the smoke was observed to be as high as the tropopause (Smith, 1950; Wexler, 1950).Satellite observations show that smoke produced by large forest fires in European Russia in August 1972 was transported eastward over the Ural mountains for distances of 5600 km in the middle troposphere (Grigoriev and Lipatov, 1978). Similar observations of long range transport of Australian forest fire smoke to New Zealand are common (D. Lowe, private communication). Such evidence indicates that the smoke produced by forest fires generally escapes prompt precipitation scavenging and disperses through the atmosphere.

In general, large oil fires appear to produce soot plumes in which little condensation or precipitation due to fire-induced convection occurs (Davies, 1959). Such plumes typically rise to several kilometers altitude. Radke et al. (1980a) observed soot coagulation in the plume of the Meteotron (a soot-generating oil-fired 1000 megawatt artificial heat source in France). They also noted that the smoke was capable of dissipating ambient clouds above the Meteotron. Following dozens of tests, precipitation associated with the operation of the Meteotron was observed only once (in an unstable air mass).

Radke and coworkers (Radke, private communication) observed smoke "processing" by a condensation cloud over a prescribed forest fire in 1978 (processing refers to the scavenging of smoke particles by water droplets, with re-emission through cloud evaporation). They measured the size distribution and visible backscatter coefficient of the smoke in two air parcels in the plume; one parcel passed through the capping condensation cloud, and one passed beneath (and clear of) the capping cloud. The cloud-processed smoke had considerably fewer particles of very small size (a factor of about 10 less at sizes less than 0.05 μm radius), had a somewhat larger number of particles in the intermediate size range (approximately 0.1 to 1.0 μm), and had fewer particles in the supermicron size range. The scattering coefficients for the two smoke samples were about equal (absorption was not measured), possibly suggesting only limited impact of processing on the overall optical properties of the smoke. However, the experiment involved a relatively small cloud formation, and was not strictly controlled (i.e., by sampling in the same smoke parcel before and after the cloud). Accordingly, the results are only suggestive of potential effects.

3.5.3.2 Nucleation

Nucleation occurs when water vapor in excess of the saturation vapor pressure condenses onto aerosol surfaces, forming water droplets or ice crystals. For ambient hygroscopic aerosols, nucleation in clouds is probably the dominant scavenging and removal mechanism (Pruppacher and Klett, 1980; Twomey, 1977). On the other hand, experimental evidence from fire plumes suggests that smoke, particularly sooty smoke, is less susceptible to nucleation scavenging.

In typical convective clouds, water vapor supersaturations seldom exceed 1%, because of the abundance of nuclei and particle surfaces to absorb the excess moisture (Pruppacher and Klett, 1980). However, the larger the convective velocity, the greater the supersaturation that could theoretically be achieved. In a fire column, high vertical velocities would be associated with enhanced concentrations of windblown debris such as ash, char and dust. Accordingly, the enhanced surface area for condensation may limit the supersaturation to normal values. Cloud condensation nuclei (CCN) are defined as those particles that can be nucleated into water droplets at supersaturations of a few percent or less.

Forest fires are potentially major sources of cloud condensation nuclei. Eagan et al. (1974) observed the production of as many as 6×10^{10} CCN active at 0.5% supersaturation for each gram of forest fuel consumed. The CCN activity may be due to the chemical nature of the smoke particles, which have been determined to consist of complex organic compounds with little amorphous carbon, or soot. Bigg (1985) reported other measurements in forest fire plumes in which about 5% of the total number of particles were active as CCN at 1% supersaturation, and 0.5% at 0.25% supersaturation. Moreover, the proportion of CCN did not appear to increase as the smoke aged. The data of Eagan et al. (1974) and Bigg (1985) are generally consistent

if a smoke emission factor of about 1% and a mean particle size of 0.1 μm—
values commensurate with observations—are assumed.

If a production rate of 6×10^{10} CCN/g is assumed for the burning of
cellulosic materials in urban centers, a total of 1 to 4×10^{26} CCN active
at 0.5% supersaturation could be produced in the nuclear war scenarios
of NRC (1985) and Crutzen et al. (1984). This is of the same order of
magnitude as the total global abundance of background CCN (Pruppacher
and Klett, 1980).

Soot particles tend to be hydrophobic (i.e., water repellent), particularly
fresh soot that has not had a chance to collect hygroscopic compounds. For
the atmospheric conditions that prevail in Western Europe, observations
by Ogren and Charlson (1984) show that soot particles are removed at a
slower rate than sulfate aerosol for the first few days following emission.
Radke et al. (1980a) measured CCN abundances in the large sooty plume
generated by the Meteotron device. They found concentrations very close to
background levels (approximately 500 to 1200/cm^3 for supersaturations of
0.5 to 1%, respectively). At the same time, the total smoke particle concentrations exceeded 10^4/cm^3, suggesting that only a small percentage of the
soot particles were active as CCN. Similarly, in carefully designed laboratory
experiments currently underway, Hallett and coworkers (personal communication) have noted that some fresh and aged soot particles can be active as
CCN at approximately 1% supersaturation; these CCN typically comprise a
small percentage of total soot particle population.

Little information is available on the ice-nucleating properties of smoke
and soot. Such particles should be poor ice nuclei (Pruppacher and Klett,
1980). Bigg (1985) reports that sampling in forest fire convective columns
yields ice nuclei concentrations of approximately 0.01/cm^3 (while this is
roughly 100 times greater than ambient ice nuclei concentrations, it is obviously much smaller than the total smoke particle concentration). A substantial increase in ice nuclei abundances could affect the microphysical
development of fire-induced clouds, and should be considered in future
studies.

Several factors could enhance smoke nucleation rates in fire plumes. Large
aggregated smoke or soot particles might nucleate more readily than the
smaller particles sampled in the experiments cited above. Chemical transformation of smoke particles—e.g., coating by sulfates generated from sulfur
in the fire fuels—make the particles more susceptible to water condensation.
Turbulence in the plume could also create local zones of considerably higher
supersaturation.

On the other hand, in the larger fire plumes, characterized by intense convection, the time available for agglomeration and chemical transformation
prior to condensation is only a minute or so, which seems insufficient for
major physical or chemical changes to occur. Moreover, as already noted,

the supersaturations in the plume are likely to be suppressed by the presence of windblown fire debris particles.

If most of the smoke particles in the plume were to nucleate, and the particle concentrations were as high as 10^4 to $10^5/cm^3$, the clouds formed could become overseeded, i.e., composed of a large number of very small droplets. Such clouds are less likely to produce precipitation because droplet coalescence is less efficient (Twomey, 1977). However, ultragiant nuclei raised by fire winds would continue to provide a source of precipitation-sized water particles. As these fell through the cloud, smaller smoke and dust particles could be scavenged and washed out (Hobbs et al., 1984). The most efficient removal would occur if the smoke particles had absorbed water and reached a size of several microns radius. These expanded smoke particles could then be collected relatively efficiently by inertial impaction on the precipitation drops nucleated on ultragiant aerosols, provided that an adequate supply of these latter particles existed.

Because observational data suggest that only a small fraction of all the smoke particles would be active as CCN, most of the smoke would have to be scavenged by processes other than nucleation. These are discussed below.

3.5.3.3 Brownian, Inertial and Phoretic Scavenging

For smoke particles with a radius on the order of 0.1 μm, Brownian and inertial collection by cloud and precipitation drops can generally be ignored (Pruppacher and Klett, 1980). For aerosols with radii much less than 0.1 μm, Brownian diffusion is important, while for aerosols with radii much greater than 0.1 μm, inertial impaction is important (particularly for those particles with radii greater than several microns). The limited fire plume and microphysics modeling accomplished to date also indicates that phoretic scavenging processes are dominant over Brownian and inertial processes (Cotton, 1985; see also Chapter 4).

Phoretic scavenging occurs when aerosols, primarily in the submicron size range, are brought into contact with a water droplet or ice crystal through motions induced by fluxes of heat and mass. Thermophoresis represents aerosol motion induced by the flux of heat to an evaporating droplet or ice crystal. As the droplet evaporates, heat is absorbed from the immediate vicinity of the droplet, producing a local thermal gradient. The corresponding gradient in the kinetic energy of air molecules then drives the aerosol in the direction of the heat flux (Pruppacher and Klett, 1980). At the same time, an outward diffusive flux of water molecules is associated with an evaporating droplet. This flux establishes a weak hydrodynamic Stephan flow of air away from the droplet. Collisions between the aerosol and the flowing air molecules cause the particles to drift away from the evaporating droplet. The resulting diffusiophoretic force opposes, but is generally less than, the thermophoretic force. As a result, evaporating droplets or ice crystals are effective in collecting aerosols; likewise, growing droplets tend to repel aerosols by this mechanism. Phoretic scavenging is generally much weaker for ice crystals than water droplets because the evaporation rates of ice crystals in clouds are normally much lower (although the time scales can be longer, particularly in the cloud anvil).

The relative importance of Brownian diffusion, inertial impaction and phoretic forces in the scavenging of an aerosol by precipitation, based on theoretical calculations, is illustrated in Figure 3.2 (Slinn and Hales, 1971). Of particular interest is the regime corresponding to aerosol radii of 0.1 to 1.0 μm where all scavenging mechanisms are relatively inefficient. Within this region, known as the Greenfield gap (after Greenfield, 1957), phoretic effects are the most important (NRC, 1985).

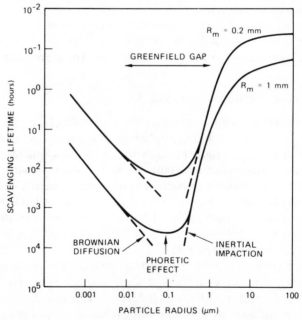

Figure 3.2. Calculated e-folding lifetimes of aerosols against removal by precipitation of 10 mm/hr, for precipitation drop spectrums with characteristic drop radii of R_m = 0.2 and 1.0 mm. The dominant scavenging mechanism in each aerosol size range is indicated (from Slinn and Hales, 1971, reproduced by permission of American Meteorological Society)

It should be noted that there is still disagreement over the magnitude of phoretic scavenging rates in clouds, and indeed over the sign of the net force (thermophoretic minus diffusiophoretic) (Vittori, 1984). It also has been suggested that other forces (e.g., electrical) may act to fill the

Greenfield gap, thereby reducing the atmospheric lifetime of sub-micron particles. Current experimental evidence bearing on this issue is mixed. Accordingly, theoretical calculations of aerosol scavenging rates should be treated as tentative estimates at this time.

3.5.3.4 Smoke Aging

Smoke particles and chemical vapors scavenged by cloud drops and ice crystals that later evaporate above or below the fire plume are released as an "aged" smoke. Particles that are not subject to water condensation can also age by coagulation with other smoke particles, agglomeration with fire debris and ambient aerosols, and deposition of chemical vapors. Particles that are aged for several days in the background atmosphere, and those passing through cloud condensation/evaporation cycles are expected to be fairly compact and hygroscopic in nature. Large soot agglomerations, for example, might collapse under surface tension if wetted, and thus could become denser and more spherical.

As mentioned earlier, the process of aging by water condensation has been observed in a forest fire plume (Radke, private communication), and aging by coagulation, in an oil fire plume (Radke et al., 1980a). Nevertheless, except for the general facts already described, extensive data on smoke aging in various atmospheric environments are not available.

3.5.3.5 Overall Scavenging Efficiency

Theoretical models of precipitation scavenging generally underestimate the aerosol removal rates actually observed in clouds (e.g., Radke et al., 1980b). There are several apparent reasons for this discrepancy. The theoretical models are not yet sophisticated enough to account for all of the possible simultaneous interactions of aerosols with water droplets and ice crystals, including the effects of electrical charge, turbulence, and transient phenomena. The physical properties of the aerosols are also important. Radke et al. (1980b) attributed the larger than predicted precipitation scavenging efficiencies of aerosols from power plant plumes to the hygroscopic nature of the particles, which they proposed could swell in size by absorption of water vapor, thereby filling the Greenfield gap. Prodi (1983) observed that ice crystals growing in the presence of supercooled water droplets readily collected submicron hygroscopic salt aerosols, but not submicron hydrophobic wax particles. Most atmospheric scavenging observations involve aerosols that are readily nucleated in clouds, or on which water readily condenses. Smoke, on the other hand, has different physical characteristics and, one might expect, lower scavenging and washout efficiencies.

An hour or less of steady rainfall (of up to 10 mm of water) is generally capable of removing most aerosol pollutants from the atmosphere (Prup-pacher and Klett, 1980). The induced precipitation in a fire plume would (for a particular air parcel) probably be much shorter in duration, but more intense.

Because the prompt scavenging of soot particles in fire-induced convective columns depends on many factors that are poorly known and extremely difficult to predict, the overall scavenging efficiency can only be crudely estimated. Clearly, a much better understanding of individual scavenging processes is required in order to make a reliable estimate. In previous studies, assumptions of 30 to 50% prompt removal of smoke (from fire plumes) have been made (Turco et al., 1983a,b; NRC, 1985; CGB, 1984). These values seem to be reasonable given the current state of knowledge (Hobbs et al., 1984).

In determining the synoptic-scale scavenging of smoke from the background atmosphere, three factors play an important role: the injection height of the smoke, the composition and morphology of the smoke particles that survive prompt scavenging, and the possible large-scale meteorological perturbations of the atmosphere. One plausible approach is to assume that the smoke particles, once processed through a condensation cloud over a large fire or a natural cloud system, can be efficiently removed during subsequent encounters with clouds and precipitation in synoptic systems. Thus, Malone et al. (1986), in their climate study using a general circulation model, assumed that smoke was essentially completely removed whenever entrained into a precipitating cloud system. This may in fact overestimate the removal rate because, in such models, precipitation occurs simultaneously over an entire grid cell, which is typically on the order of 10^5 km^2, whereas precipitating clouds are generally confined to only a fraction of this area (see Chapters 4 and 5).

While the physical characteristics of the smoke particles that escape the fire plumes have not been determined, they would presumably vary widely. Some particles would be in a relatively unaltered state, while others would be well "aged". Hence, the initial efficiency for subsequent scavenging by mesoscale and synoptic scale cloud systems could also vary widely.

3.5.4 Smoke Lifetimes

The residence times of atmospheric aerosols depend on their chemical composition, morphology and sizes. Soot generated from oil combustion is generally hydrophobic and resistant to water nucleation (Radke et al., 1980a). However, if the soot coagulates into larger particles, it may interact more strongly with water (Bigg, 1985). The smoke produced in urban and industrial fires would be coated with hygroscopic materials as it aged, making it more susceptible to removal by clouds and precipitation.

Under normal atmospheric conditions, the e-folding lifetime of atmospheric aerosols with radii less than 0.1 μm is shorter than a few days (Jaenicke, 1981). Particles with radii larger than a few microns are likewise removed rather rapidly from the ambient troposphere by precipitation scavenging and gravitational settling (although fluffy aggregates would settle out much more slowly). Aerosols with radii between about 0.1 and 1.0 μm have the longest lifetimes, generally on the order of a few days to a week in the lower troposphere, a month in the upper troposphere, and 1–2 years in the lower stratosphere (Jaenicke, 1981). Particles in this Greenfield gap size range also happen to affect sunlight most effectively (see the following section and Chapter 4).

It is important to remember that, if atmospheric stability and precipitation rates were greatly perturbed following a nuclear war, aerosol removal rates and lifetimes would be altered accordingly. In particular, if increased stability and reduced precipitation occurred on a hemispheric scale, as now seems likely (see Chapter 5), the atmospheric lifetime of smoke could be lengthened considerably. The stabilization process involves the absorption of solar radiation by smoke, which is described in the next section.

3.6 OPTICAL PROPERTIES

The primary means by which the smoke and dust injected into the atmosphere by nuclear explosions and fires affects atmospheric processes is through interaction with solar and thermal radiation. Thus, the determination of the optical properties of the aerosol is a critical factor in assessing the climatic impact. The optical properties are functions of the composition of the aerosol, of the morphology (shape) of individual particles and of the size distribution of the aerosol. As noted in the preceding sections, the composition of the smoke is determined by a complex interaction of fire intensity, fuel type and loading, particulate emission factors, and fire duration. Particle morphology and size distributions are determined by formation processes and subsequent microphysical processes. Because the physical properties of the particles are changing with time, the optical properties are also subject to change with time.

Given the complexity of the particles, it is not possible to derive exact expressions for their optical properties. Measurements of the optical properties of smoke from large fires are extremely limited. In order to arrive at an estimate for the optical properties, two approaches will be followed. First, the optical properties of idealized spherical particles will be discussed and then applied to smoke. Secondly, a simple extrapolation of laboratory measurements of the properties of elemental carbon will be carried out to infer the optical properties of smoke. Finally, some observations that bear on the problem will be considered.

3.6.1 Optical Coefficients

The optical coefficients of atmospheric particles are usually computed using Mie theory for homogeneous (i.e., uniform composition) spheres. Rigorous theoretical models are also available for some symmetric shapes such as spheroids (Asano and Sato, 1980). In order to carry out these calculations, the size of the sphere and the index of refraction of the material of which it is made must be known. The index of refraction is a physical property of the material related to its ability to reflect and absorb electro-magnetic radiation and can be measured by a variety of techniques. In addition to the exact theories, some approximate theories have been developed for irregularly shaped particles (Pollack and Cuzzi, 1980).

For homogeneous spheres, a convenient quantity used to describe the interaction between aerosol particles and electromagnetic radiation is the Mie size parameter x, defined as

$$x = \frac{2\pi r}{\lambda} \qquad (3.1)$$

where r is the particle radius and λ is the wavelength of the radiation. The particle-field interaction is often expressed in terms of the extinction efficiency, Q_e, which is defined as the ratio of the cross-section for extinction (i.e., the effective total cross-section of the particle as seen by the electromagnetic radiation) to the geometric cross-section. For a given material and, hence, a known index of refraction, Q_e is only a function of x. Typically, the particle-field interaction (and Q_e) has a maximum value when x is of order 1. For large x, Q_e tends asymptotically to a value of 2 due to diffraction effects; for small x and no absorption, Q_e decreases rapidly as x^{-4} (the Rayleigh regime). Thus, the interaction is greatest when the particle size and the photon wavelength of the electromagnetic radiation are comparable in size. Intuitively, this means that the maximum effect per unit mass of material on a radiation field is achieved by subdividing that material into particles with diameters approximately equal to the photon wavelength. Particles much smaller than the photon wavelength have only a minimal effect on the photon.

The extinction efficiency of a particle Q_e is the sum of its scattering efficiency, Q_s, and its absorption efficiency, Q_a. Radiation which is scattered by the particle is simply re-directed from its original direction of propagation to some other direction (although the majority of the scattered radiation continues on in nearly the same direction of propagation). Radiation which is absorbed by the particle, on the other hand, is removed from the propagating beam and converted to some other form of energy, usually heat. Obviously, for nonabsorbing particles, $Q_s = Q_e$. For absorbing particles, the partitioning between scattering and absorption depends on the index of refraction (i.e., the material of which the particle is composed) and the size parameter. At small values of x, absorption dominates scattering and $Q_a \sim Q_e$. At large values of x and for strongly-absorbing material, Q_a and Q_s are roughly equal. (Detailed treatments of Mie theory and efficiency factors are available in a number of texts such as Kerker, 1969 or van de Hulst, 1957).

Solar energy is emitted predominantly at wavelengths between 0.3 and $2 \mu m$, with a maximum at about $0.55 \mu m$. Wavelengths of the thermal infrared radiation produced by bodies with temperatures around 0°C or 273 K are in the range of 5 to $50 \mu m$. The representative wavelength for the thermal infrared radiation is usually chosen to be $10 \mu m$, both because peak emission occurs at about this wavelength and because the Earth's atmosphere is essentially transparent to radiation at this wavelength, which means maximum cooling of the Earth's surface occurs due to radiation in this spectral region. (see Chapter 4 for a more extended treatment of the radiation budget of the Earth and atmosphere). Thus, the ratio of the wavelength of maximum thermal emission ($10 \mu m$) to the wavelength of maximum solar energy is about 20. For a material with approximately equal values of refractive index both wavelengths, solid spheres with radii less than $0.5 \mu m$ absorb about 20 times more energy at visible solar wavelengths than at thermal infrared wavelengths, and are on the order of 10^4 times more efficient at scattering visible light than thermal infrared radiation. Since a typical distribution of atmospheric aerosols produced by combustion processes has a mean, or average, particle radius on the order of a few tenths of a μm, the distribution typically has a maximum extinction efficiency at wavelengths on the order of 0.5 to $0.6 \mu m$, which coincides with the maximum energy emission of solar radiation. At a wavelength of $10 \mu m$, $x \sim 0.05$ and the value of Q_e is substantially less than at a wavelength of $0.5 \mu m$.

Several cautionary notes should be added to the discussion in the preceding paragraph. First of all, because atmospheric aerosols exist in a range of particle sizes, the extinction efficiency for the distribution at a particular wavelength is a weighted average of the efficiencies of the individual particles at that same wavelength. In general, this averaging tends to reduce the variation in the values of the efficiency factors with wavelength. Secondly, many materials have larger indices of refraction (and, hence, larger extinction efficiencies for a given value of x) at thermal infrared wavelengths than at visible wavelengths. This also tends to increase the value of Q_e at thermal wavelengths relative to visible wavelengths. As a rough rule, Q_e for atmospheric aerosols at $10 \mu m$ is about 1/10 the value of Q_e at $0.5 \mu m$.

To illustrate these points, values of Q_e and Q_a have been computed from Mie theory for spheres as a function of wavelength from 0.2 to $30 \mu m$ (Figure 3.3). The spheres are assumed to have a complex index of refraction of 1.55–0.1i at all visible wavelengths, as suggested in the NRC report (1985).

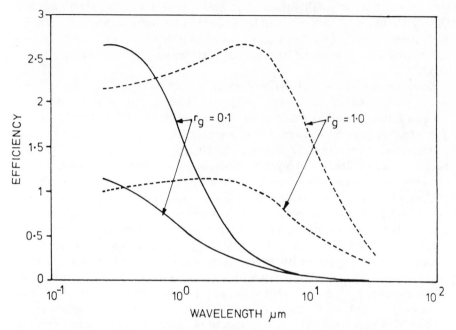

Figure 3.3. Extinction efficiency (upper curve) and absorption efficiency (lower curve) as a function of wavelength for absorbing aerosols (index of refraction = $1.55 - 0.1i$). Solid curves are for a size distribution with a number mean radius, r_g, of 0.1 μm, dashed curves for a number mean radius of 1.0 μm

While there are some indications that the imaginary part of the refractive index, and thus the absorption efficiency, of sooty material may increase somewhat at infrared wavelengths (Tomaselli et al., 1981), it was held constant to better illustrate the effects of particle size. Calculations were made for two log-normal size distributions, one with a geometric mean radius of 0.1 μm, and the other with a mean radius of 1.0 μm. The smaller value was chosen as typical of aerosol distributions produced by anthropogenic activity (Lenoble and Brogniez, 1984). The larger radius was chosen as typical of coarse aerosol distributions, typically produced by mechanical processes such as wind blowing across sand or soil; it also represents an approximate upper limit for a size distribution of climatic interest, since aerosols of larger sizes have atmospheric lifetimes of several days or less. In both cases the geometric standard deviation of the size distribution (a measure of the dispersion of particle sizes about the mean) was taken to be 2, which is typical of atmospheric aerosol distributions. For the smaller size distribution, absorption is fairly constant,although decreasing slightly with increasing wavelength, throughout the solar spectrum. At a wavelength of 1 μm, Q_c and Q_a both decrease abruptly, and are reduced by more than an order of magnitude at 10μm. For the larger particles, the peak absorption occurs at a wavelength of about 2μm, although to a good first approximation, the absorption is fairly uniform from 0.3 to 5μm. At a wavelength of 10μm, the absorption efficiency is about half its value at a wavelength of 0.5μm.

Another useful quantity, particularly from an experimental point of view, is the specific absorption, s_a. It is defined as the absorption cross-section per unit mass of absorber and is usually given in units of m^2/g. For homogeneous spheres, it is related to the absorption efficiency by the expression

$$s_a = \frac{(3/4)Q_a}{r\rho} \tag{3.2}$$

where r is the radius of the sphere and ρ is the density of the sphere. For x much less than 1 (i.e., small particles), absorption is directly proportional to the mass of the particle, which means that for a fixed wavelength, Q_a increases in direct proportion to r, the particle radius (because Q_a is equal to the absorption cross-section—which is increasing as r^3—divided by the geometric cross-section—which is increasing as r^2). Thus, from equation 3.2 above, s_a is independent of particle size for small absorbing spheres.

3.6.2 Absorption by Soot Agglomerates

Typical carbon aerosol agglomerates produced by combustion processes have mean radii on the order of a few tenths of a micron (Janzen, 1980; Borghesi et al., 1983). If these agglomerates were solid spheres of the same dimension and, thus, Mie theory were applicable, then, as illustrated in Figure 3.3, the particles would have a maximum extinction efficiency at visible wavelengths (see also Bergstrom, 1973; Faxvog and Roessler, 1978), and a very much lower efficiency at infrared wavelengths. In high density smoke plumes, the particles can agglomerate to sizes comparable to the wavelengths of infrared radiation or larger. Again applying Mie theory, these agglomerates would have reduced extinction efficiencies at visible wavelengths and increased efficiencies at infrared wavelengths.

However, Mie theory does not apply to agglomerated particles in general, and to fluffy or chained agglomerates composed of highly absorbing material such as soot in particular. While the scattering from these fluffy agglomerates would be similar to that from a solid object with the same dimensions, the absorption would be very different. Intuitively this may be understood by realizing that absorption of electromagnetic radiation is related to both the mass of the absorbing material and the amount of that mass which can be "seen" by an individual photon. Thus, for a fluffy object, which has a much greater surface to volume ratio than does a sphere of the same size, much more of the absorbing material is available for interaction with the radiation. This effect may enhance the absorption of solar radiation by the agglomerated particle since each small carbon spheroid composing the agglomerate may act as an independent absorbing particle. Furthermore, because the spheroids in the agglomerate have only a minimal area of contact with each other, they cannot act as a volume absorber for radiation whose wavelength is much greater than their size.

   This intuitive picture of absorption is supported to some extent by measurements of smoke absorption by laboratory smokes, but it needs further verification. Laboratory measurements of the specific absorption of soot agglomerates (e.g. Janzen, 1980; Wolff and Klimisch, 1982; Jennings and Pinnick, 1980; Roessler and Faxvong, 1980; see also Gerber and Hindman, 1982, for a detailed treatment of various measurement techniques and a report of an intercomparison experiment) consistently fall in the range of 8 to 10 m^2/g regardless of measurement technique or agglomerate size, supporting the picture of absorption outlined above. Lee (1983) carried out an extensive set of measurements of the specific absorption of a variety of carbon soot agglomerates. He also simultaneously obtained electron micrographs of the agglomerates. His results show that the specific absorption is independent of agglomerate size or shape until the agglomerates become very compact, i.e., they begin to resemble solid objects. Even in this case, however, the absorption is greater than that of a solid sphere of the same size. The results of measurements of spectral transmission between 0.5 and 2.2 μm reported by O'Sullivan and Ghosh (1973) are mixed. In experiments designed to study coagulation, they found a small decrease in optical density at 0.5 μm relative to 2.2 μm after 10 minutes of aging. This would be the expected result if larger particles were created through coagulation and Mie theory were applicable to these particles. On the other hand, their measurements of optical density for several different smoke concentrations indicate no change in the ratio of transmission at 0.5 to 2.2 μm, which may be due to insufficient time for coagulation to occur. Since these measurements do not distinguish between absorption and scattering, the change in absorption with aging cannot be directly deduced. Similarly, the results of Bruce and Richardson (1983) are inconclusive. They found that the specific absorption of soot at 10 μm was the same whether or not large aggregates were excluded, i.e., the large aggregates had the same absorption per unit mass as did small chains of carbon spheroids. Unfortunately, the relative abundance of large and small agglomerates was not carefully controlled in the experiments. Based on estimates from electron microscopy, they concluded that the large particles may have had insufficient mass to affect the reflectivity of their samples. Thus, they may also have been insufficient to have a detectable impact on the absorption.

   Theoretical treatments of the spheroid problem (Jones, 1979; Berry and

Percival, 1986) support the picture that spheroids in chained or fluffy agglomerates act essentially as independent absorbers. However, these treatments are only approximate and further development is needed. Additionally, the problem of chains or fluffy agglomerates surrounded by approximately transparent liquids needs to be addressed (see discussion below). These latter particles actually may be dominant in the smoke from large fires. Clearly, there is a pressing need for further measurements of the optical characteristics, both absorption and scattering, of particles produced in smoke plumes.

3.6.3 Scattering and Absorption by Smoke

The discussion thus far has concentrated on particles consisting only of a single absorbing material. However, particles emitted from fires are composed of a variety of materials with varying optical properties. Several approaches for determining the optical properties of the composite smoke have been tried. The simplest is to assume that the optical properties of the composite are mass-weighted averages of the optical properties of the individual components. This approach is particularly attractive for smokes where the emitted materials can be broadly separated into amorphous elemental carbon, which dominates the absorption, and all other materials, which scatter light only. (This approach is *not* the same as taking a mass-weighted average of the indices of refraction of the various materials and then computing the specific absorption of the mixture from the average index of refraction. Such an approach is almost certainly incorrect for mixtures of highly-absorbing and weakly-absorbing materials.)

As noted above, the specific absorption of elemental carbon is well represented by the figure of $10 \, \text{m}^2/\text{g}$. A number of other studies (e.g., Waggoner et al., 1981; Tangren, 1982) suggest that the specific scattering, s_s of submicron particles is about $3.5 \, \text{m}^2/\text{g}$ at visible wavelengths between 0.5 and 0.6 μm. Thus the specific scattering and absorption of smoke may be approximated by the expressions

$$s_s = 3.5 \, \text{m}^2/\text{g smoke} \tag{3.3a}$$

$$s_a = 10 f_{\text{EC}} \, \text{m}^2/\text{g smoke} \tag{3.3b}$$

where f_{EC} denotes the mass fraction of elemental carbon in the smoke. Given the total mass of smoke emitted, the mass fraction of elemental carbon, and the area covered by the smoke, these expressions can be used to deduce the optical depth of the smoke and the attenuation of the solar radiation impinging on the smoke (see section 3.7)

The expressions (3.3a) and (3.3b) are based primarily on measurements of submicron aerosols. Observational studies of smoke from forest fires (e.g. Radke et al., 1978; Tangren, 1982; Vines et al., 1971; Patterson and McMahon, 1984), small flaming sample fires (e.g. Bankston et al., 1981), and large 1000 MW fuel oil burners (Radke et al., 1980a) show that particulate matter in the submicron size range was produced in these fires. The production of much larger particles, however, has also been reported. Hobbs et al. (1984) recently measured the presence of a substantial mass fraction of particles with radii larger than 1 μm from prescribed burns of forest products. The measured number concentration peak was, however, at 0.1 μm. Similar observations of large particles produced by forest fires were made by Bigg (1985). Large agglomerates of soot particles can also be produced from surface oil fires (Day et al., 1979) and have been observed in urban environments (Russell, 1979). The burning of synthetic polymers can also produce large, supermicron sized, branched soot agglomerates (W.D. Woolley, J.E. Snell, personal communications). As discussed in the preceding section, (3.3b) may well be applicable to almost pure soot agglomerates in this case, provided that they are fluffy and not tightly packed. The applicability of (3.3a) for large, supermicron sized particles has not been justified.

   If the smoke particles actually consisted of agglomerates surrounded by oil or water shells, as is typical of wood smoke for example, the optical properties of the particle could be modified by the presence of the shell, although the exact nature of this modification is not known. It is possible that the surface tension of the liquid would collapse the fluffy agglomerate into a more compact, roughly spherical particle. Calculations of the effect of a homogeneous, spherical shell of nonabsorbing material surrounding a concentric core of absorbing material show that the presence of the shell increases the absorption per unit mass of the core material by as much as a factor of 2 or 3, depending on the relative sizes of the core and the shell (Ackerman and Toon, 1981). A second possibility is that the agglomerate would break up into its component spheroids and that they are dispersed more or less uniformly through the nonabsorbing liquid. Calculations carried out for this case show an even more dramatic increase in the specific absorption (Chylek et al., 1984).

   It is not entirely clear which of these models is correct either for the fresh smoke plume, where organic liquids may be condensing on the soot agglomerates, or for the plume at somewhat later stages where water may be condensing on the agglomerates. Limited experimental evidence exists for both (Chylek et al., 1984; Z. Levin, personal communication). To some extent, the appropriate model may be dependent on the material comprising the agglomerate and the forces holding it together, as well as the amount of water and/or organic liquids available to condense on it. However, according to the theoretical calculations, in either case the presence of a liquid deposit on an agglomerate would act to increase the effective absorption of the carbon, so that (3.3b) could strongly underestimate the absorption of absorbing and non-absorbing aerosol. On the other hand, if the liquid were to evaporate, the residual aerosol particle would be more compact than the original fluffy agglomerate. It would also likely be increased in size, since more than one aerosol particle could be scavenged by the drop. Both these effects would tend to reduce the specific absorption of the elemental carbon as well as the lifetime of the particle. The magnitude of the reduction would depend on the final shape and size of the "processed" aerosol.

Experimental evidence bearing on the problem is ambiguous. The results of Patterson and McMahon (1985, 1986), which were discussed in section 3.4, show no enhancement of specific absorption in wood smoke, suggesting that the theoretical models may be incorrect. However, no particle sizing or electron microscopy was performed, so the particle morphology in their smoke samples is unknown. On the other hand, comparisons of measurements of inferred elemental carbon concentrations (Rosen and Hansen, 1984) and of solar absorption (Ackerman and Valero, 1984) in Arctic haze events do suggest an enhanced specific absorption. Again, however, a complete description of particle size and composition was not obtained.

3.6.4 Wavelength Dependence

In order to assess the climatic impact of the smoke (as is considered in Chapters 4 and 5), it is necessary to know the value of the extinction and absorption at thermal infrared wavelengths as well as at solar wavelengths. Unfortunately, it is difficult to measure these properties at infrared wavelengths, so much of the available information is inferential or qualitative.

The calculations of Turco et al. (1983) and Ramaswamy and Kiehl (1985) for equivalent spheres give ratios of the extinction efficiency at 10 μm to that at 0.5 μm of about 1 to 10 or 15, and ratios of the absorption of about 1 to 5. The latter ratio is somewhat larger because a greater fraction of the extinction is due to absorption at infrared wavelengths. The transmission measurements of O'Sullivan and Ghosh (1973) and Randhawa and Van der Laan (1980) suggest this ratio may be as low as 1 to 100 for some smoke. Since aerosol extinction optical depth is directly proportional to the extinction efficiency, these values indicate that the optical depth of smoke at solar wavelengths is substantially greater than the optical depth at infrared wavelengths (which is true in general for atmospheric aerosols).

Qualitative information on the wavelength dependence of optical depth, and extinction efficiency, in actual fire plumes can be inferred from satellite imagery. Multiple views of the same scene taken with different spectral bandpasses show clearly visible smoke plumes from wildfires and agricultural burning at a wavelength of 0.5 μm, barely visible plumes at a wavelength of 3 μm, and no plume at all at a wavelength of 10 μm (Matson et al., 1984; J. Brass, personal communication). In fact, several research projects currently underway are attempting to take advantage of the transparency of the plume at infrared and near infrared wavelengths to locate and monitor wildfires.

There are essentially no data available on the thermal infrared properties of urban smoke. Further research on both laboratory aerosols and fire plumes is urgently needed to quantitatively define the wavelength dependence of smoke extinction and absorption.

Some comments (e.g., Bigg, 1985) on the "nuclear winter" hypothesis have suggested that the infrared optical depth of smoke could be equal to or greater than the solar optical depth as a result of the production of large particles by coagulation. As evidence of this, the observation of the blue Sun in Europe in 1950 (Bull, 1951) is often cited. This effect was produced by the presence of atmospheric aerosols from Canadian forest fires (Wexler, 1950).

Typically, the Sun is red when viewed through fresh smoke plumes because the relatively small particles formed in the combustion process are more effective at scattering shorter wavelengths (blue light) than longer wavelengths (red light), while their absorption is roughly constant. Thus the Sun seen in transmission appears red. As the plume ages, the particles coagulate up to larger sizes which are approximately equally efficient at scattering all visible wavelengths. When viewed through this more aged smoke, the Sun appears white, or perhaps light grey, depending on the optical thickness of the plume. By extension, it has been suggested that the blue color of the Sun was due to the presence of very large particles (with radii on the order of 1 to 10 μm or larger (Bigg, 1985) formed by further coagulation in the plume as it travelled from Canada to Europe. Furthermore, measurements of atmospheric turbidity taken in Edinburgh showed that the plume had a somewhat larger optical depth at 0.6 μm than at 0.4 μm (Wilson, 1950). However, Porch et al, (1973) and E.M. Patterson (personal communication) has pointed out that the blue Sun and the measurements can be explained by assuming that the plume was composed of a very narrow size distribution of particles having a number mean radius of 0.5 μm. Considering the long distance which these particles travelled without experiencing gravitational settling, it seems more plausible that they were particles of this size rather than particles with radii on the order of 10 μm, as has been suggested by others.

3.7 ATTENUATION OF VISIBLE LIGHT

The emissions given in Section 3.3 and the optical coefficients given in Section 3.6 can be combined to provide an estimate of the effect of the smoke on sunlight reaching the ground. The average column density, D_c, (defined as the total mass of smoke in a vertical column with a cross-sectional area of 1 m^2) of the smoke can be found by dividing the total smoke emission by the area over which it is assumed to spread. If the smoke is assumed to spread over half of the Northern Hemisphere (an area of about 1.28×10^{14} m^2), the average column density for the NRC emission estimate is 1.2 g/m^2, and for the CGB emission estimate is 0.4 g/m^2. In both cases, the average column density of amorphous elemental carbon would equal 0.23 g/m^2.

The extinction optical depth, τ, which is a dimensionless measure of the opacity of an atmospheric column, is the sum of the scattering optical depth, τ_s, and the absorption optical depth, τ_a. It can be computed from the expression

$$\tau = s_e D_c = (3.5 + 10.0 f_{EC}) D_c \qquad (3.4)$$

where the right-hand side is found from the sum of (3.3a) and (3.3b). By definition, the first term on the right is τ_s and the second τ_a. Substituting the values of f_{EC} from section 3.3 and the values of D_c gives values of τ_s and τ_a of 4.1 and 2.3, respectively, for the NRC scenario and 1.5 and 2.3, respectively, for the CGB scenario.

The transmission of the direct solar beam through a column with optical depth τ is found from the expression $e^{-\tau}$, assuming the Sun is directly overhead, i.e., at the zenith position. (For the Sun at an angle θ from the zenith, the optical depth must be multiplied by secant(θ)). Under these conditions, even for the Sun at the zenith, the total sunlight, both direct and scattered, reaching the Earth's surface would be reduced to less than $e^{-2.3}$, i.e., less than 10%, of its normal value in both scenarios.

The actual amount of sunlight reaching the ground would be even less than that given above due to scattering by the aerosols. Sagan and Pollack (1967) derived an approximate formula for an effective absorption optical depth, τ_{eff}, that accounts for the combined effects of scattering and absorption:

$$\tau_{eff} = 1.7(\tau_a + 0.15\tau_s) \qquad (3.5)$$

Although this expression was derived for optically thick atmospheres, it serves as a useful approximation in this context and was applied by CGB. Equation (3.5) gives the values $\tau_{eff} = 5.0$ for the NRC scenario and $\tau_{eff} = 4.3$ for the CGB scenario. These values imply a transmission of at most 1% of sunlight to one quarter of the Earth's surface due only to the smoke emissions.

This analysis of solar transmission through the atmosphere is only approximate and mainly descriptive. A more rigorous treatment of radiative transfer in smoke clouds can be found in Chapter 4, as well as in Turco et al., (1983a) and Ramaswamy and Kiehl (1985). In addition, it would take

for the smoke to be distributed over one quarter of the Earth.
is time, coagulation, rainout, and other microphysical processes
uce the smoke levels in the atmosphere. These factors and their
implications for the climatic impact of the smoke are discussed in detail in
Chapters 4 and 5.

3.8 DUST

3.8.1 Formation Mechanisms

Ever since the first nuclear test explosion in the desert of New Mexico on
July 16, 1945 (the Trinity test), scientists have realized that nuclear explosions can raise large quantities of soil dust and debris to high altitudes. The
dust forming mechanisms are manifold (Glasstone and Dolan, 1977):

1. The thermal radiance of the fireball causes rapid steam expansion and
   blowoff of surface soil over a large area.
2. The blast winds and turbulence churn up additional soil and dust in the
   region adjacent to the burst.
3. Detonations on land surfaces eject large amounts of soil at high velocity
   during crater formation.
4. The high temperatures and pressures of the fireball in contact with the
   surface cause soil and rock to vaporize and liquefy; some of the material
   later solidifies into fine glassy aerosols.
5. The ascending fireball lifts entrained materials to high altitudes.
6. The suction and afterwinds created by the rising fireball draw additional
   dust and debris up the stem of the mushroom cloud.

3.8.2 Quantities and Properties

Based on analyses of dust samples collected in nuclear explosion clouds
during the test series of the 1950s and 1960s, it has been estimated that,
on average, 100,000 to 300,000 tonne of soil debris can be lofted into the
stabilized cloud of a 1 Mt surface explosion (Rosenblatt et al., 1978; Gutmacher et al., 1983; NRC, 1985). Although most of the debris consists of
particles exceeding 10 μm in radius, up to 5 to 10% (by mass) may consist
of submicron particles (Nathans et al., 1970a; Yoon et al., 1985).

Information on the sizes of dust particles is sparse. For continental land
surface explosions, data from the Johnny Boy near-surface test (Nevada Test
Site, 0.5 kt, July 11, 1962) provide the most complete description of size distributions (Nathans et al., 1970a; Yoon et al., 1985). In this case, the size
characteristics of the particles were carefully analyzed in the laboratory from
filter samples collected in the stabilized explosion cloud. In this regard, it

should be noted that size distributions derived from fallout samples are not characteristic of the dust in the clouds aloft, particularly in the particle size range below several microns in radius. Data on the size distributions of dust raised by large nuclear tests on Pacific coral atolls are also of limited usefulness because of the small extent of the land masses and lack of continental soils at these sites. The Pacific tests seem to place a lower limit on the submicron dust mass fraction of about 1% (Heft, 1970). The uncertainty range in the submicron particle fraction for bursts on continental soils is at least a factor of three.

Following the largest atmospheric nuclear tests of the 1950s and 1960s, no obvious long-term effects from aerosol injection into the atmosphere were noted (e.g., Machta and Harris, 1955). This is not unexpected for several reasons:

1. In total, the principal tests amounted to about 450 Mt distributed over the decade from 1952 to 1962.
2. The largest tests occurred well above the surface, or on barren atolls, where minimal quantities of fine dust and essentially no smoke were produced.
3. The debris clouds were not carefully tracked and characterized, which precludes a present-day calibration of the expected effects.

The quantity of dust lofted by a nuclear explosion decreases steadily as the height-of-burst increases. As long as the fireball is in close contact with the surface, more than 100,000 tonne of debris can be lifted per Mt of explosive (the mass raised per unit yield decreases slowly as the yield increases above approximately 1 Mt). For a near-surface burst, in which the fireball is barely in contact with the surface, the amount of dust lofted is much smaller; in this case there is little vaporized material in the fireball and most of the dust is swept up by afterwinds. At even greater heights-of-burst, only the refractory materials used in bomb construction are available to condense as a fine aerosol (Nathans et al., 1970b).

For subsurface explosions, the amount of soil excavated from the crater at first increases with the depth-of-burst, then decreases again. However, while the quantity of soil displaced by a subsurface explosion may be greater (at some burst depths) than the quantity displaced by a surface explosion, the height of the dust cloud in the former case is lower because the fireball rise is damped in the denser medium. This also occurs in subsurface water bursts.

The heights of stabilization of nuclear dust clouds depend on the explosion yield, height-of-burst, season, and meteorological state of the atmosphere (Glasstone and Dolan, 1977). For low-altitude explosions of less than 100 kt, the stabilization height of the cloud depends to a large degree on the thermal stability of the lower atmosphere; the clouds can rise as high as the tropopause, but generally cannot penetrate into the stratosphere. For surface and low-altitude explosions on the order of or greater than 100–200 kt, the cloud stabilization height is determined almost entirely by the thermal structure of the stratosphere. The cloud of a 1 Mt explosion at middle latitudes stabilizes wholly within the lower stratosphere; larger bursts stabilize at higher altitudes. In this dynamical regime, the height of stabilization scales approximately as $Y^{0.2}$, where Y is the yield in megatons (NRC, 1985). For long-term climatological studies, primary interest centers on the quantity of fine dust injected into the stratosphere (and perhaps the upper troposphere when the atmosphere is disturbed). Hence, surface and near-surface detonations on the order of or greater than 100–200 kt should be considered.

The morphology of nuclear-generated dust particles is diverse. The smallest particles (micron to submicron sizes) can be either spherical glassy (or metallic) beads or equidimensional soil and rock mineral grains. Spherical particles are produced by the condensation of vaporized refractory compounds and by the atomization of jets of liquefied minerals. Fine soil grains are produced by the crushing, disaggregation and entrainment of earth and rock. For optical calculations, the fine dust particles may be treated as equivalent-volume spheres. In the Johnny Boy dust sample, the specific extinction of the submicron particle fraction (at a wavelength of 550 nm) was about 3 m^2/g (NRC, 1985); the absorption is usually assumed to have accounted for 1–3% of the extinction (the remaining extinction being due to scattering).

There is conflicting evidence from the inspection of filter samples concerning the agglomeration of dust particles in nuclear clouds. While the clouds do not appear to be strongly electrified, they are highly turbulent and can hold substantial masses of ice (up to several hundred thousand tonne of ice per megaton of yield, from ground water and air moisture). Thus, turbulent coagulation and collection on ice crystal surfaces are possible aggregation mechanisms. Unfortunately, reliable quantitative information on dust particle clustering is unavailable. An early study pointed to the absence of agglomeration (Nathans et al., 1970a). However, a preliminary visual reanalysis of several high-altitude filter samples reveals occasional clusters of impacted particles (G. Rawson, personal communication). Whether these clusters are related to the breakup of true dust agglomerates, or are caused by the natural "shedding" of small particles by large soil grains upon impact (Rosinski and Langer, 1974), and what fraction of the total fine particle load is associated with agglomerates are unanswered questions. It is particularly noteworthy that the local fallout from surface bursts contains negligible quantities of submicron dust; indeed, 40–60 percent of the total radioactivity carried by the finest dust grains escapes into the global atmosphere (Glasstone and Dolan, 1977). In view of the existing evidence, submicron particle agglomeration in the early stabilized clouds of nuclear surface detonations may be treated as a secondary effect, although long-term coagulation and removal of the dust must be accounted for (Turco et al., 1983a,b).

Since most of the nuclear test explosions were conducted on barren soils, little is known about the impact of soil organic matter on dust cloud optical properties. In fact, many potential targets of surface nuclear explosions, particularly missile silos, are based in regions of highly organic soil (e.g., the chernozems of the U.S. Great Plains and the peat soils of the Siberian forests). In some locales, the soils are black. The aerosols formed from these organically rich soils could strongly absorb sunlight. Any of this organic material engulfed in the fireball would be largely oxidized (burned), but some of the organic material scoured up by the blast and afterwinds would not be burned and could potentially absorb a significant fraction of the incident sunlight. By contrast, aerosols generated from barren soils are unlikely to absorb sunlight efficiently (although, occasionally, the finest glassy particles collected in nuclear clouds are black due to dissolved iron compounds). In the climate calculations carried out to date, the aerosols have been assumed to be only weakly absorbing.

Most of the nuclear tests were also conducted over coarse soils (e.g., coral atolls), whereas most nuclear targets are located in soils with substantial clay (fine particle) fractions. (The Johnny Boy test, however, occurred on a desert alluvium with a substantial fine particle abundance; G. Rawson, personal communication.) Finer parent grain sizes imply that greater quantities of submicron dust can be generated when the soil is dried and pulverized by a nuclear burst.

A number of factors could reduce the quantity of dust lofted by a surface or near-surface nuclear detonation:

1. Soil moisture, which increases soil cohesion.
2. Vegetative cover, which blocks thermal radiation and holds soil down.
3. Surface layers of hardpan, rocks, snow, or ice, which suppress dust formation.

On the other hand, vaporization and liquefaction of the soil, and pulverization of surface materials within the high-overpressure "sweep up" zone, should not be greatly affected by these factors.

3.8.3 Multiburst Effects

Nuclear attack strategies may call for multiple targeting of key military facilities (for example, double or triple detonations over missile silos). Explosions that are proximate in both space and time will interact strongly. However, there are no nuclear test data bearing on "multiburst" processes. The following effects might be expected:

1. An initial explosion would dry, excavate, and pulverize soil, which then could be more easily swept up by subsequent explosions.
2. Overlapping fireballs would reinforce buoyant motions, carrying dust to greater altitudes than might otherwise be expected (in order for reinforcement to occur, the weapons would have to be detonated within seconds of each other at nearly the same location, a feat that might be difficult to achieve operationally).

Only preliminary hydrodynamic model calculations are available to estimate the effects of interacting nuclear bursts (e.g., Filipelli, 1980; NRC, 1985). The calculations suggest a potential enhancement in dust lofting. One analysis argues on physical grounds that the dust mass raised (per megaton of yield) could be larger by a factor of 10 in multiburst environments (NRC, 1985), although detailed quantitative demonstrations of this point are lacking.

3.8.4 Integrated Dust Injections

Figure 3.4 shows a simulated dust pall that could be generated in a counterforce nuclear exchange between the superpowers (Yoon et al., 1985). The predicted total quantity of submicron dust in the upper atmosphere after 5 days resulting from 2500 Mt of land surface bursts is 40 million tonne. This is about twice the quantity computed by the NRC (1985), about one-half the baseline quantity of Turco et al. (1983), and roughly the quantity expected from the scenario outlined in Chapter 2. The differences in dust injections can be attributed mainly to differences in the assumed total yield of surface bursts. The simulation in Figure 3.4 assumes a dust mass lofting (for surface explosions) of 0.27 million tonne per megaton of yield, with about 8% of that amount in the submicron size range. Roughly one-half of the dust is generated by explosions in the yield range of 2–20 Mt, and one-half by explosions in the range of 0.3–2 Mt. Given the current evolution of weapon yields toward the smaller range, the amount of dust and the height of injection may be somewhat too large in this simulation, but the general features of the simulation would still be appropriate.

The dust simulation of Yoon et al. (1985) suggests that, after just 5 days, the initial, stabilized detonation clouds would have been displaced and sheared by the prevailing wind systems, and would blanket most of the northern mid-latitude zone under a pall of soil debris. The extinction optical depths at visible wavelengths, most of which contributes to scattering, are greater than 8 in some regions, although, if the dust were distributed uniformly over the Northern Hemisphere, the optical depth would be 0.5. While, for a given optical depth, the radiative effects of dust are considerably smaller than the radiative effects of smoke, an optical depth of 8 can reduce the average solar energy reaching the ground by 80 percent (see also Chapters 4 and 5).

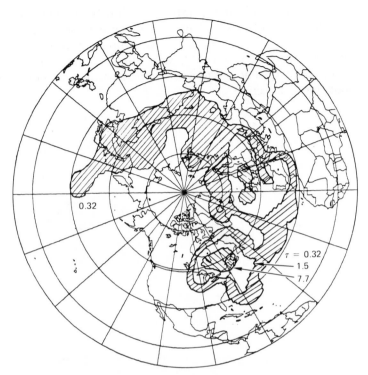

Figure 3.4.   Geographical distribution of nuclear dust clouds five days after a July counterforce exchange of 4000 Mt against missile silos and air bases. A three dimensional tracer model was used to follow the dispersion of the dust clouds. Winds for July were obtained from the 2.5° grid data of the National Meteorological Center, Washington, DC, and were updated every 12 hours in the simulation. The (zenith) extinction optical depth contours (for $\tau = 0.32$, 1.5, and 7.7) at a wavelength of 0.55μm are given. Essentialy all of the dust in the figure resides in the upper troposphere and stratosphere. The particle physics treated in the model is described by Yoon et al. (1985). (Figure supplied by B. Yoon.)

The data in Figure 3.4 may represent a reasonably conservative picture of the dust environment after a major nuclear exchange. For example, the calculations could also take into account a broader range of military targets, higher absorption by the aerosol due to the organic component of the soil, multiburst effects, and possible dust injections by powerful updrafts over intense fires. Factors that could limit the injection include early agglomeration and rainout and weather conditions favoring soil cohesion.

It should be noted that the injection of dust into the stratosphere is more important for the development of climatic effects than injection into the troposphere, because stratospheric dust has a much longer lifetime (in the unperturbed atmosphere). Owing to the uncertainties in the number of surface bursts, the dust mass lofted, particle size distributions, and multiburst interactions, the total quantity and impacts of nuclear-generated dust will remain ambiguous. However, the quantities of optically-active dust raised by a full-scale nuclear exchange could, within the parameter ranges defined by observations, be large enough to cause some environmental disturbances even without smoke injection.

APPENDIX 3A

Urban Fire Development

3A.1 INTRODUCTION

No comprehensive analysis exists of large urban fires, either nuclear-initiated or of conventional origin. In previous global estimates of the extent of urban fires and the amount of smoke generated from a nuclear attack, it has been assumed that about 250 km^2 could be ignited and burned by a 1 Mt detonation (Turco et al., 1983a,b; Crutzen et al., 1984; NRC, 1985). This assumption can be checked by analyzing fire-development processes on local urban scales. In this section, three distinct types of urban areas are studied to assess fire ignition and spread characteristics:

1. An idealized "uniform" city, representing a continuous residential area with wooden, two-storey structures;
2. A predominantly suburban/residential area (with many vacant lots serving as firebreaks), represented by San Jose, California in the late 1960s;
3. A major industrial/urban area, represented by Detroit, Michigan in the late 1960s.

The uniform city is useful for extensive parametric studies and for possible applications to urban areas in which fuel distributions may be relatively uniform. Data from the 1960s are used for San Jose and Detroit because current data are not available.

In analyzing fire history in each individual urban area, the initiation and spread of fires is considered in detail. Firestorms and rubble-zone fires are not treated because of the very limited understanding of these phenomena. Nevertheless, such fires could be important after a nuclear war, and they are discussed in Section 3A.6.

3A.2 FIRE DEVELOPMENT IN A SINGLE URBAN AREA

To determine the amount of smoke generated in an urban area following a nuclear attack, a characterization is needed of the area burned, the rate of burning, and the fuel consumption, among other factors. For precise simulation, specific information such as the yield and burst point of the nuclear weapon, weather conditions, fuel distribution patterns, etc., is also

87

required. The physical factors and the chain of events required in fire spread modelling are shown in Figure 3A.1. A theoretical treatment should consider all of these factors. However, in the context of studying the after-effects of a nuclear war, it is possible that specific details of the fire development would be less important than gross factors such as the total fuel impacted.

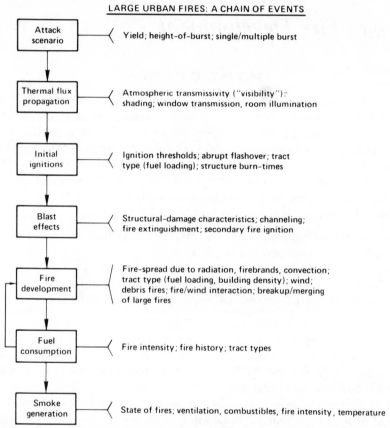

LARGE URBAN FIRES: A CHAIN OF EVENTS

Attack scenario — Yield; height-of-burst; single/multiple burst

Thermal flux propagation — Atmospheric transmissivity ("visibility"): shading; window transmission, room illumination

Initial ignitions — Ignition thresholds; abrupt flashover; tract type (fuel loading); structure burn-times

Blast effects — Structural-damage characteristics; channeling; fire extinguishment; secondary fire ignition

Fire development — Fire-spread due to radiation, firebrands, convection; tract type (fuel loading, building density); wind; debris fires; fire/wind interaction; breakup/merging of large fires

Fuel consumption — Fire intensity; fire history; tract types

Smoke generation — State of fires; ventilation, combustibles, fire intensity, temperature

Figure 3A.1.   Chain of events for characterizing urban fires and smoke generation

Physical models utilized to make urban fire calculations are described in detail by Kang et al. (1985), and references therein. Briefly, the models are based on empirical relationships between characteristics of urban buildings and fire development and spread. Clearly, such a model requires an enormous amount of data and physical knowledge of fires; not all of the needed information is adequately defined at this time. Nevertheless, the urban-fire model used by Kang et al. (1985) has been employed to provide insights into the kinds and scales of effects that might be expected in the aftermath of a nuclear explosion over a city. The results presented here should not be interpreted as literal descriptions of post-nuclear-war fire conditions. For example, the model does not include the important components of stored fossil fuels (petroleum, gasoline, natural gas distribution systems, etc.) or asphalt in its fuel loading estimates, nor an itemization of fuel types (e.g., plastics, organochemicals, etc.). Many of the assumptions and caveats pertaining to these calculations are discussed by Kang et al. (1985). More general descriptions of the problem are given by Horiuchi (1972), Takata (1972), Wiersma and Martin (1975), Aoki (1978), Sasaki and Jin (1979), Takayama (1982), and Reitter et al. (1985).

The computational procedure used by Kang et al. (1985) can be summarized as follows. The urban area is divided into uniform, square tracts that are each relatively homogeneous with regard to type and density of structures; the tracts are separated by natural or man-made firebreaks (e.g., streets, rivers, or parks). Following the initial ignition of the area by a nuclear

TABLE 3A.1.
SOME MAJOR SIMPLIFICATIONS IN THE URBAN FIRE
MODEL OF KANG ET AL. (1985)

- Tracts are small enough so that their built-up areas can be treated as homogeneous in a statistical sense.
- Tracts are large enough that ignition and fire spread can be treated probabilistically.
- Tracts can be idealized as squares, all the same size.
- Firebreaks between tracts are sufficiently large (at least 30.5 m) to prevent spread between tracts by radiation.
- Only one wall of a building is exposed to the fireball.
- Only interior fuels are important in ignition.
- Frequency of secondary (blast-caused) fires is proportional to floor area: one per 10^4 m^2 of floor space is assumed.
- Blast can extinguish primary fires at overpressures of 2 psi or greater.
- "Abrupt flashover" is neglected. ("Abrupt flashover" refers to the rapid ignition of an entire room exposed to large amounts of thermal radiation from a fireball.)
- Flashover of one room in a building leads to a sustained building fire.
- Building burning history is entirely based on the ignition of a single room, which occurs with equal probability on any floor. ["Building burning history" is the time a particular building type spends in active (flaming) combustion.]
- Building burning history is independent of moderate blast damage.
- There is no fire interaction between the debris and the non-debris regions.
- Ambient wind is constant throughout the fire area for total time of interest.
- Fire-induced aerodynamics are neglected.
- Wind effects upon the building fires and radiant fire spread are neglected.

detonation, a time-marching computational routine is employed to follow the fire-spread history. At each time step, empirically-derived probabilities are used to calculate the expected numbers of buildings ignited by radiation or by firebrands in each tract. Ignited buildings progress through several stages of burning, leading to the phase during which spread to other buildings can occur. The basic numerical model is based on work carried out by Takata and Salzberg (1968) and Takata (1972). Table 3A.1 summarizes the most important assumptions.

The overall results of many simulations (assuming a 1 Mt detonation) suggest that the dominant factors in urban fire ignition and spread in a nuclear attack are: the distance the thermal pulse can propagate and ignite fuels, fuel loadings, ambient windspeed, and firebrand production rate. The simulations for San Jose and Detroit showed additional dependence on the yield and point of detonation and the fuel-distribution patterns (as well as on the weapon yield). Details are given in Kang et al. (1985).

3A.3 UNIFORM-CITY CASE STUDY

A "uniform" city is assumed to be characterized by a single building type, a constant building density and fuel loading, and constant firebreak dimensions; each structure is assumed to have the same ignition and fire-spread probabilities. Table 3A.2 summarizes pertinent input conditions used for uniform city calculations.

Baseline calculation results are shown in Figures 3A.2 to 3A.5 for a 1-Mt burst at 3 km altitude above Ground Zero (denoted GZ). In the severely blast-damaged area (here assumed to be the area exposed to an overpressure greater than 3.5 psi = 24 kPa), fires could burn actively, smolder, or be entirely extinguished (if initially ignited), depending upon a number of complex physical processes, including the fuel to non-fuel debris ratio and the mixing characteristics of the fuels. For the current simulation, the fuel in the debris region is assumed to be "affected" by fire, i.e., the fuel could burn. Typically, in large fires, the fuel-consumption fraction is assumed to be 50% (Takata and Salzberg, 1968; Chandler et al., 1963); however, this value has not been firmly established. Any fuel not consumed in the first wave of burning could smolder for a longer period, sometimes for days if not extinguished. Because the fraction of fuel which burns rapidly (as against that which smolders or does not burn) is not known, figures for fuel are given in terms of the total fuel within the fire and the debris zones, representing the maximum available fuel in these zones (except for fuels not accounted for, as noted above, and subject to the uncertainties in the fuel loading estimates themselves).

In the non-debris area, the ignited structures serve as a source of subsequent fires within a tract through radiation and firebrands, as well as a

TABLE 3A.2
BASELINE CASE PARAMETERS FOR UNIFORM CITY

Attack Scenario:	Yield = 1 Mt, HOB = 3 km,	
Atmosphere:	Visibility = 19.3 km, Wind = 2.68 m/s (6 mph)	
Tracts:	Tract types	= 1 (uniform),
	Tract dimension	= 0.8 km × 0.8 km
	Building density	= 15%
	Density of built-up Areas	= 100%
Structures:	Wooden, residential, 2 stories	
	Height	= 5.9 m
	Window area/Wall area	= 0.1
	Window transmittance	= 0.7
Fuel:	Specific fuel loading	= 100 kg/m^2 of floor area
	Areal fuel loading	= 30 kg/m^2
Fire:	Lowest critical ignition energy	= 7.7 cal/m^2
	Secondary ignitions	= 1 fire per 10^4 m^2 of floor area
	Brand generation rate	= 18 per m^2 of roof area[a]
	Brand transport range	= 460 m
Blast effects:	Severe blast damage (debris) above 3.5 psi overpressure	
	Moderate blast damage between 2 psi and 3.5 psi	
	No blast damage below 2 psi	
	Secondary fires above 2 psi overpressure	
	Some primary fires extinguished above 2 psi	

[a] Calculated as total brands above a minimum size coming from an entire burning structure divided by the roof area.

source of fires to neighboring tracts by firebrands. Generally after several hours, the peak-burning rate is reached, involving areas initially ignited by the fireball and tracts subsequently ignited by firebrands. Figure 3A.2 shows the fire area at t = 25 hr after the explosion. Note that the fires leave behind a burned-out annular fire "ring" between the debris region and the fire front (which may in fact include the debris region).

The predicted area affected by fires as a function of time is given in Figure 3A.3. The initial affected area was approximately 510 km^2, of which about 40% was in the debris region. During the course of 25 hours of conflagration, the fire area outside the debris region increased to about 760 km^2. Figure 3A.4 illustrates the fire intensity, which reached a peak at approximately 5 hours after the burst and subsided to a steady, moderate level thereafter as the fire continued to spread. The fuel consumption was rapid in the first few

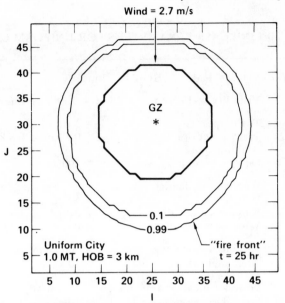

Figure 3A.2.   Fires in uniform city at 25 hr after a 1-Mt detonation at 3 km altitude, baseline case (constant windspeed at 2.7 m/s). Each I, J coordinate refers to a 0.8 km × 0.8 km tract; the numbers on the contours denote the fraction of buildings still unignited

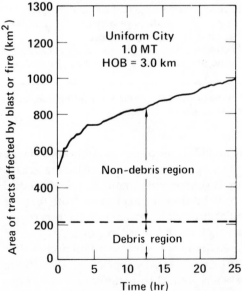

Figure 3A.3.   Fire-affected area vs. time for uniform city, baseline case (1 Mt, 3 km HOB). For simplicity, the area of the entire tract is considered to be affected by the fire when at least one structure within it is on fire. Over time, virtually all structures in a tract will burn

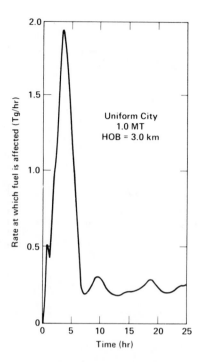

Figure 3A.4. Rate that fuel is engulfed by fire in the non-debris region in a uniform city, baseline case. A typical assumption is that 50 percent of this amount burns; this assumption is made in Kang et al.'s model

hours and then adjusts to a relatively constant rate, extending even beyond 25 hours due to spreading of the fires. The total cumulative quantity of fuel in ignited structures within the fire zone as a function of time is shown in Figure 3A.5; the fuel in the debris region is also included to emphasize the possible effect of fires there.

The results of an extensive sensitivity analysis of nuclear-induced fires in a "uniform" city are summarized in Table 3A.3. The dominant factors affecting the fire outcome are: ambient wind, atmospheric visibility, firebrand production rate, fuel loading, thermal-pulse ignition thresholds, and secondary (blast-induced) ignition frequency. All of these factors are subject to considerable uncertainty. The simulated dependence of fire behavior on each of them was, however, consistent with qualitative physical reasoning.

Fire characteristics for simultaneous bursts have also been calculated, but the detailed results were not presented here. When two well-separated 0.5 Mt bursts were detonated over the uniform city, and no interactions between the

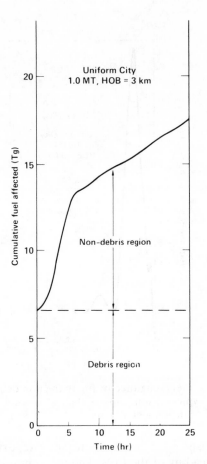

Figure 3A.5.   Cumulative fuel in structures actually on fire as a function of time for uniform city baseline case (1 Mt. 3 km HOB). The total fuel in the fire zone at any time, including structures not yet on fire, would be greater. Of the total fuel for each structure, only a certain fraction would usually be consumed in active flaming. For the non-debris firespread region, it is usually assumed that 50% of the fuel is consumed in active flaming; the rest is usually assumed to be consumed in later smoldering. For the debris, burnout estimates vary from zero to 100 percent

fires were assumed, the total fuel affected increased by about 30% over the baseline 1-Mt case after 25 hours (including the debris areas). This result suggests a modest increase in fuel consumption for simultaneous smaller bursts, depending on their sizes and relative placement. (See also Chapter 1 for a discussion of incendiary efficiency for various yield weapons.)

TABLE 3A.3.
PARAMETER SENSITIVITY STUDY FOR IDEALIZED UNIFORM
CITY (NON-DEBRIS REGION)

Parameter varied	Modification from baseline case	Fuel consumed normalized to baseline case		Normalized relative area[a]
		$t = 5$ hr	25 hr	$t = 25$ hr
Windspeed	x 2	1.30	1.23	1.18
	x 3	1.40	1.34	1.24
Firebrand	x 1/2	0.76	0.91	0.92
generation rate	x 2	1.40	2.08	1.11
Blast extent	2 psi	0.72	1.00	1.00
Visibility	12.9 km	0.67	1.00	1.00
	32.0 km	2.14	1.57	0.89
Secondary	x 2	1.42	1.02	1.01
ignitions	x 1/2	0.73	0.98	0.99
	0	0.43	0.85	1.04
Building	x 2	1.45	1.19	1.13
density	x 1/3	0.58	0.76	0,84
HOB	4.0 km	1.19	1.20	0.97
	3.5 km	0.96	1.01	1.00
	2.5 km	0.97	0.87	1.04
	2.0 km	0.97	0.82	0.99
Lowest critical	10.4 cal/cm^2	0.74	1.00	1.00
ignition energy	5.0 cal/cm^2	1.52	1.33	0.93
Window	1.0	1.76	1.18	0.97
transmittance	0.4	0.67	1.00	1.00
Specific fuel	x 2	1.41	1.08	1.04
loading	x 1/2	0.42	0.88	0.92
Window	x 2	1.42	1.27	1.14
area	x 1/2	0.65	0.72	0.86

Baseline case results: Fuel consumed[b]

outside debris region = 4.8 Tg (5 hr)
= 10.9 Tg (25 hr)
Total fuel affected = 30.0 Tg (25 hr)
Area affected = 226 km^2 (debris region)
= 534 km^2 (non-debris; 5 hr)
= 764 km^2 (non-debris; 25 hr)

[a] This column presents the ratio of two ratios; i.e., the area affected at 25 hr divided by the area affected at 0 hr for the case with a varied parameter divided by the same ratio for the baseline case.

[b] 50% fuel consumption rate is assumed.

3A.4 SAN JOSE, CALIFORNIA (1968) CASE STUDY

San Jose, California of the mid-1960s represents a typical, predominantly suburban residential area. Extensive information is available from that time on the firebreaks, building types and fuel distributions (Takata, 1969, 1972). As a case study, a 1-Mt burst was assumed to be detonated at 2.4 km altitude over the southern tip of San Francisco Bay, north of San Jose. The choice of this GZ was made in the Five-Cities study, presumably to optimize blast damage to the military-industrial complex on shore nearby. Table 3A.4 describes the baseline parameters for this case.

TABLE 3A.4.
BASELINE CASE PARAMETERS FOR SAN JOSE AREA (1968)

Attack Scenario:	Yield $= 1$ Mt, HOB $= 2.4$ km, GZ at $(I = 16, J = 33)$	
Atmosphere:	Visibility $= 19.3$ km, Wind $= 2.68$ m/s (6 mph) from the west	
Tracts:	Tract types Tract dimension Total number	$= 14^{\text{a}}$ $= 0.8$ km $\times$ 0.8 km $= 1428$ (occupied $= 699$) tracts
Blast effects:	Severe blast damage (debris) above 6 psi overpressure Moderate blast damage between 2 psi and 6 psi No blast damage below 2 psi Secondary fires above 2 psi overpressure Some primary fires extinguished above 2 psi	
Fuel:	Total mass of fuel in area Fuel in residential tracts Fuel in industrial tracts	$= 3.05$ Tg $= 5\text{–}11$ kg/m^2 $= 21\text{–}88$ kg/m^2
After attack:	Number of tracts initially involved $= 240$ (i.e., 55 blast-destroyed) (185 ignited) Fuel available in debris area	 $= 0.31$ Tg

[a] The tract types vary according to the built-upness of the area, the building density, and the building height and floor area. There are 8 residential tract types with a specified fuel loading of 50 kg/m^2-floor 2 industrial tract types with 88 kg/m^2, 2 school tract types with 24 kg/m^2, and 1 commercial tract type with 24 kg/m^2. The residential fuel loading may be low by a factor of two (Issen, 1980).

Fire initiation and spread in San Jose area (1968)

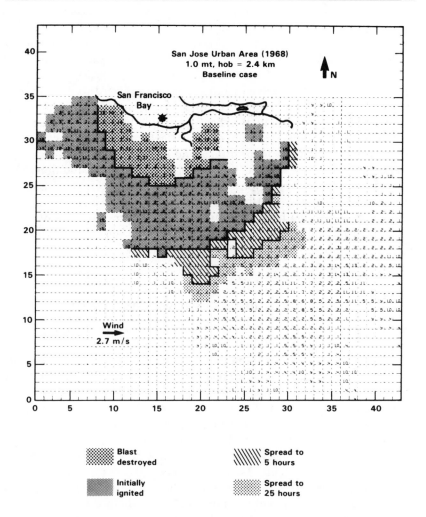

Figure 3A.6. Fires in the San Jose, California area assuming 1968 fuel load distributions (1 Mt at 2.4 km HOB)

Figure 3A.6 shows the ground zero, debris area, the fire ignition and spread patterns $t = 5$ hours and $t = 25$ hours after detonation. After 25 hours, the total area damaged and burned was calculated to be about 200 km^2 (even though the initial detonation occurred over the Bay and did not ignite downtown San Jose). The majority of structures in San Jose are primarily residential, holding about 50 kg of fuel per square meter of floor space. (While this value may be low for current residential structures (Issen, 1980), the present calculation may be considered as a sample case study.) Under these conditions, the total fuel affected in the non-debris fire-spread region was about 0.5 million tonne (Tg) after 25 hours; in the adjacent debris region, the amount of available fuel was about 0.3 Tg.

When ground zero was moved from the Bay southward to include more of the developed area of San Jose (i.e., to 9.6 km southward of ground zero shown in Figure 3A.6), the total urban area initially affected increased to about 250 km^2 (91 for the debris region, and 156 for the ignited region), then slowly increased to about 260 km^2 after 25 hours, and to about 370 km^2 after 50 hours. The total fuel affected was about 1.5 Tg (1 Tg in the non-debris region; 0.5 Tg in the debris region). These results suggest that the fire history is "city-specific", i.e., the distribution patterns of a particular city and the specific nuclear targeting near or within the city influence the fire outcome. Thus, a reasonably detailed survey, especially of the fuel distributions, may be a prerequisite for predicting the potential time-dependent fire-spread in a specific urban area.

3A.5 DETROIT, MICHIGAN (1968) CASE STUDY

Many single-burst cases have been calculated for various GZ locations over Detroit, involving 1 Mt and 0.5 Mt weapons. For the 1-Mt cases, the height-of-burst (HOB) was assumed to be 2.6 km in order to maximize blast damage, while for the 0.5-Mt cases, a HOB of 2.1 km was assumed. The wind velocity was taken to be westerly at 4 m/s. Note that fuel loadings varied according to tract type, with 50 kg/m^2 (floor space) assumed for residential structures, as in the San Jose case, and 88 kg/m^2 for industrial tracts. Figure 3A.7 is a map of the Detroit area displaying various tract types.

The results for these single-burst cases varied widely according to the GZ location. For example, Figure 3A.8 shows the fire area for a 1-Mt blast at $t = 25$ hr after detonation, encompassing about 840 km^2 (which included 110 km^2 of debris region), and involving 14 Tg of combustibles (with about 2 Tg in the debris region). In the model, the fires continued to burn in the windward direction until firebreaks were reached, e.g., Lake St. Clair. No fuel data were available to consider fires in adjacent areas in Canada, and therefore ignition and firespread in that region were not considered. The results are summarized in Table 3A.5.

Multiple, near simultaneous burst effects were also considered as a possible attack scenario, but the precise results are not presented here. Figure 3A.9 illustrates the fire zones one-day after two 0.5 Mt weapons were exploded at different locations over Detroit. Interactions between the fires were

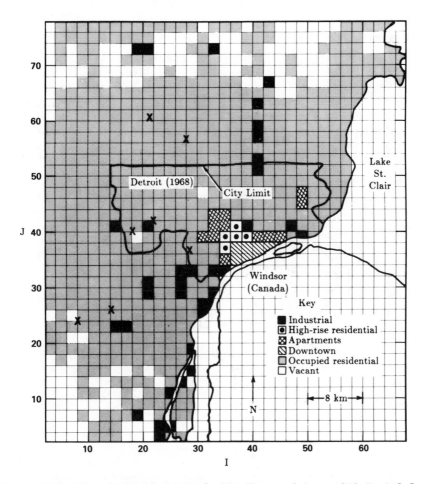

Figure 3A.7. Detroit, Michigan area fuel loading as of about 1968. Each *I, J* coordinate denotes a 0.8 km × 0.8 km tract; each **X** refers to a location that served as ground zero in a set of tests of the effects of a single detonation

ignored. The total area affected was about 1,370 km² (including 420 km² of debris region), a sizeable increase over the single-burst, 1-Mt case. On the other hand, the total fuel involved was somewhat lower at about 11 Tg (with 1.3 Tg in the debris region), because less-dense areas were burned up to this time. This again demonstrates the importance of the attack scenario and the fuel loading patterns. Note that the 11 Tg translates into an average fuel loading of about 8 kg/m², which is about the same as was assumed in the San Jose residential areas (Table 3A.4) and is also probably a low estimate.

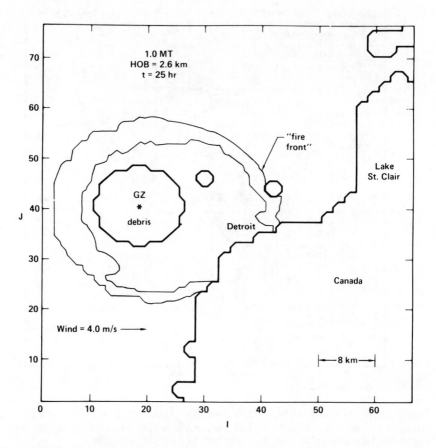

Figure 3A.8.   Fires in the Detroit, Michigan area, assuming fuel loading characteristic of about 1968, 25 hours after a 1 Mt burst. Bold lines outside the debris region
indicate fire breaks (e.g., water bodies, open areas)

Because the fraction of the total fuel in a city that becomes involved in
the fire zone is also of interest, these have been calculated and are shown in
Table 3A.5. The fractions of fuel affected by fires depend on many factors
which are specific to the city in question. Nevertheless, it appears that a
sizeable fraction of all the fuels in a major city, plus its suburbs, could be
ignited by about 1 Mt of nuclear explosives (Table 3A.5).

Table 3A.5 Urban fire areas and fuel burdens for single detonations.

| Case | Yield (MT) | Area of fire zones (km²) | | | Fuel available in fire zones (Tg) | | | Fraction of total fuel in fire zones (Percent)[a] | | |
		Debris Region	Non-debris 25 hr	Non-debris 50 hr	Debris Region	Non-debris 25 hr	Non-debris 50 hr	Debris Region	Non-debris 25 hr	Non-debris 50 hr
Uniform City	1.0	226	764	1016	6.3	21.8	35.0		(b)	
	0.5	143	516	775	4.0	12.9	23.2			
San Jose[c]	1.0	91	156	275	0.5	1.0	1.1	15%	32%	36%
	0.5	59	148	251	0.3	0.7	0.9	8	22	31
Detroit[d]	1.0	91–109	610–733	714–911	1.7–4.4	11.5–13.1	14.2–18.0	6–17	45–51	56–70
	0.5	63–65	404–582	548–755	0.3–1.6	1.4–11.4	4.5–16.3	1–6	5–44	18–64

[a] When at least one structure is on fire, the entire tract is considered to be in the fire zone.
[b] For the uniform-city case, the fraction is not given because the total urban area is not specified.
[c] For the case with a ground-zero located 9.6 km south of the ground-zero shown in Figure 3A.6.
[d] The upper and lower figures give the range of values obtained for a number of simulated burstpoints.

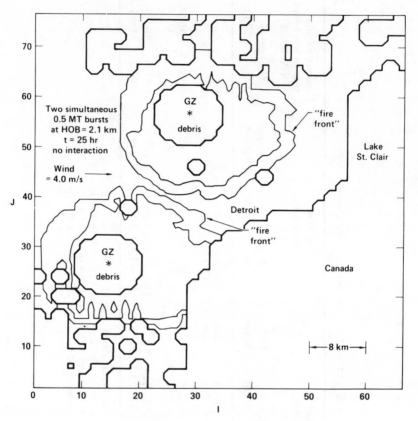

Figure 3A.9. Fires in the Detroit area, assuming fuel distribution characteristic of 1968 and for two simultaneous 0.5-Mt bursts, 25 hours after detonation. Bold lines outside debris regions indicate fire-breaks

3A.6 DISCUSSION

From the foregoing analysis, it is obvious that variability exists in the fire areas and the amounts of fuel consumed with differences between cities, and with burst location within a city, due mainly to variability in the fuel load and distribution. Nonetheless, based on these simulations several tentative conclusions may be reached. First, the dominant factors in fire spread are: fuel loading, reach of the thermal-pulse, ambient windspeed, and firebrand production rate. In addition, the cities with irregular fuel distributions display fire histories that are distinct from an idealized "uniform" city, with dependence upon the attack scenario and the fuel distribution patterns. Second, the potential burnout areas in cities under nuclear attack can be as large as 1000 km^2 after one day, when fire spread by the ambient winds is included.

These estimates do not take into account the likely significant role of liquid fossil fuels in stationary and spreading fires in urban/industrial zones. In the larger cities, the total fuel potentially involved in fires can range up to 20 Tg or more (Table 3A.5) after a day, although in sparse residential zones the value may be closer to one-tenth this figure. Indeed, a large-scale nuclear exchange could impact an enormous quantity of fuel, and presumably create large amounts of smoke (see Section 3.3).

A number of potentially important physical factors have not been included in the present calculations: synergistic effects of simultaneous fires, abrupt flashover (the Encore effect; see Chapter 1), merging and breakup of fires and plumes, fire-wind interactions, fire phenomena in the debris region, and firestorms. Fires in the debris region may burn actively, smolder, or be entirely extinguished, depending on the debris formation process, the fuel-to-nonfuel ratio, and the mixing characteristics, among other things. Taking into account the various uncertainties, it is still quite possible that fires would exist in the debris region and produce smoke in the aftermath of a nuclear attack. Thus, the debris region fires represent an important area of study.

Firestorms in particular are an important aspect of mass fires because of their potential impact on smoke transport to high altitudes. A firestorm may be defined as a stationary fire with an intense heat release and strong inflowing winds over a large area. Although the criteria necessary to initiate a firestorm are not well understood, two of the major factors seem to be that both the intensity and the extent of the fire must be quite large. For example, one of the 1943 Hamburg fires took about 2 hours to reach firestorm conditions over an urban area greater than 12 km^2. The heat intensity was on the order of 2.5×10^5 Wm^2 for more than six hours. While the present model does predict the heat release rate and area of urban fires, it does not predict the complicated interaction between the fires and fire-induced winds. Since firestorms probably develop by means of these interactions, more research is needed before reliable predictions of firestorms could be made. The convective storms that could be triggered by high fuel loads and high heat-release rates are discussed in Chapter 4.

Environmental Consequences of Nuclear War Volume I:
Physical and Atmospheric Effects
A. B. Pittock, T. P. Ackerman, P. J. Crutzen,
M. C. MacCracken, C. S. Shapiro and R. P. Turco
© 1986 SCOPE. Published by John Wiley & Sons Ltd

CHAPTER 4
Atmospheric Processes

4.1 INTRODUCTION

As an introduction to Chapters 5 and 6, which present the atmospheric response to the smoke and dust injections described in Chapter 3, this chapter focuses on the physical processes in the atmosphere that interact with the injected aerosol. The first section is concerned with plume rise and mesoscale processes that affect the early dispersion of the injected material. The next section treats physical processes such as radiative transfer and atmospheric transport, which interact with aerosol on the longer time scales of weeks to months. The last section discusses briefly geophysical analogues and their relevance to the problem of climatic disturbance following a nuclear war. This chapter should be seen as a bridge between the preceeding chapters, which are primarily concerned with determining the quantity and type of material injected into the atmosphere, and the following chapters, which are primarily concerned with the longer time scale response to the injection. In addition, it attempts to provide some perspective on the importance of various physical processes in determining the climatic response and on the degree to which these processes can be simulated in current climate models.

4.2 SHORT TERM ATMOSPHERIC RESPONSE
TO SURFACE FIRES

Fires of the intensity and areal extent of those likely to occur in the aftermath of a modern thermonuclear exchange, and the plumes associated with them, are beyond normal experience. The wildfires that occur periodically in forests and grasslands are fundamentally unlike those that would result from a nuclear detonation. Whereas usual wildfires are set from a single or limited number of ignition points, a nuclear detonation can simultaneously ignite tens or hundreds of square kilometers of forest, brush, and grass, thereby inducing fires of areal extent much larger than ever observed before (NRC, 1985). Some closer approximations to what might occur are the fires, firestorms, and associated smoke plumes caused by conventional and nuclear bombings of cities during the Second World War. The conventional

Physical and Atmospheric Effects

incendiary bombings of Dresden and Hamburg both resulted in intense fire storms. Although documented observations of the resulting fires and fire plumes are sketchy, anecdotal evidence indicates that smoke plumes reached 6–12 km in height (NRC, 1985). The Hamburg fire, although covering an area of only 12 km², had an estimated heat output of 1.7×10^6 MW and produced a smoke plume that reached altitudes ranging from 9–12 km (Ebert, 1963; Carrier et al., 1983).

Following the nuclear detonations over both Hiroshima and Nagasaki, observations of smoke were again limited. The early morning bombings occurred in early August, 1945, at which time the local maritime atmosphere was conditionally unstable. Because of topographical influences, the resulting fire in Nagasaki was less extensive than that in Hiroshima (see Chapter 1). In both cities, a large cumulonimbus cloud formed over the the fire that produced a "black rain" at the surface; at Hiroshima, the rain began within 20 minutes of the blast. Molenkamp (1980) estimates that 5–10 cm of rain fell in parts of Hiroshima in a 1–3 hour period, while the amount was somewhat less in Nagasaki (Ishikawa and Swain, 1981).

These historical fire events are small compared to those that could occur today with modern weapons. Whereas the Hamburg firestorm covered 12 km² with an average heat flux of about 14×10^4 W/m² (Ebert, 1963; Carrier et al., 1983), a current strategic nuclear weapon is capable of simultaneously igniting a city of several hundred square kilometers. Larson and Small (1982) have estimated that fuel loadings in some city centers can approach that of the pre-firestorm Hamburg fuel density of 470 kg/m² over a region of 3–13 km². They also estimated fuel loadings of 110 kg/m² over a substantially larger region of 47–100 km² surrounding the city center. Because a nuclear detonation could ignite the entire region simultaneously, total heat fluxes in a modern city could reach as high as 10^5 W/m² (equivalent to about 10^7 MW over the region) if a large fraction of the combustible material were to burn in 3–6 hours. (It is not certain that all of the material would burn, especially in the regions of heaviest debris; see Chapter 3, and Appendix 3A.) Such strong heat sources, which are roughly 10 to 100 times greater than the amount of solar radiation absorbed at the ground, could produce deep, often precipitating, atmospheric plumes that would inject smoke, dust, and moisture aloft.

4.2.1 Fire Induced Convective Plumes

The majority of nuclear weapon strategic targets lie in regions where, during the months of May–September, the atmosphere is likely to be conditionally unstable for moist convection most of the time. During March–April, and October–November, the atmosphere above the majority of the targets probably has sufficient moisture to feed convection if large and unusual surface heat sources occur. In the remaining months of December–February, at least 50% of the targets may be in a conditionally unstable environment or possess enough moisture to supply unstable moist convection with a large surface heat flux. Therefore, over much of the year and in many locations, smoke resulting from large surface fires would be likely to enter a cumulonimbus cloud that is supported or triggered by the fire induced heat flux.

Fire induced moist convection could be somewhat more intense than natural moist convection for a given time and locality since it is augmented by the surface heat source. In fact, the heat flux of an intense, massive fire could be comparable to that which drives an intense cumulonimbus storm (which may have latent heat fluxes aloft exceeding 10^5–10^6 W/m^2 at mid levels over a region of 10–25 km^2). If heat fluxes are less than about 10^4 W/m^2, strong convection might not ocur. Smoldering fires would tend to inject smoke only into the boundary layer. In general, natural deep convective systems are found preferentially to detrain (release to the enviroment) most of their mass near the tropopause level (Yanai et al., 1973; Knupp, 1985). Hence, as a first approximation, it can be expected that deep moist convection induced by flaming fires would deposit much of the smoke carried up by the cloud into the region just below, or possibly just above, the tropopause (excepting, of course, for that smoke which is removed from the cloud by scavenging processes; see Chapter 3 for a discussion of scavenging mechanisms).

As with natural convective plumes, dynamic interactions with local winds and vertical wind shear can result in a helical and inertially stabilized plume updraft (Lilly and Gal Chen, 1983). For fires that persist for a period of several hours, a strong inward spiraling surface vortex might develop that would fan and further enhance the fire. There was evidence of such a whirlwind in the Hamburg firestorm, although not in the other mass fire cases.

Unlike natural moist convection, cumulonimbus clouds occurring in conjunction with surface fires would contain extremely high levels of pollutants (dust, smoke, debris, etc.) that could alter their microphysical development, which in turn could affect the plume development. Much higher than normal CCN (cloud condensation nuclei) concentrations might be produced, resulting in the formation of many very small droplets and preventing the formation of raindrops by autoconversion (growth to large drops by coalescence of smaller cloud droplets). This is probably not an important modification since raindrop formation by autoconversion is rare over continental regions, even with present background levels of pollutants and CCN. Other cloud microphysical effects may be anticipated as well. If the pollutants act effectively as ice nuclei (silicate is a primary ice nucleus) in addition to CCN, ice nucleation by condensation freezing could be augmented. This may lead to larger than normal ice crystal production rates in portions of the cloud above $-20°C$ (where depositional nucleation rates are normally small) and perhaps smaller than normal ice crystal sizes overall. It is unclear, however, whether this would augment or weaken the precipitation growth process.

Because of the complex interactions among the production of heat and smoke by the fire, atmospheric motions, and the microphysical processes affecting the particles, modelling fire plume behavior is extremely difficult. A number of two-dimensional simulations of fire plumes have been conducted. These have been used to study the relationships among heat addition from the fire, production of buoyancy, generation of pressure gradients, and induction of fire winds, as well as the role of sub-grid turbulence and fire-plume radiation (Luti, 1981; Small et al., 1984a, 1984b; Proctor and Bacon, 1984). While these models have the advantage of high spatial resolution and give reasonable agreement with observations and experiments, they are not able to simulate the complex relationships between plumes and the ambient atmosphere which occurs within a three-dimensional framework.

A more realistic simulation of the plume development in the ambient atmosphere can be performed using three-dimensional cumulonimbus models, although these models suffer from a coarser resolution and do not include fire-plume radiation effects. Over the past decade, three-dimensional models have been quite successful in simulating the observed characteristics of cumulus and cumulonimbus clouds (Miller and Pearce, 1974; Cotton and Tripoli, 1978; Klemp and Wilhelmson, 1978a,b; Schlesinger, 1978; Clark, 1979; Tripoli and Cotton, 1980, 1986). Because of their superior representation of cloud dynamical processes, three-dimensional models have greater fidelity in simulating important interactions between the cloud and the surrounding environment than do numerical models in one- or two-dimensional frameworks. The three-dimensional models used to simulate fire plumes to date are the Colorado State University Regional Atmospheric Modeling System (RAMS) (Tripoli and Cotton, 1982; Cotton et al., 1985) and the model of Penner et al., 1985; (see also Haselman, 1980). Both models include the predicted effects of latent heating. However, RAMS also predicts the growth and effects of ice and rain precipitation, which allows it to simulate the effects of buoyancy changes resulting from precipitation movement relative to the air parcel. This effect could be equivalent to as much as a 1–3°C temperature perturbation. For these studies, both models also predicted the evolution of a passive tracer representing smoke. The RAMS model also calculates the effects of thermophoretic, diffusiophoretic, and Brownian-induced scavenging by the simulated cloud and precipitation elements; a single-sized smoke (soot) particle with a radius of 0.1 μm is assumed. Because experimental evidence (Section 3.5) indicates that soot particles are generally poor CCN, nucleation scavanging was neglected by RAMS. However, some studies have found that smoke particles from forest fires do act as CCN (Radke et al., 1980b), implying that nucleation scavenging could prove to be an important removal mechanism.

RAMS was employed by Cotton (1985) to simulate a hypothetical post-nuclear-attack urban fire occurring in the early morning in Denver, Colorado on 4 June, 1983. On that day, sufficient moisture was available to trigger intense local thunderstorms. (In fact, strong thunderstorms actually were observed later that day.) The simulation was performed using three surface heat fluxes: 8×10^4 W/m^2 (intense), 4×10^4 W/m^2 (medium), and 0.8×10^4 W/m^2

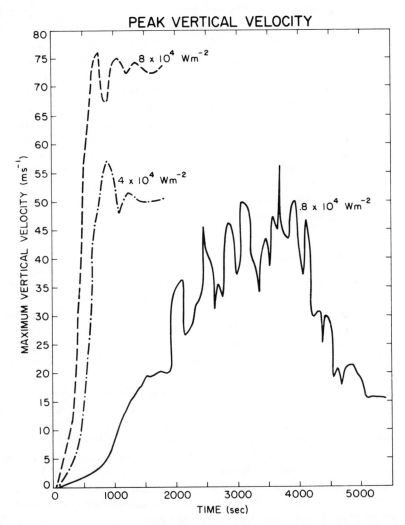

Figure 4.1.   Peak vertical velocity in the simulated cumulonimbus cloud as a function of time for three different values of the fire heat source assumed by Cotton (1985). The curves are labeled with the source intensity and all fires cover a circular region with a radius of 4 km

(weak) applied over a circular region of 50 km^2. In each case, soot was injected along with the heat, assuming that 0.2% of the fuel is converted into soot. (In such tracer studies, the absolute amount injected is not critical since the tracer does not feed back into the rest of the model physics.) In addition to the water vapor entrained with the ambient air, water vapor released by fuel combustion was also included.

The intense case was found to produce a very strong updraft (Figure 4.1) that reached velocities of over 75 m/s and became nearly steady in time after 15 minutes of simulation. Within the updraft, temperature perturbations from ambient conditions of over 23°C were predicted. Because of the strong vertical motion, air parcels spent less than five minutes in the updraft before flowing into the anvil at 8–12 km above ground level. Since such a relatively short time was spent in regions of high condensation and large amounts of cloud water, less than 2% of the tracer aerosol was predicted to be removed by phoretic and Brownian scavenging (neglecting nucleation scavenging). The simulated plume produced an anvil that extended below and above the tropopause (see Figure 4.2) leading to maximum smoke and water detrainment in this region. The simulation was terminated after half an hour. By that time, no significant precipitation had yet reached the ground. Ice was deposited into the spreading anvil with concentrations reaching 6 g/kg of air, compared to normal amounts of 1–3 g/kg.

4.2.2 Smoke Injection Heights

The vertical profile of smoke injection resulting from the simulation is displayed in Figure 4.3. For the case with a heating rate of 8×10^4 W/m^2, over 50% of the injected tracer aerosol is deposited above the tropopause level, which is located at about 10 km above ground level. Another maxima occurs below about 2 km because some smoke is detrained and trapped at low levels. This appears in Figure 4.2 as the low-level, flange-like structure to the plume.

Reducing the fire heat source to half of the intense heat source produced little change in the percent injection of smoke with height. However, when the rate was reduced to one-tenth of the intense case, the resulting convection was noticably altered. In this case, the surface heat flux was insufficient to loft the tracer aerosol high enough to initiate strong condensational plume growth. After some time, precipitation did develop at the top of the plume which produced occasional strong updrafts in excess of 50 m/s. As a result, a weak maximum of aerosol injection eventually developed around the tropopause level (see Figure 4.3). There was also considerably more smoke deposited in the boundary layer in the first half hour (as is evidenced in Figures 4.2 and 4.3) for this weak intensity case than in the full and one-half intensity cases.

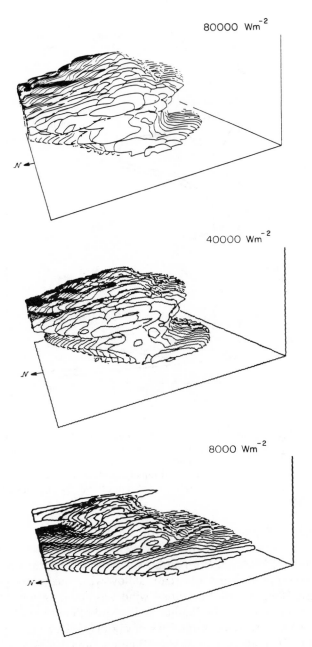

Figure 4.2. A three-dimensional depiction of the the exterior surface of the tracer aerosol cloud for the three values of the fire heat source assumed by Cotton (1985). The surface is the 10^{-8} g/m^3 contour level and the time is 30 minutes after the start of the calculation

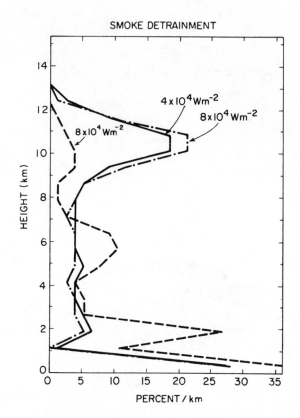

Figure 4.3. Vertical distribution of the tracer aerosol injection for the three different values of the fire heat source assumed by Cotton (1985). The horizontal axis is in units of percent of mass injected per vertical kilometer of atmosphere. The curves are labelled with the source intensity and all fires cover a circular region with a radius of 4 km. These numbers are calculated 30 minutes after the start of the calculation

Predictions of plume injection for spring/fall midlatitude standard atmospheric conditions were made by Penner et al. (1985) and by Banta (1985), who used RAMS. Both authors assumed a vertical shear with a unidirectional component and a horizontal flow with maximum speeds near the tropopause. Banta's peak horizontal wind was about 14 m/s, compared to a value of 18 m/s assumed by Penner et al. While Penner et al. assumed standard atmosphere relative humidities of about 77% at the surface decreasing to about 10% at the tropopause, Banta assumed 50% relative humidity throughout the entire depth of the atmosphere.

Penner et al. (1985) performed three simulations based on this thermodynamic profile. The first simulation considered a heat source of 8.9×10^4

W/m^2 (intense), the second simulation used a heat source of 2.3×10^4 W/m^2 (medium) while the third simulation used a heat source of only 0.23×10^4 W/m^2 (low). Each flux was specified over a circular region covering 78 km^2. Banta (1985) assumed a heat source of 9.4×10^4 W/m^2 over the same area. Using only that heat source, Banta (1985) performed three simulations. The first simulation considered moisture flux from the fire due to the release of water vapor by combustion in addition to the latent heat contained within the atmospheric water vapor in much the same manner as Cotton (1985). In the second simulation, Banta assumed no background atmospheric humidity in order to demonstrate the relative contribution of moisture released by the fire to plume penetration. Finally, in a third simulation, Banta removed all moisture effects in order to demonstrate the effects of only the fire heat to the plume rise.

The predicted smoke injections from all three simulations are displayed in Figure 4.4. Note that Banta (1985), who used RAMS, included only phoretic scavenging (which was less than 2% of the injected aerosol), while Penner et al. (1985) neglected all scavenging. Thus, both simulations treat the particles essentially as inert tracers. The predictions of Banta (1985) show the deepest smoke penetration with a maximum above the 11 km tropopause level for the case including natural and fire-induced moisture. The other two RAMS simulations demonstrate that the moisture from combustion is of only minor importance, but that the ambient moisture is of major importance. Even with the intense heat source considered, the majority of the smoke particles were detrained between 4 and 5 km when atmospheric moisture was omitted.

The most intense case simulated by Penner et al. (1985) resulted in a lofting maximum some 2 km in altitude lower than that predicted by Banta (1985). The most probable reasons for the difference between the results are that Banta initialized with greater moisture and that Penner et al. neglected the weight loss to the air parcel resulting from precipitation. A third factor which may bear on the difference is the diffusion of energy and aerosols within the models. In RAMS, the aerosol tends to be detrained at the highest level to which it is carried; in the Penner et al. model, the aerosol tends to overshoot its equilibrium level and then subside to the level at which it subsequently detrains. The cause of these differences is not understood. The Penner et al. results for the medium and low intensity cases demonstrate that, as the heating is reduced, the detrainment maximum quickly drops down to the 1–3 km level. This is consistent with the previously discussed simulations of Cotton (1985).

The results of these plume simulations suggest strong sensitivity of the height of smoke injection to the natural environment and to the intensity of the fire itself. This implies that atmospheric smoke injection resulting from a nuclear exchange would necessarily depend on the details of the at-mospheric environments at the times and places that the fires occur. The

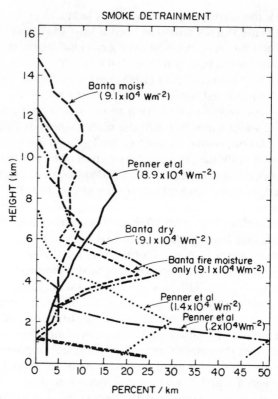

Figure 4.4. Vertical distribution of the tracer aerosol injection for the calculations of Banta (1985) and Penner et al. (1984) for the values of the fire heat source and for the moisture assumptions indicated. The horizontal axis is in units of percent of mass injected per vertical kilometer of atmosphere. These values are based on distributions 30 minutes after the start of the calculations

smoke injection profile that would arise statistically would most strongly depend on the time of the year. During the summer months, many of the likely strategic targets would be within conditionally unstable or moist (and thus potentially conditionally unstable) environments. For such targets, even the relatively small fires (less than 10^6 MW) could potentially lead to intense cumulonimbus clouds that could loft as much as 50% of the smoke into the upper troposphere and lower stratosphere, assuming minimal precipitation removal due to nucleation scavenging or hydrodynamic capture.

In less conditionally unstable but moist environments, intense cumulonimbus clouds would occur only over the larger fires. However, even in these environments, some smoke injection into the stratosphere is possible. In general, the ambient surface temperature would be less important than the available ambient moisture and the temperature aloft, because the heat necessary to initiate a plume would come from the fire. Therefore, during the summer months, the average smoke injection profile for urban fires with heat releases greater than 10^6 MW will likely show a relative maximum near the average tropopause level.

During spring the atmosphere locally can be more conditionally unstable, but statistically is probably less favorable to deep convection than in the summer because of frequent stabilizing temperature inversions, lingering stable cold air regions, and drier air masses. The fall season would be even less favorable for the development of deep convection because of the existence of more warm air aloft and drier surface conditions. The winter season would be least favorable to deep penetration by fire-induced cumulonimbus clouds because of generally stable and dry conditions.

Overall, the model results suggest that the immediate lofting by plumes could inject, on the average, 50% of the smoke produced by intense post-nuclear-attack fires to within 3 km of the original tropopause during the summer months. The remainder would be injected at lower altitudes and, in part, removed by precipitation. The fraction lofted during spring and fall could be substantially less, and little would be injected at the tropopause level during the winter months. It is possible, that if nucleation scavenging or hydrodynamic capture were effective, the above estimates of lofting could be reduced, although it is impossible at this point to say by how much. The various studies of nuclear war effects (e.g., Turco et al., 1983a; NRC, 1985; Crutzen, et al., 1984) have assumed that the precipitation scavenging fraction is on the order of 30 to 50%. This estimate takes into account precipitation scavenging from all fires, including those that do not create strong convection.

4.3 MESOSCALE RESPONSE

Beyond the scale of individual fire plumes, the response of the atmosphere would involve weather systems on the scale of 100–1000 km. This scale is referred to by meteorologists as the meso-alpha scale (Orlanski, 1975). Following a major nuclear exchange, large amounts of smoke and, for moist convection, ice would be deposited in the upper levels of the atmosphere. Because smoke and ice are radiatively active in the visible and infrared parts of the spectrum, their presence could trigger meso-alpha circulations. Some preliminary meteorological evidence suggests that the radiative effects of cumulus anvils and the cooling effects of melting ice might drive meso-alpha convective activity. The effects appear to be stronger at night, perhaps due to lack of warming by sunlight at the anvil top, which may partially compensate for the longwave cooling at cloud top. Within soot-filled anvils, the diurnal variation of heating would probably be enhanced as a result of strongly increased solar absorption.

Observational evidence for the existence of meso-alpha systems of this type is found in a class of systems called Mesoscale Convective Complexes (MCCs) (Maddox, 1980). Occurring frequently over the Great Plains of the United States on summer nights, they seem to begin often as intense afternoon convective systems that initially produce a large ice anvil. Usually about 1–3 hours after sunset, these systems undergo a dramatic transformation from cellular convection into a large anvil system in which steady stratiform rain is the primary precipitation. Current theory suggests that long wave radiation and precipitation melting, in conjunction with a strong relative flow of low level moisture into the region, help organize the system. A similiar mesoscale weather system found in the tropics is called a tropical cloud cluster. W. Gray (personal communication) has also found that these systems have a similar strong diurnal variation.

As a result of geostrophic adjustment to anvil outflow, it has been shown that strong alteration of the upper level flow occurs in the region of an MCC (Maddox et al., 1981; Fritsch and Maddox, 1981). The result is the formation of mesoscale high pressure aloft in conjuction with a divergent anticyclonic outflow pattern. In the Northern Hemisphere, upper level jet streaks of up to 50 m/s form to the north west of the MCC at the 20 kPa (200 mb) pressure level. As low level moisture supplies are exhausted and the MCC weakens, elements of the upper level flow pattern can remain. The residual flow pattern seems to be capable of restarting the system when it again encounters favorable conditions (Cotton et al., 1982).

Because fire induced cumulus plumes have been shown by model results to cause significant anvil outflow of ice, smoke, and air mass, a meso-alpha system similiar to a MCC could result. In regions where urban strategic targets are clustered, the effect of merging anvils may be stronger. The large volume of ice deposited in these anvils must eventually precipitate or evaporate and would be likely to induce some motion due to the cooling at the melting or evaporating level. Large cooling rates due to infrared radiation divergence would be expected in the upper anvil, which, at night, could act to destabilize the upper anvil region. During the daytime, shortwave absorption by smoke mixed with ice would oppose such cooling, and could, in fact, dominate. The net radiative effect at this time is unknown. Given that such systems are observed and that some of the known conditions would exist in fire-induced cumulonimbus systems, a mesoscale response may be expected to occur. If so, the induced mesoscale circulation could redistribute the smoke vertically and horizontally from the regions predicted by the three-dimensional cloud models. However, it is not yet known whether individual fire plumes could form anvils extensive enough to trigger MCCs.

The larger mesoscale anvil systems that might occur would contain large amounts of ice, possibly leading to light rain at the surface over large regions for a day or so following the initial strong convection phase. This would provide a second opportunity for scavenging mechanisms to deplete the large smoke concentrations deposited throughout the atmosphere. The smoke would have aged so that precipitation scavenging could be more efficient. Since slow sublimation of the ice would be occurring, some ice thermophoretic scavenging could be expected. Due to the many complex physical processes involved, and the uncertain composition of the post-nuclear exchange anvil and the environment, it is impossible to estimate the amount of scavenging that might occur in this mesoscale phase.

When large ambient vertical wind shears are present over the fire zones, cohesive anvil systems would be less likely to develop. In that case, precipitation formed aloft could evaporate after falling into drier layers below. This would introduce cooling and cause some mesoscale circulation response, which could be effective in continued mixing and dispersal of lofted pollutants.

It is less likely that only small amounts of ice would be present and that moist processes could be neglected. Even so, the mesoscale distribution of soot and other pollutants could cause horizontal temperature variations. This could, in turn, induce mesoscale circulation fields, that would continue to disperse smoke horizontally and vertically.

Mesoscale circulations would already exist in conjunction with normal synoptic and mesoscale weather patterns. Vertical motions associated with such features as jet streaks and frontal zones would be effective in subsequent vertical transport and horizontal dispersion of the lofted pollutants. Certain deformation fields associated with normal weather patterns could be effective in organizing strong, local vertical transport of pollutants. In particular, the regions near frontal zones, where secondary ageostrophic circulations exist, would have the largest influence.

For smoke deposited in the lower troposphere, thermally-driven mesoscale circulations such as seabreezes (Pielke, 1974) and slope flows (Defant, 1951) could lead to the preferred transport of materials to regions of mesoscale rising and sinking motions. Often, such zones of mesoscale rising motion are associated with natural cumulonimbus systems (Pielke, 1974; Banta, 1982) that could act to loft the material into the upper troposphere.

There are several potentially important short-term effects of mesoscale circulations in redistributing smoke initially lofted by fire plumes. Overall, however, the circulations are likely to lead to increased dispersion both vertically and horizontally, as there are no circulations that can increase local aerosol concentrations. One possible exception is the creation of clear spots through precipitation scavenging, leading to an increase in "patchiness". Without better data on scavenging rates and modeling of the potential radiation effects on the circulations, it is difficult to assess the importance of such processes. Although the lower stratosphere is highly stable, radiative cooling in the ice anvil could generate sufficient instability to create large scale overturning. Some smoke and ice separation could occur at cloud top, which would leave some of the smoke behind in the stratosphere. Whether the mixture of smoke and ice could lead to net cooling or warming of the anvil during the daylight hours may have an important impact on the distribution of smoke in the upper troposphere and lower stratosphere.

Because of our limited understanding of mesoscale circulations and the lack of experience with plumes from large, intense fires, much of the preceeding discussion has been speculative. It is worth noting, as was done in Chapter 3, that there are observations of wildfire plumes which show that smoke can be transported over very long distances. Wexler (1950) described transport of smoke from fires in Alberta, Canada extending down into the U.S. and across the Atlantic to Europe. More recently, Chung and Lee (1984) used satellite imagery to track plumes from fires in this same area across Canada to the Atlantic coast and down across the Great Lakes to New York. Voice and Gauntlett (1984) also made use of satellite imagery to track plumes from the Australian bush fires of 1984 across the Tasman Sea to New Zealand. In these and in other cases, there is no evidence of induced mesoscale circulations. This may be because the fires were not sufficiently intense to generate cumulonimbus systems or because large wildfires typically burn under hot, dry, and windy conditions that do not favor the development of deep convection. In any case, it is clear that this scale of interaction deserves attention in future research programs.

4.4 SYNOPTIC SCALE RESPONSE

Scales of atmospheric motions between 1000 and 10,000 km are referred to as synoptic scales. Generally, synoptic systems are responsible for weather variations occurring over periods of 2–7 days. For example, mid-latitude cyclones and cold air masses are synoptic weather systems. These systems could potentially play an important role in redistributing, transforming, and removing the smoke particles, and, in turn, the smoke could modify these systems.

In the first few days after a nuclear exchange, cloud-scale and mesoscale processes could act to mix soot and ice throughout the troposphere and possibly into the lower stratosphere over much of the Northern Hemisphere. The fire plume models suggest that the depth of the initial lofting would be strongly dependent on local atmospheric conditions. It follows that destabilized regions on the synoptic scale, such as those having cold air advection aloft or positive vorticity advection at mid levels, would be characterized by the deepest soot and ice penetration aloft. On the other hand, very stable air masses would lead to the highest smoke concentrations below 3–4 km with little ice deposited aloft. Subsequent heating by solar radiation might then act systematically to weaken or strengthen the system relative to its natural state. Obviously, the induced changes in the strength of the synoptic system would then affect the mixing of the smoke.

As discussed previously, the generation of anticyclones and elevated jet streaks by mesoscale systems could significantly alter the normal progression of local synoptic scale systems. The possible generation of extensive regions of cumulonimbus clouds in the aftermath of a concentrated nuclear exchange might have a similiar effect. If a particularly dense arrangement of targets, e.g., missile silo fields, happened to lie under a region of strong conditional instability, it is possible that intense outflow aloft could create an upper level ridge on the scale of one or two thousand kilometers. A flow perturbation of this scale could propagate well beyond the time scale of the initial forcing. In addition, instabilities in the flow or forced flows at a later time might lead to the growth of subsequent weather systems. It is difficult to be more precise about these effects because the understanding of the relationships between mesoscale and synoptic scale systems in the natural atmosphere is very limited.

If surface fires can indeed amplify an existing disturbance or create a new disturbance on the synoptic scale, it is likely that the disturbance would propagate for at least several days. It can be anticipated that air motions associated with such systems would most likely lead to enhanced dispersion of the smoke for several days following the nuclear exchange. Precipitation processes could also act on the same time scale to remove some of the smoke. In any case, the synoptic scale effects would depend sensitively on the weather patterns at the time of the nuclear exchange. A more precise understanding of these effects may be obtained through future regional scale atmospheric modeling studies.

4.5 INTERACTION WITH SOLAR AND INFRARED RADIATION

As discussed in the preceeding section, smoke and debris clouds produced by nuclear detonations and the attendant fires could be spread by plume rise and mesoscale circulations over large areas of the Earth on spatial scales of 100 to 1000 km. These large smoke clouds would then interact with a variety of physical processes which are important in determining the weather and maintaining the current climate. The most important interaction is that with the radiation fields, including both the absorption of solar radiation and the emission of infrared radiation. This section will be concerned with the direct interaction of the injected smoke and dust particles with these fields and the implications for the radiative budget of the atmosphere.

4.5.1 Solar Radiation

The immediate result of the injection of optically-thick aerosol clouds into the atmosphere is to alter radically the pattern of absorption of solar radiation in the Earth–atmosphere system. Under normal conditions, about 30% of the incident solar radiation is reflected by clouds and the surface, 25% is absorbed by the atmosphere, and 45% is absorbed at the surface. This energy absorbed by the surface is subsequently transferred to the atmosphere by latent heat (evaporation of water), sensible heat (warming of the atmosphere by contact with the warm surface), and infrared radiation. The atmosphere convectively transports the latent and sensible heat upwards, distribut es it horizontally by atmospheric motions, and ultimately radiates it back to space as thermal or longwave radiation. A schematic of the energy deposition and transfer is shown in Figure 4.5 (taken from Liou, 1980).

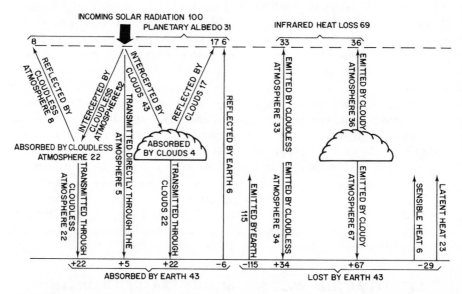

Figure 4.5. The heat balance of the Earth–atmosphere system for the present climate. Values are normalized to an incoming solar flux of 100 energy units (taken from Liou, 1980). Reproduced by permission of Academic Press

When an aerosol layer is introduced into the atmosphere, the direct solar beam is attenuated exponentially by the aerosol. The energy removed from the direct beam is partitioned into three categories: it may be absorbed by the aerosol particles, it may be scattered upward, or it may be scattered downward towards the surface. For most naturally-occurring aerosols such as cloud droplets, ice crystals, and wind-blown soil, downward-scattering dominates the other processes. Thus the basic effect of these aerosols is to convert the direct solar beam to diffuse (scattered) solar radiation (hence, the white color of clouds). At the same time, some additional energy is scattered upwards and lost to space, thereby reducing the solar energy available to the surface-atmosphere system. The fraction of solar radiation reflected to space by the atmosphere and surface and lost from the system is called the planetary albedo.

If the smoke particles are assumed to contain an amorphous elemental carbon fraction on the order of 20% and have typical dimensions on the order of 0.1 μm to 1.0 μm, about half the extinction events (defined as a photon interacting with a particle) for photons at visible wavelengths result in the photon being absorbed by the particle. In contrast, for dust aerosols of the same size, about 2 in a 100 events result in absorption, and for pure water less than 1 in a million result in absorption. In addition, most of the scattering events, perhaps 7 out of 8, result in scatter towards the ground. Thus, for dust or water, the majority of photons continue on towards the Earth's surface after a single extinction event; for the smoke particles, only about half continue on towards the surface.

A quantitative comparison of the effect of nuclear dust and smoke aerosols on solar radiation is shown in Figure 4.6 (adapted from Turco et al., 1983a) where solar transmission (defined as the fraction of the incident solar energy that penetrates the aerosol layer as either direct or diffuse radiation) is plotted as a function of optical depth. For a given value of optical depth, the transmission through the dust is considerably higher than that through smoke. The direct beam transmission is the same in both cases. For dust, the great majority of extinction events lead to scattering, while in the case of the smoke, a little less than half of the extinction events lead to absorption. The amount of transmitted radiation can be related directly to this difference (see Section 3.7). Furthermore, it is this difference which is primarily responsible for the potentially large climatic effects that are discussed in Chapter 5.

The presence of an aerosol layer also affects the amount of solar radiation reflected by the surface–atmosphere system. In the event of a large-scale nuclear war, ground bursts of nuclear weapons could inject dust layers directly into the upper troposphere and lower stratosphere, and air bursts could ignite large fires that could create black smoke layers throughout the troposphere. Thus, the most probable initial distribution of aerosols would consist of a dust cloud essentially overlying a smoke cloud, but with some mixture of the two in the upper troposphere (and above water clouds). Incoming solar radiation would then first be scattered by the dust, and then strongly absorbed by the smoke layer. For smoke extinction optical depths on the order of 3 or more, the planetary albedo would be on the order of 10 to 15% (Turco et al., 1983b; Cess, 1985; Ramaswamy and Kiehl, 1985), as opposed to the normal, averaged value of 30% (see Figure 4.5). The addition of an

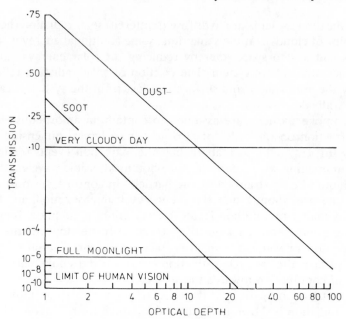

Figure 4.6.   Transmission of solar radiation at 0.55 μm as a function of extinction optical depth for soot and dust aerosols. Size distributions and indices of refraction were taken from Turco et al. (1983b). Calculations were made with a multiple scattering model and assumed a solar zenith angle of 60°

overlying dust layer of optical depth 1 would increase the planetary albedo to a value of 20 to 25%, which is still less than that of the unperturbed planetary system. Thus, the Earth–atmosphere system could actually absorb more solar energy than in normal conditions. Despite this, the change in the height of the solar absorption leads to the apparently paradoxical cooling of the surface.

An illustrative calculation of solar transmission and reflection as a function of aerosol optical depth in an atmosphere containing no water clouds is given in Figure 4.7 (from Cess, 1985). The curves labelled Case II in the figure are for a smoke only layer (the smoke having a single-scattering albedo, ω —defined as the ratio of scattering to extinction—of 0.70 at 0.55 μm) distributed through the lower 75% of the atmosphere. Case I has the same smoke layer, but has a dust layer overlying the smoke. The dust layer has an optical depth equal to one-third that of the smoke and a ratio of scatter to extinction of 0.96. Case III is the same as Case I except two-thirds of the dust is mixed with the smoke and only one-third is above the smoke. For all three cases, the transmitted flux decreases steadily as the optical depth increases. For Case II (the smoke only case), the planetary albedo decreases with increasing optical depth, while for Case I it increases due to the enhanced back-scatter by the dust. In Case III (which is not plotted) the reflected radiation is almost identical to that of Case II because only a small amount of dust overlies the soot.

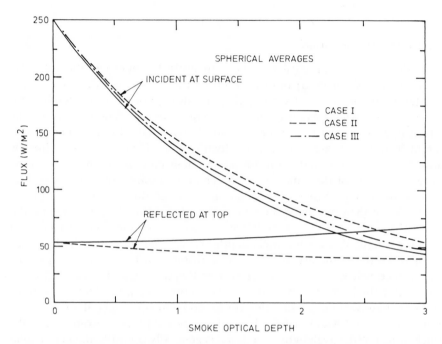

Figure 4.7. Spherically-averaged fluxes reflected at the top of the atmosphere and incident at the surface, as a function of extinction optical depth. Cases are explained in the text. The reflected fluxes for Cases II and III are virtually identical, so only Case II is shown (from Cess, 1985). Reproduced by permission of D. Reidel Publ. Company

As noted above and in Chapter 3, the transmission of solar radiation through an aerosol layer is determined in large part by the ability of the aerosols to absorb the radiation. In the calculations shown in Figure 4.7, Cess (1985) assumed that the smoke had a single-scattering albedo of 0.7. Turco et al. (1983a) and the NRC (1985) assumed a value of ω between 0.6 and 0.65, while Crutzen et al. (1984) assumed a value of about 0.4 (see section 3.6 for an extended discussion of smoke optical properties). If Cess had used a more absorbing smoke, (i.e., smoke with a lower single-scattering albedo), the transmitted solar radiation would have been reduced.

Calculation of the effects of aerosol particles on solar radiation, such as those shown in Figures 4.6 and 4.7, requires solution of the integral-differential equation of radiative transfer. Some of the multi-dimensional climate models which have been applied to the nuclear war problem use approximate solutions which do not incorporate particle scattering. In order to treat this problem correctly, those models not now treating scattering will have to be modified; alternative approximate methods may also be suitable (Slingo and Goldsmith, 1985).

4.5.2 Infrared Radiation

As discussed in the earlier section on optical properties, the solar extinction (and absorption) optical depth of the smoke and dust aerosols is greater than the infrared extinction (and absorption) optical depth. This greater solar opacity has two immediate consequences. The first is that the climatological impact of a layer of absorbing aerosols is fundamentally different from that of greenhouse gases such as CO_2. These gases are primarily absorbers of infrared radiation. Increasing their concentration increases the infrared opacity of the atmosphere, which in turn warms the Earth's surface by making it more difficult for radiation to escape through the atmosphere. Aerosols, on the other hand, make it more difficult for the solar radiation to reach the surface. If the aerosols are non-absorbing, then the surface must cool regardless of aerosol optical depth because the aerosols will always act to increase the planetary albedo. For aerosols that absorb solar radiation, the surface will warm if the layer is near the surface and optically thin; it will cool if the layer is high or if the layer is optically thick. Detailed discussions of these various possibilities are found in Ackerman et al. (1985a) and Ramaswamy and Kiehl (1985). Even if the aerosol layer becomes optically thick at infrared wavelengths, the layer is generally optically thicker at solar wavelengths. The importance of this difference will be discussed in the next section.

The second consequence of the smaller infrared optical depth is that an optically thick aerosol layer must develop a temperature gradient. Solar heating of the layer occurs primarily between the layer top and the level at which visible optical depth 1 or 2 is reached. The layer's ability to cool by emitting infrared radiation is directly proportional to its ability to absorb infrared radiation, which is related to its infrared optical depth. This cooling from the layer top also occurs primarily in the region from cloud top to an infrared optical depth of 1 or 2. However, the thickness of this "infrared cooling" layer is greater than that of the "solar heating" layer because the infrared extinction cross-section per unit mass is smaller and, therefore, it takes more mass to reach the same optical depth. It follows that much of the infrared radiation originates well below the layer of solar heating. The greater thickness of the infrared emitting layer relative to the solar absorbing layer can be reduced by the presence of other infrared emitters such as water vapor and CO_2. However, due to the spectral characteristics of gaseous emitters, (i.e., the presence of infrared "windows"), these emitters are generally not able to completely compensate for the difference in the aerosol optical depths at infrared and solar wavelengths.

As a result of this vertical distribution of heating and cooling, the smoke layer would develop a vertical temperature gradient with a maximum temperature at or near the layer top. This, in turn, would result in a stably stratified aerosol layer and inhibit mixing from below. This differential heating would only stop when the layer temperature becomes sufficiently hot that the emitted infrared energy from the layer top exactly matches the absorbed solar energy. However, long before this occurs, the heated air would mix upward into the ambient air above the layer, carrying the aerosol upwards. Obviously, the diurnally varying solar heating and relatively constant infrared cooling would result in diurnal variations in temperature at the top of the aerosol cloud. However, averaged over the daily cycle, the absorption of solar energy would dominate the emission of infrared energy.

In the preceeding discussion, the radiative concepts have been expressed in terms of the extinction optical depth of the aerosol. In reality, the more relevant quantity is the absorption optical depth, which determines the amount of incident radiation, either solar or thermal, absorbed by the layer. Because amorphous carbon is a highly absorbing material, the comparison of extinction optical depths between wavelength regimes is correct in a qualitative sense for smoke. However, for other materials such as water, which is transparent at solar wavelengths and has strong absorption features in the infrared, comparisons must be made on the basis of the absorptivity.

For absorbing particles that are small compared to the wavelength of the radiation being considered, absorption tends to dominate scattering. This fact can be used to derive an approximate method for treating the effects of aerosols on infrared radiative transfer. The aerosols are assumed to be absorbers only and scattering is neglected. In this case, the aerosols are essentially treated as a "gas" with an equivalent optical depth equal to the aerosol absorption optical depth as a function of wavelength. This approach, which simplifies the radiative calculations considerably, has been used without significant loss of accuracy in the studies by Turco et al.(1983a), Crutzen et al. (1984), and Haberle et al. (1985).

One situation that is likely to be important, and which cannot be treated by the absorption only approximation, involves the condensation of water on the smoke particles. As discussed in the section on mesoscale effects, following injection and the attendant initial precipitation, some layers might be saturated and condensation or freezing would occur on the particles, or smoke could be scavenged by water droplets or ice crystals, resulting in a polluted cloud or haze. Because of the large particle concentrations, the particles might only grow to sizes on the order of few microns in radius. However, this growth would still have an appreciable effect on the radiative properties of the aerosol, especially in the infrared. Due to the increase in particle size by water condensation, the infrared extinction cross-section and optical depth would increase substantially. At the same time, the visible extinction optical depth would increase due to the increase in total aerosol mass, although the increase would not be as great as at infrared wavelengths. In short, if sufficient water were available for condensation, the infrared optical depth of the aerosol layer could become equivalent to the visible optical depth through particle growth. Since it is unlikely that enough water would be present to produce raindrops, the aerosol layer might resemble a haze layer rather than a typical stratiform cloud.

It should be clearly pointed out that these particles would be substantially different from normal haze or cloud droplets. Because of the inclusion of absorbing material within the droplet, they would have large absorption cross-sections even at visible wavelengths. In fact, the water could enhance the ability of the elemental carbon to absorb solar radiation, as has been shown theoretically by Ackerman and Toon (1981) and Chylek et al. (1984). (For an extended discussion of this problem, see section 3.6 on optical properties.) Therefore, it should not be concluded that the addition of water would allow the solar radiation to penetrate the aerosol layer; on the contrary, both extinction and absorption would likely be increased.

The GCM studies done to date have not included the effects of aerosols on the infrared radiative transfer, although they have been included in several of the one- and two-dimensional simulations. Neglecting the infrared effects presumably would cause the model surface temperatures to decrease too quickly and too deeply in the early period when both the visible and infrared aerosol optical depths are large. Although rapid cooling would be expected in any case due to the cutoff of solar radiation, increased infrared radiation from the aerosol layer could moderate the cooling rate, depending on the layer location and the initial temperature profile. This increased infrared emission from the layer may also act to cool the bottom of the layer more rapidly, which in turn may offset the moderating influence of the enhanced infrared emission at the surface. Including the infrared effects may also reduce the magnitude of the cooling somewhat if a considerable fraction of the injected aerosol is scavenged and removed rapidly from the atmosphere. Infrared effects would have only a minimal impact on the duration of the temperature perturbations (assuming that the smoke persists for periods of weeks or more), since the return to normal temperatures would be controlled mainly by the return of normal insolation levels, i.e., by the rate at which the smoke is gradually removed from the atmosphere. One further possibility is that the infrared effects would be important in regions such as the polar latitudes where there is minimal solar radiation. Infrared cooling may be especially significant in the polar night stratosphere where cooling might enhance downward motions, thereby aiding in the ultimate removal of the smoke.

4.5.3 Radiative Equilibrium

The effect of an aerosol layer on equilibrium planetary temperatures can be examined qualitatively using a simple model consisting of a black surface and a grey atmosphere (Ackerman et al., 1985a). In this model the atmosphere is assumed to be uniform in temperature, to be transparent to solar radiation (in the absence of aerosols), and to have an infrared gaseous emissivity of 0.8, which is roughly equivalent to the average emissivity of the Earth's atmosphere. A spectrally-grey aerosol that absorbs but does not scatter radiation is introduced into the model. The infrared absorption optical depth, $\tau_a(IR)$, is specified as a simple fraction of the solar absorption optical depth, $\tau_a(S)$, and equilibrium temperatures are then computed as a function of the solar absorption optical depth.

The results of this simple model, plotted in Figure 4.8, illustrate several of the points discussed in the preceeding two sections. As $\tau_a(S)$ increases, the equilibium surface temperature decreases, first slowly, then sharply in the region of $\tau_a(S)$ approximately 1. The decrease in temperature is a result of the reduction in solar energy reaching the surface. The slight increase in temperature at large $\tau_a(S)$ is caused by the aerosol layer becoming optically thick in the infrared. No temperature increase is seen in the case where $\tau_a(IR)$ is assumed to be 0.

As the surface temperature decreases, the atmospheric temperature increases as it absorbs an increasing fraction of the solar energy. For a layer that is optically thick at all wavelengths, both surface and atmospheric temperatures tend to the radiation-to-space temperature, which is defined as the average black-body radiation temperature necessary to emit to space the total solar energy absorbed by the surface–atmosphere system. For the Earth, the normal radiation-to-space temperature is about 254 K. Because this simple model has no vertical atmospheric structure, it cannot illustrate temperature gradient effects within the aerosol layer itself or within the atmosphere.

A similar analytical model has been used by Golitzyn and Ginsburg (1985) to obtain estimates of the surface temperature and mean atmospheric temperature of a planet for several different cases. The model gives reasonable results for both clean and dusty Martian atmospheres (Ginsburg and Feigelson, 1971; Ginsburg, 1973), for clean and dusty (due to an asteroid impact) terrestrial atmospheres, and for doubled CO_2 in the Earth's atmosphere. When applied to the nuclear war scenarios, the results are similar to those described above.

Both Crutzen et al. (1984) and Ackerman et al. (1985a) have extended these analytic equilibrium models to include a three-layer atmosphere. Although these models can be used to examine some of the effects of optically-thick aerosol layers of vertical temperature gradients, they are not capable

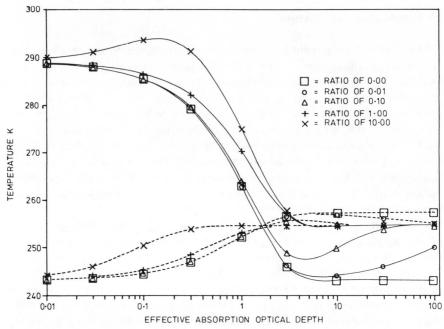

Figure 4.8. Equilibrium temperature as a function of the effective absorption optical depth for solar radiation (Ackerman et al., 1985a). The computations are from a one-layer grey atmosphere model. Solid curves indicate surface temperatures and dashed curves atmospheric (grey layer) temperatures. Symbols mark curves computed with different ratios of infrared to solar effective optical depth. Ratio values are shown in the legend

of adequately resolving temperature structure within the aerosol layer itself and therefore, still only provide approximate solutions.

While these simple models are limited in their vertical resolution and spectral detail, more realistic studies can be performed with one-dimensional radiative-convective models (RCMs). In addition to the more detailed radiative transfer calculations generally included in the RCMs, they also incorporate vertical structure and an atmospheric stability criterion. This criterion requires that when the temperature profile becomes unstable (i.e., when the vertical temperature gradient exceeds some critical value), the atmosphere is assumed to mix air upwards instantly to reduce the gradient back to its critical value. This type of climate model has been used in a variety of climate studies over the past 15 years (for a review of RCMs, see Ramanathan and Coakley, 1978). For the nuclear war climate problem, sensitivity studies of equilibrium responses have been carried out with RCMs by Ackerman et al. (1985a), Cess et al. (1985), and Ramaswamy and Kiehl (1985). The latter study presents results at 20 days after injection, which is essentially an equilibrium response of the model.

As discussed earlier in this chapter, various vertical distributions of the smoke have been postulated. Ramaswamy and Kiehl (1985) compared the effects of the same total amount of smoke distributed with a constant density between the surface and 10 km latitude (Profile D) and smoke distributed with an exponential scale height of 3 km (Profile M). The same amount of solar radiation is absorbed by both distributions, but the heating rate distributions are quite different. The "M" distribution has a broad maximum centered near 5 km, while the "D" distribution has a sharp maximum at the layer top (10 km). The equilibrium temperature profiles for these two cases, as well as for an unperturbed atmosphere, are shown in Figure 4.9. The "D" profile has a sharper and more elevated inversion and a considerably greater surface cooling (a temperature change of $-32°C$ as opposed to $-22°C$ for the "M" profile). Since the surface receives the same amount of solar radiation in both cases, the enhanced cooling is primarily the result of less downward infrared radiation reaching the surface for the "D" profile. This is in turn related to the height at which the maximum absorption of solar radiation takes place. The higher the level at which this absorption occurs, the less atmospheric gaseous infrared opacity lies above the layer. The gaseous opacity tends to trap the infrared radiation emitted by the layer (the greenhouse effect). Thus, when the layer lies above the bulk of the infrared opacity, it can more easily radiate the absorbed energy to space, and the total energy radiated by the layer, both upwards and downwards, is reduced. Ackerman et al. (1985a) essentially pointed out the same effect when they noted that for an aerosol layer of given optical depth, the higher the layer, the colder the equilibrium surface temperature.

Cess et al. (1985) used a 2-level RCM to study the sensitivity of the surface temperature to various aspects of the aerosol layer such as vertical distribution and optical depth. Their RCM incorporates the same vertical structure and boundary layer physics as are in the Oregon State University 2-level GCM (Ghan et al., 1982), but the solar radiation code has been replaced with a delta-Eddington scheme, and the hydrologic cycle has been replaced with an assumption of constant relative humidity. They found that, as opposed to results for forcing by CO_2 concentration changes or by changes in the solar constant, the sensitivity response to increasing aerosol concentrations is non-linear due to the convective coupling between surface and troposphere and to the exponential behavior of solar absorption. For tropospheric aerosol layers with small optical depths, sufficient solar radiation continues to reach the ground to convectively couple the troposphere and surface. As long as this coupling is maintained, the surface temperature and tropospheric temperatures remain essentially unchanged; both temperatures actually may increase if the planetary albedo is reduced as a result of the aerosol absorption, or may decrease if the albedo is increased due to aerosol scattering. However, under dense smoke clouds, the surface and

Physical and Atmospheric Effects

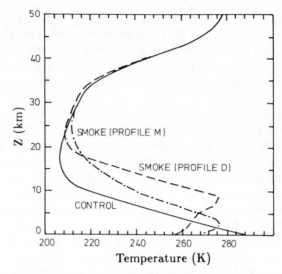

Figure 4.9. Thermal structure of the atmosphere for the unperturbed case, and for smoke profile M and smoke profile D. Temperature profiles for the smoke cases are for 20 days after the assumed injection (taken from Ramaswamy and Kiehl, 1985)

lower atmosphere can decouple due to a combination of increased stability in the middle troposphere and reduced heating at the surface, both of which suppress convection. As a result, the surface temperature can decrease dramatically because the direct solar heating is lost, mechanical transfer of heat from air to ground is suppressed, and the surface becomes very sensitive to small changes in the radiative forcing.

In addition to the RCM studies, Cess et al. (1985) were able to perform sensitivity studies with the OSU GCM, which was modified to include the delta-Eddington solar scheme. Their results show that, not unexpectedly, the surface cooling is sensitive to the layer optical depth, the aerosol single-scattering albedo, and the vertical distribution of the aerosol. The sensitivity to both the aerosol single-scattering albedo, ω, and the vertical distribution is demonstrated in Figure 4.10. As indicated, the visible extinction optical depth of the layer is 1. The changes in surface-air temperatures are computed over non-ocean areas only. The results of the change in vertical distribution are consistent with the RCM results discussed above and show that the decrease in surface–air temperature is more severe for more elevated smoke layers. As expected, decreasing the single-scattering albedo, which increases the absorption optical depth for a given extinction optical depth, also increases the severity of the cooling. The authors point out that the effect is more pronounced for the constant density distribution than for the constant mixing ratio distribution and attribute this difference to the greater effectiveness of surface–troposphere decoupling for the constant density case.

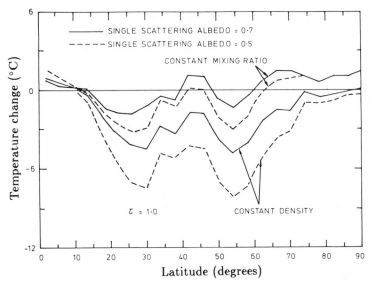

Figure 4.10.   Zonally-averaged (over land and sea-ice areas) day 10 changes in July surface-air temperatures for changes of the vertical smoke distribution and the smoke single-scattering albedo, computed with the OSU 2-level GCM. The smoke optical depth is 1.0. (taken from Cess et al., 1985)

An important limitation of the studies by Cess and his colleagues may arise because of the vertical resolution of their models. The two model layers each are assumed to have equal atmospheric mass. Therefore, for the constant mixing ratio case, the optical depth is simply equally divided between the two layers. For the constant density case, approximately two-thirds of the smoke is in the upper layer and one-third in the lower layer. At these small absorption optical depths of 0.3 and 0.5 (for ω equal to 0.7 and 0.5, respectively) and for a constant mixing ratio distribution, roughly equal amounts of solar radiation are absorbed in each layer and each layer experiences roughly the same solar heating rate. Thus, both the change in overall atmospheric stability and the change in the average temperature of the lower layer are minimized, which also minimizes the surface-troposphere decoupling. If the absorption optical depth were greater, or if more of the aerosol were placed at higher levels in the atmosphere (as occurs in the constant density case), the solar heating per unit mass of air would be greater at the layer top, which would tend to heat the top layer preferentially and to stabilize the atmosphere, thereby increasing the decoupling of atmosphere and surface. In addition, since the bulk of the water is in the lower layer, the warmer temperature may increase the water vapor content, which would in turn increase the atmospheric opacity and the downward infrared radiation. Furthermore, the surface-air temperature is determined diagnostically, in part, from the average temperature of the lower layer. All of these biases tend to reduce expected decrease in surface and surface-air temperature resulting from the attenuation of solar radiation, and may in some cases actually force a warming.

In the constant density case, the upper layer absorbs approximately twice as much radiation as the lower, and thus heats about twice as much. The temperature of the lower layer is decreased relative to the previous case due to the reduction in solar absorption. As a result, the atmospheric stability is increased. The increased absorption in the upper layer also implies a greater sensitivity at low optical depths to changes in the absorption optical depth, (i.e., changes in ω), because, as ω is decreased, an increasingly greater fraction of the absorption takes place in the upper layer due to the exponential nature of the attenuation of solar radiation.

4.5.4 Diurnal Variations and Daylengths

The majority of climate models are run with diurnally-averaged solar insolation. This means that the solar radiation is computed at a single value of the solar zenith angle chosen to approximate an exact average for insolation at the top of the atmosphere over daylight hours; this "averaged" solar radiation is then further multiplied by the fraction of the day during which the sun shines, and the result is used as the solar energy input to the climate model. While this is done primarily to increase computational speed, it has been assumed to introduce relatively little error in the values of daily average surface temperature predicted by the models. The success of this approximation is basically due to the fact that, under normal conditions, about two-thirds of the solar radiation absorbed by the surface-atmosphere system is deposited at the ground. Thus, most of the diurnal variations in the atmosphere on the scale of typical general circulation model grids are confined to the boundary layer (the lowest few kilometers of the atmosphere) and have relatively little impact on the predicted temperatures and wind fields at higher levels. However, it obviously has a substantial impact on the variables at the surface and precludes the calculation of diurnal variations in such quantities as surface temperature and evaporation rates.

In the case of optically-thick, elevated aerosol layers, this approximation would not be valid. The aerosol layer itself obviously would experience strong diurnal variations in solar heating, which presumably would lead directly to variations in the predicted model quantities. To date, the effects of such variations have not been studied in any detail, although several model runs with a diurnal cycle have been carried out using the 2-level OSU/GCM (Cess, personal communication; MacCracken and Walton, 1984). Unfortunately, however, comparisons of otherwise identical cases but with and without diurnal cycles have not been carried out.

An additional, and perhaps equally important, consequence of diurnally-averaged calculations was noted by Cess (1985). Because the solar radiation is attenuated exponentially by the aerosol layer, the effect of the solar zenith angle is very pronounced for layers with solar optical depths on the order of 1 to 3. The amount of sunlight reaching the surface on a daily basis can be substantially underestimated when an average solar zenith angle is used. Again, the climatological consequences of this underestimate have not been studied in detail, but it is assumed that this additional solar radiation reaching the surface will reduce the surface cooling somewhat. However, since heat transfer processes at the surface are highly non-linear, the magnitude of the reduction, or even if that reduction would occur, cannot be ascertained without further research.

TABLE 4.1.

DAYLENGTHS, IN HOURS, FOR A CLOUDLESS SKY AND VERTICAL AEROSOL EXTINCTION OPTICAL DEPTHS, τ, OF 0, 1.0, AND 3.0 AT A WAVELENGTH OF 0.55 μM. SUNRISE AND SUNSET WERE DEFINED TO BE THE TIME AT WHICH THE TOTAL DOWNWARD SOLAR FLUX AT THE SURFACE REACHED 10% OF NORMAL (CLOUD-FREE) NOON INSOLATION. THE AEROSOL SINGLE-SCATTERING AT 0.55 μM WAS ASSUMED TO BE 0.7

| Latitude | A. Northern Hemisphere Summer | | |
	$\tau = 0$	$\tau = 1.0$	$\tau = 3.0$
60° N	16.0	11.8	4.1
30° N	13.9	10.9	7.6
15° N	12.7	10.1	5.5
0°	11.0	8.8	4.4
	B. Spring/Autumn Equinox		
60° N	11.0	7.5	—[a]
30° N	11.0	8.3	2.1
15° N	11.0	8.7	3.8
0°	11.0	8.9	5.1
	C. Northern Hemisphere Winter		
60° N	5.5	2.5	—[a]
30° N	8.5	5.1	—[a]
15° N	10.0	6.5	—[a]
0°	11.0	8.8	4.4

[a] Solar flux never exceeded the 10% normal criterion.

The effects of the smoke on normal diurnal variations in insolation may also be important for plant communities, many of which are highly sensitive to both the total amount of sunlight received and the period (the daylength) over which it is received (see Volume II of this report). The reduction in daylength due to absorbing aerosol layers with extinction optical depths of 1 and 3 and absorption optical depths of 0.3 and 0.9, respectively, were computed as a function of season and latitude. It was assumed that no water clouds were present in the atmosphere, and sunrise and sunset were defined as the time at which the solar insolation was one tenth of its normal, clear sky, noontime value. The results are given in Table 4.1. The daylength reductions are quite large, particularly at high latitudes, where the longer slant path of the solar beam becomes more important. In several instances, the solar flux reaching the surface never exceeds the one-tenth normal flux criterion.

4.6 INTERACTIONS WITH OTHER ATMOSPHERIC PROCESSES

As a result of the interaction of the smoke and dust with the radiation, various other atmospheric processes would be affected. The primary effect would be on the dynamical motions of the atmosphere, which would in turn influence the transport of the aerosols. There would also likely be significant alterations in boundary layer processes and in the hydrologic cycle. These interactions are described briefly in this section in a qualitative manner. More extensive discussions, particularly on the dynamical interactions and modifications to the hydrologic cycle are found in Chapter 5, where the results of the general circulation model studies are described.

4.6.1 Atmospheric Transport

Prior to studies of the smoke problem, the vast majority of studies of the transport of a trace species by atmospheric motions considered the trace species to be passive, i.e., to have no effect on the atmospheric motions. However, if a tracer has a significant effect on the distribution of radiative heating and cooling in the atmosphere, its presence will affect the motions of the atmosphere and, hence, its own transport. This is clearly the situation that arises when a layer of absorbing aerosols is present in the atmosphere.

From the previous calculations of the absorption of solar radiation by the smoke cloud, it can be inferred that the top of the cloud will heat and become buoyant, except perhaps during the winter season when insolation is weakest. As a result, the smoke would be lofted both by induced, small-scale convective motions and by the generation of large-scale upward vertical velocities. The extent of lofting would be directly related to the amount of available solar radiation, and thus (as already noted) would have a strong seasonal and latitudinal dependence. In addition, the atmosphere is likely to develop a temperature inversion below the level of maximum heating which would act to reduce mixing from below. Wexler (1950) noted such an inversion in his study of the large smoke plume generated by forest fires in Alberta, Canada, in 1950 although the cause of the inversion could not be unambiguously determined. The combination of these two effects, lofting and stabilization, suggests that at least the upper part of the aerosol layer (absorption optical depth about 1) could be transported upwards and could stabilize the atmosphere below. In essence, this layer would form a stabilized "stratosphere", even if the smoke were not initially injected into the ambient stratosphere. For this reason the determination of the exact initial height of injection is perhaps less crucial than previously thought, although it still is likely to be an important factor in winter scenarios.

Inferences concerning subsequent smoke and dust transport can be drawn from observations of stratospheric winds and the transport of stratospheric aerosols. For a variety of reasons, including a reduced influence of both land-sea temperature contrast and topographical effects, stratospheric circulations tend to be strongly zonal, i.e., the flow around the Earth is along lines of constant latitude. Thus one might expect the lofted aerosols to be mixed fairly uniformly within latitudinal bands after a period of a few weeks. For example, Robock and Matson (1983) discuss the dispersion of the El Chichón volcanic debris cloud, which in the relatively weak zonal flow of the tropical stratosphere took about three weeks to form a band around the globe. In regions where no injection had occurred, atmospheric "eddies" would "diffuse" the material from other latitudes. This meridional spreading would presumably be somewhat faster in the case of absorbing smoke aerosols than in the case of non-absorbing volcanic aerosols due to the stronger temperature gradients produced by the smoke. The induced vertical motions would also lead to enhanced meridional dispersion due to strong wind-shear effects.

Obviously, the related issues of aerosol transport and modification of atmospheric dynamics and stability are a crucial aspect of the climatic impact problem being assessed here. The lifetime of the particles, which is discussed in more detail in Chapter 5, is highly dependent on atmospheric transport and thermal stability. In the current atmosphere, tropospheric particles tend to have lifetimes on the order of a few days to weeks, depending on their height and the local synoptic conditions. Stratospheric particles have typical lifetimes on the order of 6 months to 2 years due to the stability of the stratosphere and the lack of precipitation scavenging. The principal removal mechanism for stratospheric aerosols is, in fact, not direct removal, but injection of the aerosols into the troposphere, either through mid-latitude tropopause folding events or through migration to the winter poles and descent in the polar vortex, and subsequent removal from the troposphere by

scavenging. Since the particle lifetimes determine the longevity and severity of the climatological response, uncertainties in the smoke lifetimes in a
modified atmosphere need to be reduced by further research.

4.6.2 Boundary Layer Processes

The impact of variations in climatologically-important parameters is usually quantified in terms of their effects on surface variables, particularly
temperature. While this reflects primarily our bias as surface dwellers, it
also reflects the important role that surface processes play in the planetary
energy balance. Under normal conditions, the daily input of solar energy
to the surface is nearly balanced by evaporative or latent heat transfer, turbulent conductive or sensible heat transfer, and infrared radiative exchange
with the atmosphere. The relative importance of these terms, as well as the
conduction of heat into the sub-surface, varies with season and location.
Furthermore, there are obviously substantial differences between the thermal response of land and ocean surfaces. In the presence of an optically
thick aerosol cloud, very little solar radiation will reach the ground; thus the
focus of the present discussion will be on how the other terms in the energy
balance might adjust to compensate for this loss of solar radiation.

4.6.2.1 Land Surfaces

For land surfaces, the effect of a thick smoke layer on the sensible heat
flux is quite predictable. Due to the substantial reduction in insolation, the
ground temperature quickly drops below the surface air temperature and an
inversion forms. While this phenomenon occurs every night, the polar night
provides a dramatic example of what can occur on longer time scales. Here
the inversion deepens to a kilometer or more and the temperature difference
over this layer may be as much as 15–20°C. As an example, a typical average
temperature sounding for the month of February at Barrow, Alaska is plotted in Figure 4.11. Note the extreme stability of the lowest kilometer, which
inhibits downward sensible heat transfer driven by mechanical turbulence.

Mechanical turbulence is produced by frictional drag on the winds at the
Earth's surface. The drag generates vertical eddies that mix heat downward
to the surface under stable conditions. Compared to buoyancy (which only
mixes heat upwards from the surface), mechanical turbulence is a relatively
poorly understood process dependent on the roughness of the surface and
the local wind speed. As a general rule, the current generation of GCM's has
incorporated rather simple boundary layer models. The ability of these current formulations to model adequately stable boundary conditions will have
to be examined in view of the importance of these processes in determining
the surface temperature. A correct determination of the downward heat flux

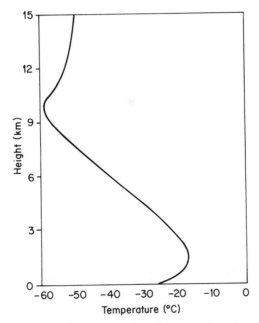

Figure 4.11.   Monthly-mean temperature profile for February, 1978 at Pt. Barrow, Alaska

is especially critical at the very early times when the rate and magnitude of the surface cooling is likely to depend largely on the amount of heat that can be extracted from the atmosphere.

The role of evaporative cooling in the land surface energy balance most probably is reduced substantially under these conditions. For low or negligible insolation, evaporation must be strongly inhibited. During the initial transition period when the boundary layer is cooling, condensation would in fact occur. At the later stages, the low saturation vapor pressure of the cold boundary layer air would prevent much evaporation, even though water or ice is available at the surface. Given the low evaporation rates and the reduced convective mixing, a large reduction in precipitation rates over land would seem to be unavoidable. While transport of moist air from over the oceans to the land would occur, it is unlikely to compensate fully for the lack of evaporation and mixing. Overall, it seems plausible to conclude that latent heat would play a relatively unimportant role in the surface energy budget over land.

These qualitative conclusions concerning the surface energy balance are reinforced by the results of Covey et al. (1985), who present a more detailed analysis of the GCM model results reported in Covey et al. (1984). They point out that for land surfaces underneath an optically-thick aerosol layer, the moisture term in the surface heat budget actually becomes slightly positive, indicating a net release of heat by condensation under these conditions. These calculations are discussed in more detail in Chapter 5.

Additional support is lent by the computations of Cess et al. (1985) with the 2-level RCM described in the section on radiative equilibrium models. The aerosol was assumed to be distributed between the surface and the model top (defined to be 200 mb) with a constant mixing ratio. The infrared opacity of the smoke was neglected. In the RCM, the surface is assumed to have no heat capacity, so an energy balance is determined by balancing absorbed solar radiation, net infrared radiation, and latent and sensible heat losses. As shown in Figure 4.12a, the solar radiation reaching the surface is reduced by the presence of an aerosol layer. The combined heat fluxes gradually decrease and eventually change sign. The small heat flux remaining for optical depths greater than 2.5 represents heat mixed to the surface by mechanical turbulence. The sum of this residual heat flux and the reduced solar radiation must be balanced by infrared losses from the surface.

Under normal conditions, the heat flux from, or into, the soil is relatively insignificant. Its primary role on seasonal time scales may be seen as controlling the amplitude of the diurnal temperature cycle. For soils which conduct heat poorly such as sand, little heat is stored in the soil and the diurnal amplitude is large. For soils such as clay or loam, the conductivity is higher, so the heat storage tends to reduce the diurnal temperature amplitude somewhat. Soil thermal conductivity is intimately related to moisture; wet soils are far more efficient heat conductors than dry soils. This suggests that should surface temperatures drop below freezing, the soil heat flux would decrease substantially. In short, the soil heat flux in soils with a high thermal conductivity may be an important factor in mitigating the surface cooling on the time scale of a few days. At longer time scales or for poorly conducting soils, it is unlikely to be significant.

The final energy component to be considered for the land energy balance is the downward infrared radiation. At first glance, one might expect this term to increase as a result of enhanced temperatures in the aerosol layer. However, the situation is more complicated. First of all, the majority of the downward infrared radiation reaching the surface under normal conditions originates in the relatively warm regions of the lower troposphere. If this layer cools as a result of an overlying aerosol layer, the downward infrared radiation reaching the surface actually decreases due to a reduction in the effective emission temperature of the lower atmosphere. Although radiation from the warm aerosol layer may partially compensate for this decrease, it is unlikely to do so completely unless the bulk of the aerosol layer is very low in the atmosphere.

This effect is demonstrated in Figure 4.12b, also taken from Cess et al. (1985). For uniformly mixed smoke at small optical depths, the net infrared radiation absorbed at the surface actually increases due to heating of

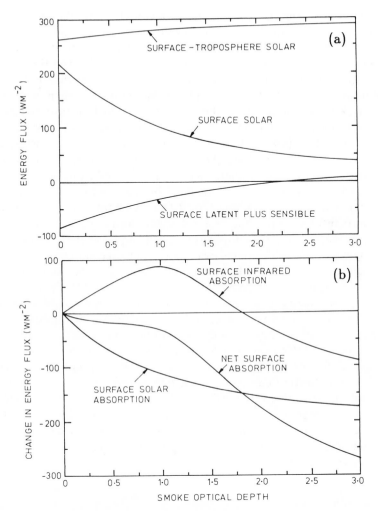

Figure 4.12.   The equilibrium response (W/m^2) of a radiative convective model as a function of smoke extinction optical depth for (a) surface–troposphere solar absorption, surface solar absorption, and latent plus sensible surface heat flux; and (b) changes in surface solar absorption, surface infrared absorption, and net surface radiation absorption (taken from Cess et al., 1985)

the lower atmosphere. As the optical depth increases the net infrared begins to decrease and becomes a deficit at about optical depth 2. Taking into consideration that the ground temperature has decreased by 15°C at this same optical depth and, therefore, is emitting much less infrared radiation (since infrared emission by a blackbody is proportionally to the temperature raised to the 4th power, T^4), it is clear that the downward infrared flux from the atmosphere has been reduced by a large degree. Because the model neglects the infrared opacity of the aerosols, the computed downward infrared radiation may be somewhat too small. However, this is probably only a small correction at these optical depths. Furthermore, as noted in the discussion on radiative equilibrium results, the uniform mixing ratio distribution of the aerosols assumed in the Cess et al. (1985) model places much of the smoke in the lowest few kilometers of the atmosphere, which in turn enhances the downward infrared radiation from clouds and the air to the surface.

Ackerman et al. (1985a) have also presented calculations illustrating this effect. Their results, in which the infrared opacity of the aerosol layer is included, show that, for radiative-convective equilibrium, a soot layer of visible extinction optical depth 3 located between 8 and 14 km can reduce the downward infrared reaching the surface by about 20% compared to the downward infrared flux under clear sky conditions. If the aerosols in this layer are assumed to have an infrared opacity equal to their visible opacity, the downward infrared remains essentially unchanged at equilibrium. Thus, while optically thin aerosol layers located near the surface may increase the downward infrared radiation reaching the surface, moderately thick layers in the middle troposphere or above are actually likely to reduce the downward infrared because they raise the effective level of solar absorption.

Model calculations that allow the aerosol heating to influence the dynamical motions (MacCracken and Walton, 1984; Aleksandrov, 1984; Haberle et al., 1985; Stenchikov, 1985; Malone et al., 1986; Thompson, 1985) show that the solar heating can produce vertical lifting of the particles and the surrounding air. This lofting reduces the temperature of the aerosol layer, through expansion and mixing, relative to the case of a fixed aerosol layer. Thus, the emission temperature of a smoke layer would be less than that computed from radiative-convective equilibrium, thereby reducing further the downward infrared flux.

4.6.2.2 Ocean Surfaces

Over the oceans, the situation would be quite different. Given the long thermal response time of the ocean mixed layer (generally the topmost 50 to 100 m of the ocean), the ocean surface will cool only very slowly when the insolation is removed. When cold air masses from land areas move out over the warmer oceans, strong air modification events with large upward fluxes of heat and moisture are to be expected. The depth of the modified air layer is difficult to specify from qualitative arguments. Although deep convection could be suppressed by the presence of an elevated inversion in the smoke layer, it seems plausible that a moderately deep layer of modified air with a thickness of a few kilometers might develop over the warm ocean.

There is, of course, considerable uncertainty in making the preceeding generalizations since the land and ocean cases have been considered as if they were unrelated. Obviously, the modification of continental air masses as they move offshore and the corresponding modification of marine air as it moves onshore need to be considered. While the uncertainties in predicting coastal effects are unlikely to be completely removed, our understanding of these effects can be enhanced through the use of appropriate numerical models and through the study of intense air modification events such as those which occur in the China Sea.

4.6.3 Hydrologic Cycle

Discussion of the qualitative effects of an aerosol layer on the hydrologic cycle is difficult because the hydrologic cycle is strongly dependent on both radiation and atmospheric dynamics. Some plausible arguments can be set forth but few can be stated definitively without further analysis. Consequently, the following discussion attempts to identify issues of importance.

The modification of condensation and precipitation processes would occur during both the transient phase and the quasi-equilibrium phase. Neglecting initial scavenging in the fire plume, which has been discussed elsewhere (Chapter 3), scavenging can occur by water vapor that has been redistributed by entrainment from the boundary layer to the middle troposphere or above. Considerable confusion has been generated on this issue because, on a gram per gram basis, more water vapor than aerosol is generated by combustion. The real issue, however, is the relative change in atmospheric water vapor concentrations produced by the injection. While background aerosol concentrations are typically small, water vapor column amounts are order of 10^4 g/m^2. In the NRC (1985) report, it was estimated that the average global water vapor concentration in the upper troposphere might increase by a maximum of 20% due to all the fires in a major nuclear war. As pointed out earlier in this chapter, this excess water vapor can be very important on a local scale in producing severe cumulonimbus storms with some attendant rainfall and particle removal. On the longer timescale and larger spatial scale, the effect of such an increase is uncertain. There will be some increase in infrared opacity and perhaps some residual ice crystals at the early times following injection. However, once the smoke begins to heat the air by absorbing solar radiation, the air will be well below saturation and condensation will cease in the aerosol layer (Covey et al., 1984; Malone et al., 1986).

A related issue that has received considerable attention by critics of the nuclear winter hypothesis is the scavenging of aerosols by condensation processes as the atmosphere cools beneath a smoke cloud (Teller, 1984; Katz, 1984). They have argued that as the atmosphere cools, condensation and scavenging will occur. This condensation, however, is not likely to affect the upper levels of the aerosol layer, which are being heated. Also, this condensation would probably be analogous to nocturnal fog formation rather than to precipitating cloud formation. The amount of water actually condensing would be limited to the water vapor already in the air mass and would not result in large amounts of precipitation or efficient particle removal, although some coagulation might occur.

Possible changes in cloud optical properties also need to be considered. In an earlier section, it was pointed out that the inclusion of absorbing material in cloud droplets actually enhances the effective absorption of the material. In an atmosphere so impregnated with aerosols, it is difficult to imagine that any clouds would be formed that did not include a significant amount of absorbing material. Thus, the clouds which did form would be much more absorbing in the visible than present clouds. Furthermore, the presence of large numbers of aerosol particles would presumably lead to clouds with more but smaller droplets, which might increase the reflectivity of clouds (Twomey et al., 1984). Since absorption within the cloud would tend to decrease the reflectivity, the exact result would depend on the amount of absorbing material present within the cloud. In any case, either process would reduce cloud transmissivities, thus further reducing the solar radiation reaching the ground. Of course, significant absorption within the cloud would also act to evaporate the cloud.

Because of the physical complexity of the processes involved, modelling cloud formation in GCMs is a very difficult problem. Current schemes tend to be idealized and empirically tuned to the present climate, as they must be since the grid spacing in the climate models is much larger than the scale of individual clouds. Interactions with radiative transfer codes are carried out in a variety of ways, none of which are entirely appropriate for simulating the actual interactions. While some models do a fair job of predicting globally-averaged cloud statistics, regional-scale cloud predictions are often less successful and the precipitation predictions are often marginal. As noted before, clouds are the dominant removal mechanism for aerosols. Modelling the microphysics of particle scavenging is a difficult problem even within the context of very detailed cloud models. The problem has hardly even been addressed within GCMs with their coarse resolution and highly parameterized clouds (MacCracken and Walton, 1984 and Malone et al., 1986 represent the first attempts within the studies of nuclear weapons effects). In short, since the current schemes are not completely satisfactory for study of the present climate, they are likely to be inadequate for studying clouds, precipitation, and washout processes of smoke generated by a nuclear war.

Interestingly, this short-coming of the models may not be as critical in the study of the short-term effects as in the long term. The low relative humidities associated with the strong heating in the aerosol layer make the issue of cloud formation relatively unimportant in the early stages. At later times as the smoke disperses and thins, cloud formation would again become a very important part of the climate simulation.

4.7 GEOPHYSICAL ANALOGUES

There are phenomena in nature where the aerosol interactions discussed in this chapter are of importance. To some extent these phenomena can, and have, been used both to validate climate models in general and to aid in understanding the climatological effects of a nuclear war. Several of these analogues will be considered briefly regarding their relevance to the present problem. One other, the possible impact of an asteroid causing the extinction at the Cretaceous–Tertiary boundary (Alvarez et al., 1980, 1984; Toon et al., 1982), will not be discussed since virtually no evidence on the atmospheric effects of such an event can be deduced without recourse to the very same models that are currently being used to study the nuclear effects problem.

4.7.1 Volcanic Eruptions

In 1783 Benjamin Franklin proposed that the volcano Laki which erupted in Iceland was responsible for the cold summer weather in Europe of that year. Humphreys (1940) was the first to examine the volcanic hypothesis quantitatively. Although his treatment was incorrect in several important aspects, it illustrated the potential for volcanic eruptions to affect global climate. The idea that large atmospheric injections of aerosol could cause extremely severe climatic perturbations was suggested by Budyko (1974), who considered the potential effects of several large eruptions occurring within a relatively short period of time. A recent review of the current status of our knowledge of volcanic effects was given by Toon and Pollack (1982). They concluded that large volcanic eruptions can affect the climate through the injection of gases leading to the formation of sulfuric acid droplets into the stratosphere. The ash particles injected by the volcanos are typically much larger in size and tend to settle rapidly out of the atmosphere.

While volcanic aerosol layers might appear to be reasonable analogues of those projected to be produced by nuclear detonations and fires, they are not. Sulfuric acid droplets have very different optical properties compared to smoke particles. Sulfuric acid droplets are essentialy transparent at visible wavelengths. Thus, their primary effect is to scatter solar radiation and thereby increase the planetary reflectivity (albedo). They do have appreciable absorption at thermal infrared wavelengths and can actually heat the stratosphere locally by absorbing upwelling infrared radiation emitted by the surface and lower atmosphere (Labitzke et al., 1983).

Furthermore, while the model predictions of the effect of volcanic aerosols are quite consistent in predicting a small decrease in surface temperatures over large spatial scales, this decrease has never been observed directly, although considerable inferential evidence is available that cool summers have followed large volcanic eruptions (Stommel and Stommel, 1983). Kelly and Sear (1984) claim that volcanic eruptions also may be responsible for declines in monthly mean land surface temperatures on the order of a few tenths of a degree in the several months following the eruption. On the local scale, Mass and Robock (1982) showed that diurnal variations in surface temperature were damped under the volcanic ash plume following the eruption of Mount St. Helens. Apparently the ash cloud cooled the surface by reflecting and absorbing solar radiation during the day and warmed the surface by increased emission of infrared radiation at night. This was, however, only documented for a few days following the eruption. Attempts to isolate temperature signals at later times both at the surface and aloft were unsuccessful, apparently because most of the large ash particles fell out within the timespan of a few days and the remaining sulfuric acid and ash particles dispersed too rapidly.

4.7.2 Dust Storms

Regional scale dust storms are a frequent phenomena on Earth, and can reach the global scale on Mars. Saharan dust storms containing as much as 8 million tonne of dust have been observed (NRC, 1985). These storms move out of west Africa, westward across the Atlantic as far as South America. They can generate optical depths at visible wavelengths as large as 1. Observations show that convection is suppressed underneath the dust cloud and direct atmospheric heating occurs within the cloud. Brinkman and McGregor (1983) reported on dust clouds over Nigeria with extinction optical depths up to 2, reductions of daily mean total solar radiation of up to 30%, and corresponding temperature decreases of up to 6°C.

Although the dust particles involved are usually considerably larger than typical smoke particles and much less absorbing, the available evidence shows a climatic response very similar to that proposed for optically-thick smoke clouds. The rapid decrease in surface temperatures, the heating of the atmosphere, and the suppression of convection are all consistent with the physical processes discussed in the preceeding sections.

A similar phenomena is observed on a much larger scale on Mars where dust storms originate in the Southern Hemisphere during its summer and occasionally spread globally in a matter of a week or two. Extinction optical depths can reach values on the order of 5 and surface cooling of as much as 10 to 15°C occurs (Zurek, 1982). The Martian dust is apparently more absorbing than typical desert sand on Earth and thus produces strong heating in the Martian atmosphere. The somewhat smaller temperature reduction than would be predicted for the Earth under the same conditions is likely due to both the large infrared optical depth of the Martian dust and the much weaker greenhouse effect and colder surface in the unperturbed Martian atmosphere.

The effects of Martian dust storms on the dynamics of the atmosphere have also been observed. Data from the Viking Landers show that the normal cyclonic activity of the Martian atmosphere is suppressed when the dust is present (Ryan and Henry, 1979). Boubnov and Golitsyn (1985) proposed an explanation based on the suppression of baroclinic instability in atmospheric flows when the vertical stability of the atmosphere is increased. A similar suppression of synoptic-scale fluctuations was observed in the NCAR GCM in experiments with injected smoke layers performed by Thompson (1985).

While Mars lacks an ocean to moderate temperature changes, the periodic, large Martian dust storms offer perhaps the most convincing, although still imperfect, analogue to the projected atmospheric effects of a major nuclear war. The large surface coolings, the heating of the atmosphere, the global transport of the dust, and the suppression of baroclinic activity are all similar to, and consistent with, the computed effects of optically-thick smoke layers.

4.7.3 Smoke from Forest Fires

As noted previously, forest fires occasionally produce large amounts of smoke that can be transported over large areas. Examples of this include the plume from the Alberta fires reported by Wexler (1950), plumes from the large Australian bush fires of 1984 (Voice and Gauntlett, 1984), smoke from peat fires in the U.S.S.R. in 1972 which were reported to travel over 5500 km (Grigoryev and Lipatov, 1978), and plumes from recent fires in Alberta seen in satellite photographs (Chung and Le, 1984). In several of these events, temperatures under the smoke plumes were observed to be several degrees lower than forecast, presumably due to the obscuration of sunlight by the plume (Wexler, 1950). An early review by Plummer (1912) gives some very interesting historical information on forest fires that led to smoke transport over long distances in North America at the beginning of this century. For example, he reported that a large forest fire in Idaho in August, 1910, covering an area of 10^4 km^2, caused "dark days" to occur over a total area of more than one million square kilometers, so that artificial light had to be used even during daytime. He also wrote: "In connection with the 1910 phenomenon it was noted that a cool wave followed, passing eastwardly over the same area, but spreading further southward, which gave the lowest temperatures, with frosts, for the month of August". Because much of the available information on fire plumes is anecdotal, however, it is difficult to extract a reliable, quantitative picture of smoke behavior. While the smoke produced by forest fires is generally less absorbing than that produced by urban fires, plumes from large fires clearly present research opportunities that should be exploited in the future.

4.8 SUMMARY

The introduction of an optically thick aerosol layer into the atmosphere would have a significant effect on most of the important physical processes in the atmosphere. Severe storms induced by the strong heating of the fires could produce local effects such as the Japanese black rain. These severe storms, particularly in the case of adjacent targets, could lead to mesoscale and synoptic-scale disturbances. The most immediate and obvious effect on longer timescales would be on the deposition of solar energy both aloft and at the Earth's surface. Owing to the size of typical smoke particles, the effect on infrared radiative transfer would be in general considerably less important. However, the infrared effects could be important in layers which are optically thick at infrared wavelengths, as well as in locations where condensation and transient cloud formation produced high infrared opacities. Modifications in solar energy absorption patterns could both increase the buoyancy at the top of smoke layers and the stability at the bottom. Altered heating patterns also would force a response in the atmospheric dynamics, which would, in turn, alter the vertical and horizontal distribution of the smoke. The effect of this coupling is difficult to anticipate from qualitative arguments and will have to be explored using three-dimensional models (see Chapter 5).

The changes in solar heating would also have large effects on the surface energy budget. If most of the incoming solar radiation were absorbed by the smoke, the boundary layer over land would change from being weakly unstable on average to being very stable. In this case, sensible and latent heat fluxes would be considerably less important than normally, and the dominant term in the surface energy budget would be infrared exchange between the atmosphere and surface. Under these conditions, the surface would cool dramatically. Over the ocean, large upward fluxes of heat and moisture would probably occur, causing intense modification of cold air masses moving offshore. Questions regarding the warming of land surfaces by marine air masses, and the generation of mesoscale storms along coastal margins cannot yet be answered.

The effect of smoke on cloud properties is similarly important but difficult to quantify. Beneath the aerosol the formation of ground fogs would be probable. However, these fogs, like those found in such locations as Fairbanks, Alaska during the winter, would be unlikely to have strong effects on surface temperatures. Cloud formation above the aerosol seems unlikely but, if it occurs, might increase the downward infrared energy reaching the surface. There is the possibility of some cloud formation in the stratosphere because of the much larger than normal accumulation of water vapor. Water cloud optical properties would certainly change because of both the greater availability of cloud condensation nuclei and the inclusion of absorbing material in and between the cloud droplets. If the near-surface air cooled below freezing, damaging frost and ice could form on vegetation.

While some of the effects of smoke can be understood using simple models and straightforward physical arguments, many of the effects are highly non-linear and cannot be easily quantified. These effects must be determined through the use of sophisticated general circulation models or specialized meteorological models. The present generation of GCMs lack the ability to simulate a number of the important physical processes. In particular, radiative transfer codes and boundary layer parameterizations must be improved. The models must also incorporate interactive aerosol transport, as is the case now with several models. In addition, treatments of the hydrologic cycle and cloud formation must be examined to see if they are adequate for study of the nuclear war problem. This model evaluation and development cannot be accomplished on a short time scale. If the results are to provide reliable answers to the questions that have been raised, careful and systematic studies will need to be carried out in future years.

Environmental Consequences of Nuclear War Volume I:
Physical and Atmospheric Effects
A. B. Pittock, T. P. Ackerman, P. J. Crutzen,
M. C. MacCracken, C. S. Shapiro and R. P. Turco
© 1986 SCOPE. Published by John Wiley & Sons Ltd

CHAPTER 5
Meteorological and Climatic Effects

5.1 INTRODUCTION

Models are a principal tool for studying natural systems such as planetary atmospheres and oceans. Conceptually, the construction of these models is relatively simple. The important physical processes within the system are identified; mathematical equations that describe the processes and their interactions with other processes are written down; and then the equations are solved, usually using computational aids, which may range from hand calculaters to the largest, fastest computers available. When studying the Earth's atmosphere, these mathematical models are used to identify the relative importance of many individual factors that affect weather and climate, as well as to simulate the overall response of weather and climate to outside forces such as the clouds of smoke and dust that might be generated by a nuclear war.

The components of the overall climatic system that need to be incorporated into a given model depend on the time scale and purpose of the model. For example, to model ice ages it would be necessary to include factors for the atmosphere, oceans, glaciers, and even the solid earth, since all of these sub-components can change on the hundred thousand year time scales appropriate to ice age/interglacial cycles. Yet only a limited description of the average properties and effects of individual weather disturbances may be necessary in such a model. On the other hand, to forecast the weather over a week's time, more details of the atmospheric behavior are needed, while other components of the weather-climate system (e.g., sea surface temperature) may be held as fixed "external" or boundary conditions, since they change very little in a week.

The "resolution" of a model refers to spatial separation of the points at which computations are made. A one-dimensional model might treat the vertical dimension (i.e., altitude) in detail, but average out all variations in the horizontal. On the other hand, a three-dimensional model would also resolve north-south and east-west dimensions, in addition to the vertical. Time may also be included, and the most comprehensive models explicitly include three spatial dimensions and time variations. Modelers speak of a "hierarchy of models" that ranges from simple models, which predict average surface temperature for the whole Earth, up to high resolution, three-dimensional, time dependent models, which explicitly resolve atmospheric motions, temperatures, precipitation, cloudiness and other atmospheric constituents, including smoke.

While the most highly resolved models are more physically comprehensive, they are much more complicated to build and interpret, and they consume vastly greater human and computational resources than simpler models. Choosing the "optimum" or necessary minimum combination of factors is an intuitive procedure that trades off completeness and hoped-for accuracy for tractability and economy. Such a trade-off is not "scientific" *per se*, but rather is a value judgement, weighing many factors. However, the value judgement is still subject to scientific test and validation. Making this judgement depends strongly on the problem the climate model is designed to address. The best strategy is often to use a hierarchy of approaches, where models of various complexity and resolution are all applied to the same questions, with the simpler ones helping to illustrate basic physical principles and the relative importance of individual factors, while the more comprehensive models are used to provide geographic detail or insight into the outcome of many simultaneously interacting processes or "feedback mechanisms".

Feedback mechanisms are important controls on the climate system, which may act either to enhance initial changes ("positive" feedback) or to oppose them ("negative" feedback). As an example of a climate feedback, consider the simple phenomenon of ice forming on a lake or on a sea coast as the weather turns cold. The ice is brighter than the unfrozen water, and thus reflects more sunlight upward than the liquid water. This leads to a positive feedback, because the increased reflectivity further decreases the amount of solar heat absorbed by the lake, thus allowing the original cooling to accelerate.

Many such feedback processes in the climatic system, both positive and negative, have been identified. Some of these are explicitly treated in the more comprehensive models. Indeed, such models are already able to produce many of the major features of the Earth's climate reasonably well. Such features include the seasonal cycle of temperature and winds and the broad geographic distribution of climatic variables such as temperature and precipitation. In addition, such models can reproduce the radically different climatic conditions of our neighboring planets Mars and Venus when the physical parameters in the model are changed to those of these planets.

Despite these important successes, models cannot yet provide credible, detailed predictions of how any arbitrary perturbation to the surface-atmosphere system would perturb the weather and climate. First of all, no model can resolve every important atmospheric process on all relevant scales. That is, smoke particles, clouds, and even small-scale storm complexes, cannot be individually and accurately treated, even in the highest resolution weather and climate models. One-dimensional vertical models, by definition, do not resolve land and sea differences, winds or any other horizontal variations, although some can treat aerosol physics and radiative transfer in considerable detail. The inability of any model to treat explicitly every physical feature necessitates the development of procedures to account collectively for the effects of these neglected features on the processes retained in the model. This procedure is known as "parameterization", a contraction for parametric representation. Instead of solving for sub-resolution scale details explicitly, a search is made for a relationship between variables on time and space scales that are resolved and what is happening on scales that are not resolved. While it is not possible to find a perfect correspondence between these averaged variables and what is actually experienced at a point, reasonably accurate relationships have been found that are valid in a variety of circumstances. Whether the parameterizations are accurate enough for each application is a principal issue of debate among climate modelers and others.

Verification experiments (including the simulation successes mentioned above) have confirmed that the present generation of climate models are powerful tools for analyzing how the surface-atmosphere system behaves, but these verification exercises—on, say, the seasonal cycle—do not guarantee the model's accuracy on completely different problems, such as the climatic response to 100 million tonne of smoke being injected into the atmosphere of the Northern Hemisphere. Therefore, modelers perform so-called "sensitivity experiments" in which an external forcing, such as a nuclear smoke cloud, is imposed and the climatic response studied for a variety of internal assumptions—such as the height distribution of the injected smoke, the vertical resolution of the model, the cloudiness parameterization, etc. Through this procedure, modelers can determine if those model characteristics which are most uncertain have a significant influence on the potential climatic response of interest, such as the resulting surface temperature variation under a thick smoke cloud.

The most reliable procedure is to repeat these climatic sensitivity experiments across a hierarchy of models, constantly comparing the results both across the hierarchy and with observational data for appropriate climatic variables, where such data are available. Indeed, it is this approach that has been chosen by the various independent climatic modeling groups around the world in studying the possible effects of nuclear war on weather and climate. As a result of the extensiveness of these efforts, some qualified statements about such effects now can be made with some confidence.

Recognition of the potentially serious consequences that nuclear war could inflict on weather and climate commenced with the identification and crude initial quantification of smoke injections into the atmosphere (Crutzen and

Birks, 1982). This led to "back-of-the-envelope" calculations of the optical effects, to simple analytical studies, and to "first-generation" one-
dimensional, radiative-convective models (for example Turco et al., 1983a,b;
MacCracken, 1983; Crutzen et al., 1984; Golitsyn and Ginsburg, 1985; Ackerman et al., 1985a; Ramaswamy and Kiehl, 1985). These models were applied to the sequence of events indicated in area 1 of Figure 5.1, and have
been useful in gaining a semi-quantitative understanding of the dependence
of the results on such variables as the amount and height distribution of absorbing and scattering particles, their scattering coefficients, visible absorption coefficients, infrared absorption coefficients, particle size distribution,
and coagulation processes.

The next stage utilized two- and three-dimensional atmospheric models,
in which smoke and dust, in some cases varying with time in a prescribed
manner, were inserted as radiatively active components, modifying the

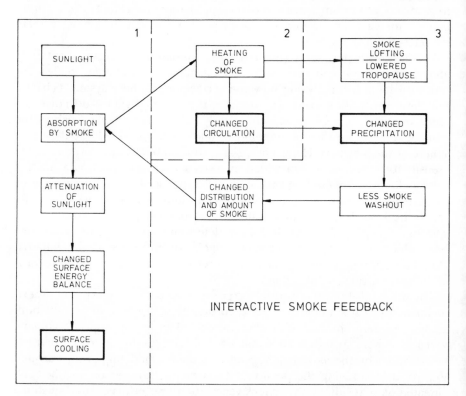

Figure 5.1.   Schematic diagram of smoke-weather/climate effects. The area labelled
1 shows interactions included in first generation models. Second generation models
added calculations of effects in area 2. Third-generation, or fully interactive models
complete the feedback loops as shown in area 3. Changed weather/climate effects
are indicated by thicker lines. Adapted from Malone et al. (1986)

temperature structure and the circulation, but not being transported around within the model. These effectively added the boxes labelled 2 in Figure 5.1. MacCracken (1983) used a two-dimensional model, while Aleksandrov and Stenchikov (1983), Covey et al. (1984), Thompson et al. (1984) and Cess et al. (1985) used three-dimensional models. Although some of these models divided the atmosphere into only two layers, which have, in some ways, limited their ability to simulate the atmospheric response, others using nine-layer models, which allowed more detailed calculation of the changes in vertical structure of the atmosphere, calculated similar results. All of these models clearly demonstrated that large amounts of smoke would change the temperature structure, including, especially, the surface temperatures. These results also demonstrated that significant changes in atmospheric circulation would occur. Such changes would, if the smoke had been allowed to move, have rapidly transported the smoke to latitudes other than those where it was injected. The realism of the results was severely limited, however, by the lack of the feedback mechanism allowing the changed circulation to alter the distribution in height and horizontal position of the smoke (and dust).

A third generation of studies is now well underway, in which "fully interactive" smoke has been placed in two-dimensional (Haberle et al., 1985), and more realistically, in three-dimensional atmospheric circulation models (MacCracken and Walton, 1984; Aleksandrov, 1984; Stenchikov, 1985; Malone et al., 1986; Thompson, 1985). These models include the processes indicated in area 3 of Figure 5.1, i.e., the smoke can be heated by solar radiation, which warms the air and leads to changes in atmospheric circulation, which, in turn, alter the vertical and horizontal distribution of the smoke and precipitation. These changes, in turn, affect the amount of smoke remaining and its ability to absorb sunlight and start the cycle again. The strong coupling between the heating, atmospheric stability, and the induced motions requires realistic treatment, which is difficult in models with low vertical resolution.

In the following sections, the principal conclusions that can be drawn from these studies will be summarized. Also discussed, with the aid of inferences drawn from physical reasoning and present model results, are possible effects on the oceans, monsoonal and coastal perturbations, and the possibility that effects might last several or more years. The findings are also summarized in terms of the range of possible surface temperature changes that could be experienced following a major nuclear war. This is attempted as a function of locality, season, and magnitude of the smoke inputs. At this stage such quantitative interpretations must be regarded as subject to a wide range of uncertainty for a variety of reasons, including uncertainties in targeting and scenarios (see Chapter 2), uncertainties in the extent of fires and the amount and characteristics of smoke emissions from those fires (see Chapter 3), and uncertainties in plume processes and local precipitation and other scavenging and microphysical processes (see Chapter 4). Nevertheless, such estimates have general qualitative validity and have been prepared in order to be helpful in the further investigation of the possible biological impacts (see Volume II of this report).

In the climate studies discussed below, certain quantities of smoke (e.g., 150 million tonne) are usually specified, along with an assumed set of physical properties that determine the optical effects of the smoke. In many of the studies, a specific absorptivity of about 2 m^2/g has been assumed, based on an elemental carbon content in the smoke of about 20% (see Chapter 3). In other studies, more detailed analytic approaches have been taken to determine the absorptivity. Thus, it should be kept in mind that when total smoke amounts are quoted, usually only about one-fifth of the mass consists of strongly light-absorbing soot, which is the critical component.

The optical depth, or thickness, of the smoke after it has been spread over a specified area of the globe, is usually specified as an absorption optical depth or as an extinction optical depth (which also includes the effect of scattering in reducing the intensity of a direct beam of light). The absorption optical depth is mainly due to the soot component of the smoke, while the non-soot aerosol component is the major contributor to the scattering. Optical depth is referred to with reference to overhead, or zenith, viewing. In the literature, "extinction optical depth" and "optical depth" are often used interchangeably. Because values of the various smoke parameters (smoke mass, soot content, area of distribution, absorptivity, scattering coefficient, etc.) are not standardized in the studies, caution must be exercised in making comparisons and interpretations of computed effects.

5.2 RESULTS OF ONE-DIMENSIONAL STUDIES

In their seminal paper, which was based on the Ambio war scenario (Ambio Advisors, 1982), Crutzen and Birks (1982) calculated that the average smoke loading resulting from the burning of a million square kilometers of forest and wildlands would be about 0.1 to 0.5 g/m^2, when spread over half of the Northern Hemisphere and assuming an average particle residence time of 5 to 10 days. They concluded that this could lead to an average reduction in sunlight reaching the ground by a factor of 2 to 150 at noon in summer. They suggested that there could be marked climatic effects, including suppression of rainfall due to the setting up of a temperature inversion in the lower atmosphere, but they did not attempt to estimate the possible effects on surface temperature. They also suggested that the burning of oil and gas wells, cities, and fossil fuel stockpiles could contribute comparable amounts of smoke.

Turco et al. (1983a) considered the effects of a variety of smoke and dust loadings, based on some three dozen different war scenarios ranging from a 3000 Mt pure counter-force war (i.e., one in which only military targets outside cities would be hit) to a massive cities and counterforce war of 10,000 Mt, and a 100 Mt cities-only case using 1,000 warheads, each of 100 kt. Table 5.1 summarizes a selection of these cases.

TABLE 5.1.
MAIN FEATURES OF THREE OF THE NUCLEAR WAR SCENARIOS
CONSIDERED BY TURCO ET AL. (1983A). SEE TEXT FOR DISCUSSION

Case	Total yield (Mt)	Pct. yield on surface burst	Pct. yield urban or industrial targets	Warhead yield range (Mt)	Total number of explosions
1. Baseline case, countervalue and counter-force[a]	5,000	57	20	0.1–10	10,400
11. 3,000 Mt nominal, counterforce only[b]	3,000	50	0	1.0–10	2,250
14. 100 Mt nominal, countervalue only[c]	100	0	100	0.1	1,000

[a] In the baseline case, 12,000 km² of inner cities are burned; a fuel loading of 10 kg/m² of combustibles are assumed to be burned, and 1.1% of the burned material is assumed to rise as smoke. Also, 230,000 km² of suburban areas are assumed to burn, 15 kg/m², with 3.6% rising as smoke.

[b] In this highly conservative case, it is assumed that no smoke emission occurs and 25,000 tonne/Mt of fine dust are injected into the upper atmosphere.

[c] In contrast to the baseline case, 20,000 km² of inner cities are assumed to burn, but with 3.3% injected as smoke into the atmosphere.

This large range of cases was used in a sensitivity analysis to delimit the influence of different variables on the atmospheric effects. Variables considered included the total yield, the percentage of detonations that were surface bursts (which were assumed to generate dust, but not smoke), the percentage yield on urban and industrial targets (which were assumed to generate large amounts smoke), the warhead yield range (which affects the height to which the nuclear fireball is assumed to rise in the atmosphere), and the total number of explosions. (Chapters 1 and 2 provide a more extensive discussion of these various quantities). None of these cases was considered to be necessarily the most probable, although case 1, with 5,000 Mt detonated on military and urban/industrial targets, was designated the "baseline" case. For each case, smoke emissions from urban and wildland fires and dust injections from surface bursts were estimated, microphysical calculations of the time evolution of the aerosol were carried out, and a time-dependent, radiative-convective calculation of atmospheric temperatures was performed, including the effect of the smoke on both solar and infrared radiation.

Figure 5.2 shows the resulting land surface temperatures for the cases shown in Table 5.1 as calculated using a one-dimensional radiative-convective model that assumed zero heat capacity at the surface, in effect a land-only planet (Turco et al., 1983a,b). Since this model takes no account of possible horizontal transport of heat from the oceans, which are a vast store of heat, the resulting temperature changes are not representative of what would occur in coastal areas, but can be used to estimate what might occur in mid-continental areas far removed from oceanic influences. Also, mean annual solar insolation was used in the model so seasonal effects are not included.

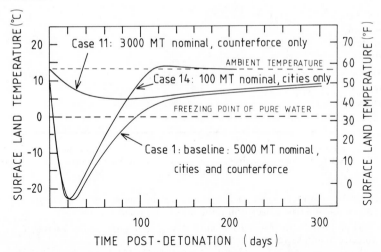

Figure 5.2. Surface land temperatures as a function of time after detonation, as calculated by Turco et al. (1983a) for the nuclear war scenarios listed in Table 5.1. Values apply to mid-continental regions and do not take into account buffering of temperature changes by oceanic heat capacity

Note that in the "baseline" case (case 1) the surface temperature drops from an assumed normal global annual mean temperature of about $+13°C$ to about $-23°C$ in about three weeks, and stays below freezing for some three months. In this model, the surface temperature was taken as the average temperature of the bottom two kilometers of the atmosphere, which tended to slow the rate of cooling somewhat. Infrared absorption by the smoke is included in this calculation, but does not substantially slow the rate of cooling, because the smoke has a lower optical depth in the infrared and because the smoke has been thinned by spreading it over the Northern Hemisphere. Thus, the smoke does not have an effect similar to that of the much stronger infrared absorption of water clouds in reducing the rate of surface cooling on a cloudy night. In their baseline case, the total smoke emission is estimated at 225 million tonne, after prompt removal of up to 50% of the smoke particles by scavenging processes in the fire plume. Five percent of the smoke is assumed to be injected into the stratosphere. They assume 960 million tonne of dust are generated, 80% going into the stratosphere, of which 8% is in the sub-micron size range with a long atmospheric lifetime. (The remaining 92% is removed relatively rapidly from the atmosphere by gravitational settling, see Chapter 3.) Smoke and dust in the troposphere are assumed to be removed at the rate observed in the normal atmosphere, i.e., about 50% in a week or so (Ogren, 1982).

In the pure counterforce case (case 11), all of the warheads are assumed to be of a size that would put dust into the stratosphere, where the sub-micron fraction would have a long residence time. However, there is assumed to be no smoke. Dust is far less absorbing of solar radiation than is smoke (see Chapter 3) and consequently the surface cooling is much less, although it is still estimated to be about 8°C in mid-continental regions. While this represents a large change in a climatological sense (it is, for instance, greater than the global average difference between a glacial and an inter-glacial period), the climatological significance of the change is uncertain because this temperature decrease might occur only in continental interiors. Major volcanic eruptions injecting almost as much material into the stratosphere have, for example, reduced large-scale time-averaged temperatures by at most a few degrees, although they may have induced anomalous weather events exhibiting larger changes (see Section 4.7.1). These dust injections would last over a much shorter time period than climate changes associated with glacial events, or even with changes associated with processes such as the buildup of the atmospheric CO_2 concentration; thus, the perturbation is not likely to induce a permanent change. The cooling might, however, last for a year or more because of the long residence time for dust in the stratosphere. In case 1 there is also a long-lasting cooling due to the stratospheric part of the smoke and dust injection.

Case 14 is of great interest because it involves the generation of an amount of smoke similar to the baseline case (but no dust) from an attack employing 100 Mt detonated on urban centers. Turco et al. (1983a,b) estimated an emission of 150 million tonne of smoke from the detonation of 1000 100-kt weapons on large urban areas. All the smoke is assumed to be deposited in the troposphere where it would normally have a relatively short residence time. Note that the initial cooling is almost as great as in case 1, but that in case 14 the recovery is much faster, with the temperature returning to almost normal within three months. The faster recovery is the result of the smoke being deposited in the troposphere, where precipitation scavenging is assumed to be efficient, rather than in the stratosphere, where the lifetime could be months to years.

Turco et al. (1983a,b) included treatment of detailed microphysical processes for smoke and dust in their calculations and performed a number of sensitivity tests to assess the importance of these parameters, as well as for variations in optical parameters. These tests identified the importance of smoke optical constants, injection heights, and particle lifetimes in determining the degree and duration of the surface cooling. They were unable, however, to include detailed calculation of smoke removal processes, so their estimate of the duration of the cooling is highly uncertain. They noted, however, the strong heating of the upper troposphere and suggested that this could lead to stabilization of the smoke cloud that could reduce the normal scavenging rates they assumed and accelerate transport of the smoke to the Southern Hemisphere. However, because of the inability of one-dimensional models to simulate horizontal transport, they were unable to provide any quantitative estimates of the degree of stabilization that would occur in the actual atmosphere or to estimate transport times.

MacCracken (1983) also carried out studies using a one-dimensional radiative-convective model including land surface heat capacity. For injection of smoke, dust, and nitrogen oxide amounts similar to those used by Turco et al. (1983a), the model projected up to a 30°C cooling within two weeks. This result is consistent with the findings of Turco et al. (1983a). In one important sensitivity case, MacCracken (1983) removed the assumed cloud cover (Turco et al., 1983, had held cloud cover constant) on the presumption that warming and stabilization of the smoke layer might reduce the relative humidity. This change led to an even greater cooling as a result of the loss of the clouds that had been moderating the cooling by trapping upwelling infrared and reradiating some of that radiation and some of the solar radiation back down to the surface. Accurate simulation of the potential atmospheric response thus requires interactive calculations of the hydrologic cycle and cloud cover. MacCracken (1983) also calculated that the NO_x emissions and consequent ozone perturbation alone would induce a comparatively small climatic perturbation, roughly consistent with past changes following a major volcanic eruption.

Crutzen et al. (1984) used a one-dimensional, radiative equilibrium model to estimate surface and atmospheric temperatures for revised estimates of urban/industrial and forest smoke inputs, as well as somewhat different parameterizations of particle coagulation, dispersion, and washout. Their model assumed a three-layer atmosphere with a smoke- and cloud-free layer between 1000 mb and 750 mb, and two layers of equal mass above 750 mb, each containing half the injected smoke. Characteristic smoke removal rates of 15 days in the middle layer and 30 days in the upper layer were assumed. Latent and sensible heat fluxes from land surfaces were parameterized in terms of the calculated surface temperatures, but the thermal inertia of the atmosphere and the heat transfer from the oceans were neglected. Solar energy input was computed for equinoctial conditions at about 30° N. The model also included a simple ice/snow-albedo feedback, with the surface albedo increasing from 12% to 50% when the land surface temperature dropped below 0°C. The latter figure was taken as representative of dirty snow (Chylek et al., 1983).

In their primary case, Crutzen et al. (1985) estimated that a total of 100 million tonne of smoke would be produced from the burning of 0.25 million km^2 of forests and a similar area of urban/suburban fires, and that 36% of the smoke particles would be in the form of amorphous elemental carbon. Equilibrium temperatures calculated for this case for the conditions prevailing at each time step are shown in Figure 5.3. The results show surface temperatures dropping to about $-25°C$ after a few days, which is faster than, but otherwise in agreement with, the Turco et al. (1983a,b) results. The upper smoke layer heated to about $+27°C$. The dashed curves show the evolution of temperatures with a constant surface albedo of 12%. It is apparent that the albedo feedback considerably prolongs the cooling. The sudden jump in temperature on day 80 was caused by the assumed instantaneous melting of the snow and ice, and the coincident change in the surface albedo from 50% to 12% when the surface temperature reached 0°C. This albedo effect is, of course, dependent on there being enough available moisture to cause an appreciable snow cover when surface temperatures are below freezing.

Crutzen et al. (1984) also calculated the effects of several other scenarios. One involved the burning of 1 million km^2 of forest, which, in combination with the urban fires, gave two hundred million tonne of smoke, but a lower fraction of elemental carbon (22%). In this case the temperature excursions were only slightly larger, and the return to normal took about 10 days longer. In another calculation, 100,000 million tonne of water vapor was injected into the upper two layers along with the smoke plumes. Except for a slower cooling in the first few days, the results did not differ significantly from those in Figure 5.3. Finally, a case in which "only" 25 million tonne of smoke were injected into the atmosphere was considered. This might be equivalent to a "limited nuclear war". Even in this case, very substantial cooling appeared possible in the mid-latitude continental interiors. Oceanic effects and thermal inertia would, of course, be particularly important in this case, but it does support the case 14 results obtained by Turco et al. (1983a), and again illustrates the nonlinearity of the effects.

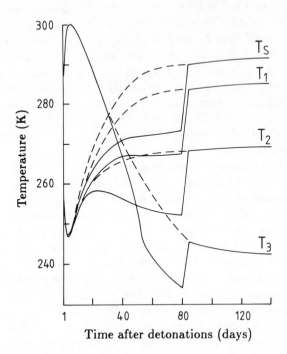

Figure 5.3.   Equilibrium temperatures at (K) the Earth's surface (T_S) and in three atmospheric layers (T_1, from the surface to 750 mb; T_2, from 750 to 375 mb; and T_3, from 375 to 0 mb) in smoke-covered continental regions from 30–60°N, as a function of time after the insertion of 50 million tonne of smoke into each of layers 2 and 3 (from Crutzen et al., 1984). Temperatures indicated with dashed lines are calculated with a surface albedo of 12%. The temperatures indicated with solid lines assume a surface albedo of 50% for ground temperatures below 0°C. Reproduced by permission of D. Reidel Publ. Company

5.3 RESULTS OF GENERAL CIRCULATION MODELS
WITH FIXED SMOKE

The first three-dimensional simulations of the effects of large quantities of smoke and dust generated by a nuclear war were carried out using general circulation models (GCMs) that were not greatly changed from those used in simulating the undisturbed atmosphere. Nevertheless, the results pointed the way to more elaborate simulations and highlighted effects that are not intuitively obvious.

Prior to these simulations, a model study was carried out by Hunt (1976), which made no reference to the possible consequences of nuclear war, but has some interesting parallels as well as very important differences. In this study, the solar energy input in a GCM was completely switched off and the resultant behavior of the atmosphere was observed. This model assumed an all-land planet, thus neglecting the heat storage in the oceans. The time variations of selected hemispheric integrals are shown in Figure 5.4. Notable features, besides the cooling rate (integrated over the whole depth of the atmosphere) of more than 1 °C per day, are a very rapid decline in water vapor content (due to reduced evaporation and cooling of the atmosphere) and in the kinetic energy of the atmosphere, with a slower rate of decrease in energy dissipation. Surface cooling rates were found to be about 4 °C per

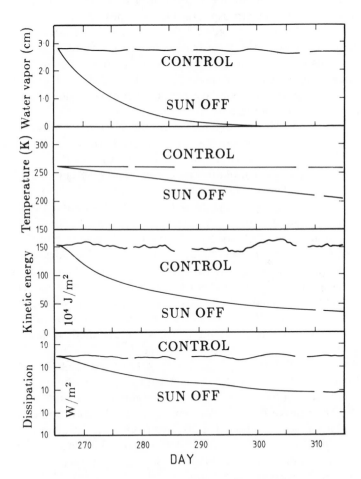

Figure 5.4. Time variations of selected hemisphere integral quantities for a control run and a run in which the solar insolation was switched off (from Hunt, 1976). (Gaps in the curves occur where printer output was lost owing to computer malfunction.)

day in the first week, decreasing to around 1 °C per day after the first month. The zonally-averaged zonal wind speed dropped by a factor of 2 by day 48 after the Sun was turned off, and the eddy kinetic energy decreased even more rapidly. These conditions might be like those which would prevail in the lower atmosphere under a uniform, optically-thick elevated smoke layer, although infrared effects could moderate the temperature and energy losses somewhat, depending on the infrared optical depth of the smoke. The presence of oceans in the real world would greatly reduce both the cooling and the loss of water vapor in the lower atmosphere. The absorption of solar radiation in the smoke layer would of course drastically change the picture in the upper atmosphere. Despite these differences, the results point to the possible importance of reductions in water content in the lower atmosphere in the post-nuclear war case, and in the vigor of the hydrologic cycle. They also suggest the possibility of a significant decline in synoptic disturbances and in mean wind speeds below the smoke layer, except perhaps in coastal areas or at the boundaries of an incomplete global smoke cover where horizontal thermal gradients could be large.

The general circulation model used at the National Center for Atmospheric Research in Colorado, known as the NCAR Community Climate Model, or NCAR CCM (Washington, 1982; Williamson, 1983), was first applied to the nuclear war simulation by Covey et al. (1984) (see also Thompson et al., 1984). The model is a nine-layer spectral model truncated at wavenumber 15, corresponding to a horizontal resolution of about 4.5° latitude and 7.5° longitude. The top layer is centered at about 30 km. Interactive clouds are predicted based on the relative humidity and the presence or absence of convection. The radiative transfer code includes absorption of sunlight by ozone, water vapor, carbon dioxide, oxygen and clouds, and cloud albedo effects. Infrared emissivities are included for water vapor, ozone, CO_2 and clouds, but neglected in the case of smoke, as is visible scattering by the smoke particles. Sea surface temperature is specified at the seasonally varying climatological value, and land surfaces are assumed to have zero heat capacity. The diurnal cycle is not considered.

Based on a draft "baseline" case for a 6,500 Mt war (NRC, 1985), Covey et al. (1984) assumed that a smoke layer with an absorption optical depth of 3 was distributed uniformly between 1 and 10 km altitude in the latitude belt 30–70° N. This smoke loading was kept fixed for the duration of the model run. Simulations were run out for 20 days from a model-generated weather situation emulating typical weather patterns for 30 June ("summer"), 27 December ("winter"), and 22 March ("spring") supplied from an earlier simulation of the unperturbed annual cycle.

For the summer case, Covey et al. (1984) found cooling below the smoke layer, strongest in inland continental regions, and strong heating of about 60–80°C near the top of the smoke layer, with some heating even above the smoke, presumably due mainly to transport of heat by the atmospheric motion. Estimated surface temperatures for this case are shown in Figure 5.5 at day 0 (the unperturbed state), day 2, and day 10 of the simulation. Areas with temperatures below $-3°C$ (i.e., 270 K) are hatched. By day 10, the temperature of land surfaces in some areas has dropped by up to 25°C, with considerable day-to-day variability in particular regions dependent on weather variations calculated by the model. For example, weather variability produces off-shore winds and below-freezing temperatures in Western Europe on day 8, but not on day 10 when the winds are on-shore. In spring, with less incident solar radiation, land average surface temperature depressions reached only about 11°C, and in winter only about 5°C (Thompson et al., 1984).

Average zonal winds show an increase in the westerlies north of the smoke and around 30-45° S, with greatly enhanced easterlies at the 20 kPa (200 mb) level from about 45° N to 20° S. Water clouds largely disappear in the middle troposphere due to reduced water vapor transport upwards through the smoke-induced, stable temperature inversion in the lower troposphere and to substantial heating in the upper layers.

The zonally-averaged meridional circulation of the atmosphere is greatly affected by the presence of the smoke. For the summer case, the normal cross-equatorial Hadley cell circulation is greatly strengthened in the first few weeks, while in spring the two tropical Hadley cells are replaced by a single cell transporting air upwards in the northern sub-tropics, southward across the equator at about 10–15 km altitude, and descending in the southern sub-tropics (Figure 5.6a,b). In the winter case (Figure 5.7a,b), there is very little change in the mean meridional circulation, although instantaneous streamlines (Covey et al., 1984) indicated that individual streamers of smoke could move as far south as the thermal equator, where it was suggested that solar heating of the smoke could cause subsequent changes in the circulation.

These changes in atmospheric circulation patterns must be accepted with some caution. Because the smoke layer is held fixed in its spatial extent, very large thermal gradients are formed at the layer boundaries. The gradients, in turn, force the development of strong wind fields that can advect the heat away from the top and southern boundary of the smoke layer. If the smoke were allowed to be transported by the winds, such very large thermal gradients would not develop and the associated wind fields would be somewhat different (see following section).

The model employed at the Computing Centre of the U.S.S.R. Academy of Sciences in Moscow was also used to simulate post-nuclear war conditions (Aleksandrov and Stenchikov, 1983; Thompson et al., 1984). This model has a horizontal resolution of 12° latitude by 15° longitude, with two layers in the vertical representing the troposphere from the surface to about

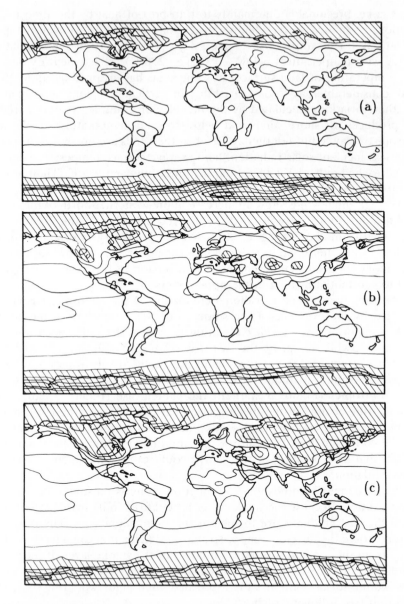

Figure 5.5. Surface air temperatures calculated with the NCAR CCM by Covey et al. (1984) for an injection at time $t = 0$ of a smoke layer having absorption optical depth 3, between 30 and 70° N. Diagrams are for the Northern Hemisphere summer case at (a) day 0, (b) day 2, and (c) day 10. The contour interval is 10°C and areas with temperatures below -3°C are hatched. Reproduced by permission from *Nature*, Macmillan Journals Limited

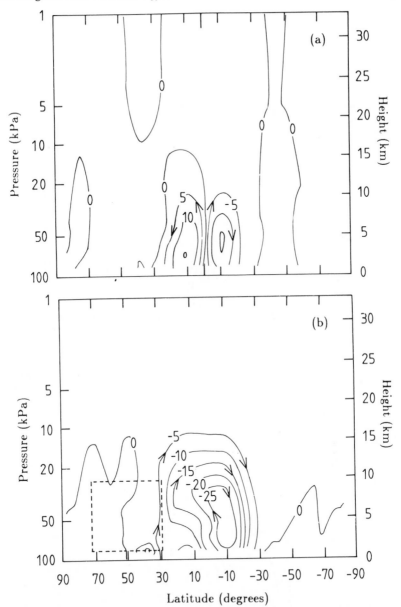

Figure 5.6. The zonally-averaged, north-south atmospheric circulation from a NCAR CCM simulation (Covey et al., 1984). Arrows indicate the direction of motion. Units: 10^{10} kg/s. Data are averaged over days 16–20 of the simulation. (a) Control run for April; (b) smoke-perturbed run for April (a smoke layer with absorption optical depth of 3 was inserted on day 0 within the area indicated by the dashed box). Reproduced by permission from *Nature*, Macmillan Journals Limited

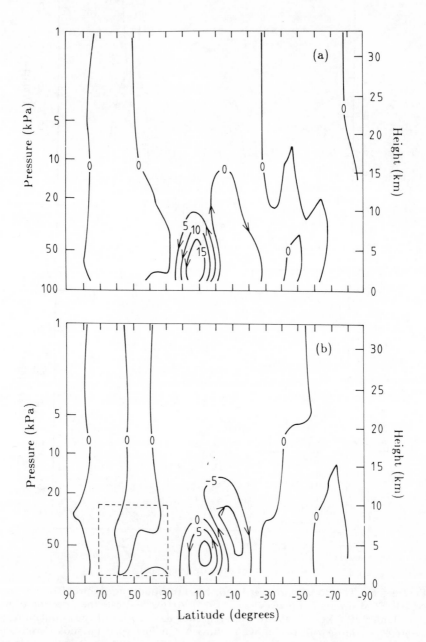

Figure 5.7.   Same as Figure 5.6, but for January. (a) control run; (b) smoke-perturbed

20 kPa (about 12 km altitude). This model was coupled to a thermodynamic model of the upper ocean, enabling changes in estimated surface ocean temperatures to be calculated. It used annually-averaged solar input, and was intended to represent annual average conditions rather than individual seasons. Cloud cover and precipitation were calculated.

In the Aleksandrov and Stenchikov (1983) simulation, smoke and dust that were supposed to correspond to the 10,000 Mt war scenario of Turco et al. (1983a) were instantaneously injected and uniformly spread into the model between latitudes 12° and 90° N. (In fact, because of the absorption and scattering properties assumed for the smoke and dust, their simulation corresponded to an injection several times larger than that in the 10,000 Mt case of Turco et al. (1983a) and was roughly equivalent to the upper range for smoke injections suggested by NRC, 1985.) Solar radiation reaching the troposphere (i.e., the model top) was assumed to have been reduced by the presence of dust in the stratosphere; the smoke was injected equally into the two model layers. The initial hemispheric-average absorption optical depth was assumed to be 6. This was reduced to 3.5 after 30 days, and further reduced in steps at later times in order to approximate the effects of coagulation and removal processes.

Calculated globally-averaged atmospheric and land surface air temperature changes during the first 60 days are shown in Figure 5.8, and the change in surface air temperature, relative to the initial conditions, on day 40 is shown in Figure 5.9 (Aleksandrov and Stenchikov, 1983). As in the NCAR model spring and summer cases (Covey et al., 1984), the mean meridional circulation in the smoke-perturbed case shows that the two normal Hadley cells, appropriate for annual mean conditions, are replaced by a single large cell with rising motion in the subtropics of the Northern Hemisphere, and sinking motion in the southern subtropics. The calculated circulation three months after smoke injection is shown in Figure 5.10b. Figure 5.10c shows results for a case intended to be similar to case 14 of Turco et al. (1983a), but which again has a considerably greater absorption optical depth.

MacCracken (1983) used a combination of the unperturbed wintertime circulation from a two-layer, three-dimensional model (Gates and Schlesinger, 1977) to disperse smoke from discrete source regions and to calculate the geographical distribution of visible optical depths, and a two-dimensional climate model (MacCracken et al., 1981) to calculate resultant surface temperatures. He also discussed the possible effects of reduced rates of removal of smoke, relative to those experienced in the unperturbed atmosphere (Ogren, 1982), in prolonging the effects. An initial 150 million tonne of smoke from urban/suburban fires was injected above the surface boundary layer on the first day, and an additional 57 million tonne from wildland fires over the first seven days, from four discrete target areas covering North

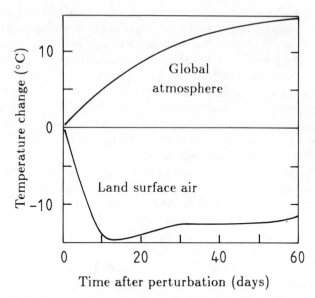

Figure 5.8.   Globally-averaged atmospheric and land surface temperature changes from initial (annual mean) conditions during the first 60 days after injection into the Northern Hemisphere of smoke with a hemispheric average absorption optical depth of 6 (from Aleksandrov and Stenchikov, 1983)

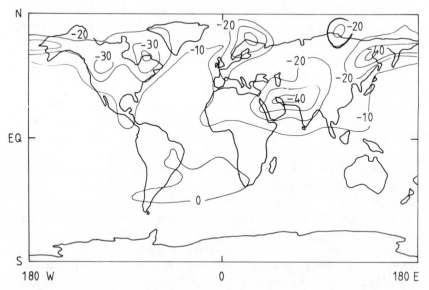

Figure 5.9.   The change, in °C, from initial (annual mean) surface temperature for the smoke-perturbed case, as in Figure 5.8, on day 40 after smoke injection (from Aleksandrov and Stenchikov, 1983)

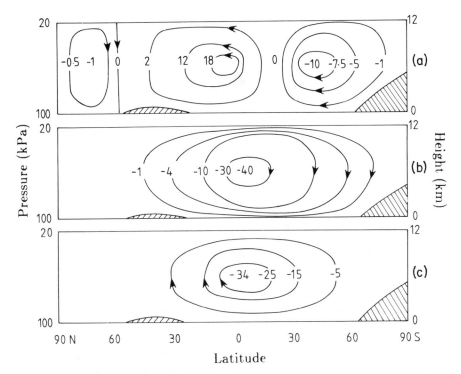

Figure 5.10   The zonally-averaged, north-south atmospheric circulation from the Computing Centre (Moscow) model (Aleksandrov and Stenchikov, 1983). Arrows indicate the direction of motion. Units: 10^{10} kg/s. (a) The normal undisturbed mean annual circulation; (b) 3 months after injection into the Northern Hemisphere of smoke with an initial absorption optical depth of 6; (c) as in (b) but for an absorption optical depth of 3

America, Europe and western Asia. After 30 days, the hemispheric-average extinction optical depth was found to be 1.1. MacCracken then reduced the scavenging rate by a factor of $\exp(-\tau/3)$, where τ was the local extinction optical depth; this factor was chosen as a plausible representation for the effect of smoke on precipitation rates, based on a *ad hoc* relationship between observations of precipitation rate (Jaeger, 1976) and solar radiation absorbed at the Earth's surface. This reduced scavenging rate gave a hemispheric-average extinction optical depth after 30 days of 4.5, with the highest values at middle and high latitudes.

Using a two-dimensional climate model that treats land and sea surface areas separately within each latitude zone, MacCracken (1983) calculated the reduction in surface temperatures over land and oceans for normal and reduced rates of scavenging. The average results for the Northern

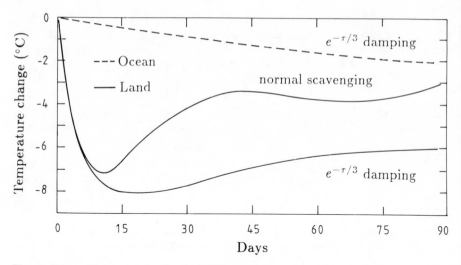

Figure 5.11.   Reduction in Northern Hemisphere average land and ocean surface temperatures as a function of time following the injection of some 200 million tonne of smoke into the Northern Hemisphere. Both "normal" smoke removal rates (scavenging) and "normal" removal rates reduced by a factor of $e^{-\tau/3}$, where τ is the extinction optical thickness, were used (from MacCracken, 1983)

Hemisphere are shown in Figure 5.11. With normal scavenging, the maximum cooling of land occurred within 2 weeks, but was much smaller than in one-dimensional models in which transport of heat from the oceans to the land was not included. Also, the one-dimensional calculations were for land under the smoke cloud, whereas MacCracken's calculated temperatures were averaged over all land in the Northern Hemisphere, including land not yet underneath the assumed spreading smoke cloud. The mid-latitude temperature changes MacCracken (1983) calculated under the smoke cloud were about twice as large as the change for the hemispheric land average (NRC, 1985), and he noted that the cooling in mid-continental areas could be much larger. In the case of reduced scavenging, the maximum cooling was not much greater, but severe cooling lasted much longer. Recovery was slowed not simply because the smoke optical depth was greater, but also because the smoke had had time to spread more uniformly, thereby intercepting a greater amount of global insolation, and because the ocean temperature decreased slowly throughout the simulation.

The precipitation rates calculated by the two-dimensional model show marked reductions. After three months in the case of assumed reduced scavenging, precipitation was 25% less than normal over land and 20% less over the oceans. The precipitation was found to be confined to lower altitudes than normal, with the reduction largely due to less precipitation in the intertropical convergence zone. This was presumably the result of the dramatic change in the global circulation pattern and an increase in the vertical stability of the lower atmosphere.

Covey et al. (1985) report further diagnostic studies on the simulations which used the NCAR CCM. They have found that the land temperatures in the perturbed case were strongly influenced and, in fact, prevented from cooling further, by diffusion of heat downward from the lower troposphere. This heat was supplied by horizontal transport from the relatively warm oceans. They conclude that the substantial downward vertical heat diffusion into the lowest layer of the model was almost certainly over-estimated in the smoke-perturbed conditions of high vertical stability by the particular parameterization scheme used in the NCAR model. Consequently, they suggest that, in the absence of other errors, use of this parameterization results in an underestimate of the cooling of the land surface for the case of optically-thick smoke layers. They warn, however, that there are other omissions and approximations in the present models that make it difficult to conclude that the model-predicted temperature changes are in fact underestimates of the actual changes which could occur.

Covey et al. (1985) also note that the thermal balance in the perturbed atmosphere as a whole would be dominated by intense solar heating of the upper troposphere smoke layer in middle latitudes, which would be balanced by dry convection and large-scale dynamical heat transport. Clouds largely disappeared in the mid to upper troposphere in smoke-affected regions of their model, due to a decrease in relative humidity resulting from the higher temperatures and, to a smaller extent, from a decrease in vertical transport of moisture. They suggested that to study the effects of nuclear war-generated smoke particles, the most important areas for improvement of general circulation models include improving representation of boundary layer processes and incorporating radiative interaction, with aerosol transport and removal processes.

Cess et al. (1985) performed a number of simulations with fixed smoke layers using a version of the two-layer, three-dimensional OSU model in which the solar radiative transfer scheme was modified to include both aerosol absorption and scattering at solar wavelengths. The model was primarily used as a tool for conducting sensitivity tests and the majority of the simulations were truncated after 10 days. The results of some of these tests are discussed in Chapter 4, and summarized only briefly here.

Analysis of the surface energy budget in their simulations illustrates the sensitivity of the surface temperature to the extinction optical depth of the aerosol. As the aerosol amount is increased, the initial response of the surface temperature is small because the loss of solar energy is compensated by an increase in downward infrared energy and a decrease in heat lost by convection and latent heat. At some point, these latter terms can no longer compensate for the loss of solar radiation and rapid surface cooling occurs.

The point at which this rapid cooling begins, and the total cooling which occurs within the first 10 days following the injection, are a function of several variables. Not unexpectedly, Cess et al. (1985) found that the surface cooled more when the smoke was mixed uniformly with height (constant density) than when it was mixed uniformly with pressure (constant mixing ratio), primarily because solar absorption occurs at a higher level in the atmosphere for the constant density case. Their results also indicate sensitivity to the absorption optical depth of the smoke as a result of both variations in the total (extinction) optical depth and variations in the assumed single-scattering albedo of the aerosol (see discussion in Chapter 4) with fixed total optical depth. In their conclusions, the authors stress the need to improve parameterizations of boundary layer and surface processes, but also point out that the treatment of infrared emission by the atmosphere may be a critical area that will require improvement to obtain more accurate simulations.

5.4 RESULTS OF GENERAL CIRCULATION MODELS WITH INTERACTIVE SMOKE

5.4.1 Two-Dimensional Models

The importance of allowing the injected smoke both to be transported by the atmospheric circulation and to interact with the circulation through radiative effects was first suggested by studies of the great Martian dust storms (Ryan and Henry, 1979; Haberle et al., 1983). These storms form on regional scales, but can grow rapidly into global-scale storms as a result of the interactions between the circulation of the Martian atmosphere and solar absorption by the dust.

Haberle et al. (1985) modified the fully interactive, zonally-symmetric, two-dimensional circulation model of Haberle et al. (1983) to approximate important aspects of the terrestrial atmosphere circulation. Ground temperatures are fixed at their mean annual values, which is roughly equivalent to having an ocean-covered planet, and surface sensible heat fluxes are calculated from a drag law formalism. Latent heat fluxes are prescribed using observed values and water cloud amounts are fixed at 50% coverage within each latitude zone. Although the model predicts winds, temperature, and the movement of trace species in the meridional plane (i.e., across latitude bands), it does not include a parameterized representation of large-scale eddy fluxes (which cannot be calculated in two-dimensional models) of these quantities. Although these eddy fluxes are an important factor in the present atmosphere, it is not certain how they would change, and thus how they should be parameterized, in the case of a highly-perturbed atmosphere. This lack of eddy fluxes and of variable surface temperature in the model is a significant limitation with regard to the prediction of changes in climatic variables such as air temperature, but they may be less important when investigating global-scale smoke transport by the mean meridional circulation.

Such a model is obviously a very crude approximation to the real atmosphere. In the unperturbed case, the model predicts two Hadley-type circulation cells, that are shallower and extend further polewards than in reality. These discrepancies are thought to be mainly due to the lack of a large-scale eddy parameterization and the crude representation of latent heating. The mid-latitude jet streams are also too strong, but the model does produce a statically-stable stratosphere, which is an important barrier to buoyantly generated vertical motion originating in the troposphere. The model includes a full radiative treatment of the smoke, including scattering and absorption of solar radiation and absorption and emission of infrared radiation. Coagulation of smoke is ignored and removal is by a fixed rainout rate giving an average tropospheric lifetime for smoke of 10–15 days. Smoke optical properties are those specified by Turco et al. (1983a).

An initial smoke layer of 265 million tonne was injected between 27.5° and latitude 62.5° N, and between either 0 and 4 km or 6 and 10 km altitude, designated the "low cloud" and "high cloud" experiments, respectively. The corresponding initial extinction optical depth is 14 (with an absorption optical depth of about 5). For each initial smoke cloud, three 20-day simulations were run: a passive tracer run in which the smoke was not allowed to affect the local heating rates or change the circulation; an interactive tracer run in which heating by the smoke was allowed to alter the circulation; and an interactive tracer run in which both heating by the smoke and upward vertical mixing of the smoke by convection were included.

The results for the passive low cloud experiment are shown in Figure 5.12a. After 20 simulated days, the southern part of the passive smoke cloud had been transported toward the equator, giving a greater total latitudinal spread than existed initially, and the rainout term had removed all but 37 million tonne of smoke. This left typical optical depths of about 2. Virtually no smoke rose above 4 km.

When solar heating of the smoke was included (Figure 5.12b), a plume of smoke rose well into the stratosphere by day 20 due to large-scale circulation changes induced by the added source of heat. In this case, some 44 million tonne of smoke remained, because smoke rising above 10 km was no longer subject to removal by rainout. Rising motion was favored at more southerly latitudes of the Northern Hemisphere because solar heating was greater nearer the equator.

When dry adiabatic mixing of the smoke was included (Figure 5.12c), more smoke was lofted, and nearly 66 million tonne remains after 20 days.

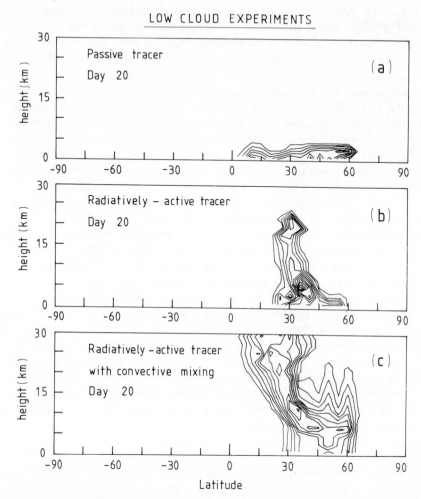

LOW CLOUD EXPERIMENTS

Figure 5.12. Isopleths of smoke density on day 20 after the injection of 265 million tonne of smoke into a two-dimensional atmospheric circulation model. Initial injection was between 27.5° and 62.5° N latitude and 0 to 4 km altitude. (a) The smoke is transported by the undisturbed circulation only; (b) the smoke is transported by and allowed to modify the large-scale circulation; (c) upward transport of smoke by sub-grid-scale convective motions, induced by heating of the smoke due to absorption of solar radiation, was also included (from Haberle et al., 1985)

This was almost double the amount remaining in the passive case. Strong cloud-top heating destabilized the atmosphere above the smoke over a range of latitudes allowing most of the smoke cloud to move upwards, and causing a much greater portion of the northern mid-latitude atmosphere to heat up. As a result a stronger direct component of motion toward the equator developed in the upper levels, transporting more smoke toward the Southern Hemisphere.

In the high cloud experiment, initial heating rates were more than double those in the low cloud experiment (due to the lower air density), inducing stronger vertical motions and convection. After eight days the smoke in the radiatively interactive case reached the top of the model, and was artificially forced to spread southward. A tendency noted particularly in the high cloud experiment, but also in the low cloud case, was the rising of the initial plume of smoke to shade the smoke below it. As a result, the plume, which became even more strongly heated at higher altitudes, broke away from the main body of smoke. A weaker plume developed at a higher latitude where the main body of smoke was not shaded. This process could limit the vertical transport of smoke into the stratosphere to a total visible optical depth in the stratosphere sufficient to suppress heating of the smoke below. Spread of lofted smoke into the Southern Hemisphere might, however, reduce the optical thickness of lofted smoke in the Northern Hemisphere, possibly leading to further lofting of the underlying smoke.

A similar interactive smoke run was also performed using the two-dimensional climate model of MacCracken et al. (1981) modified to allow movement of the smoke (Walton et al., 1983). Since this model contains a number of additional parameterizations for the Earth's atmosphere, including a treatment of eddy fluxes, variable surface temperatures, and an interactive hydrologic cycle, the results provide an interesting comparison to those of Haberle et al. (1985). In the MacCracken et al. model, the predicted lofting of the smoke was reduced, being limited to about 20 km, but the horizontal spread towards the south was much greater. These differences apparently were related predominantly to the inclusion of the horizontal eddy transport term. This transport tended to spread the smoke layer horizontally and, at the same time, to mix the heated smoke parcels with cooler ambient air. This reduced the buoyancy of the smoke and the lofting. However, it also reduced the shielding of the smoke at lower levels, leading to heating through a deeper column of the atmosphere. Because of this apparent importance of eddy transport terms, these results indicate clearly that three-dimensional studies are essential.

While two-dimensional models have obvious deficiencies as vehicles for quantitative simulations of what might happen in the real three-dimensional atmosphere, they do illustrate, however, several qualitative effects that may be very important. The most significant is that the inclusion of interactive smoke, radiation, and transport processes, including local convective mixing, may lead to rapid lofting of large quantities of smoke to heights well above levels where washout processes could remove the smoke. This implies the potential for much longer-lasting effects on surface temperatures than hitherto considered likely.

5.4.2 Three-Dimensional Models

MacCracken and Walton (1984) have performed fully interactive simulations using the OSU GCM (Gates and Schlesinger, 1977) coupled to a three-dimensional extension of their GRANTOUR model (MacCracken, 1983). In this version of GRANTOUR, the troposphere is divided into 10,000 equal volume parcels, initially in four layers, and the parcels are moved by the three-dimensional wind field calculated with the GCM. Two classes of particles are treated in the model: those with diameters less than 1 μm and those with diameters greater than 1 μm. These particles are assumed to have extinction cross-sections of 6.7 m^2/g and 2.6 m^2/g, respectively. The smoke particles are assumed to be scavenged by the precipitation calculated in the GCM at the nearest grid cell; since the precipitation rate changes as the climate evolves, the particle lifetime will also change interactively. The larger particles are assumed to be scavenged about four times as rapidly as the smaller particles.

Coagulation was ignored in their initial calculation, which may result in a potential over-estimate of extinction (and absorption) optical depth after 30 days by up to 50% (Penner and Haselman, 1985). As in the noninteractive case discussed earlier, the OSU GCM has two layers in the troposphere and does not include the stratosphere. Sea surface temperatures and the solar radiation were held fixed at July conditions. This latter assumption probably leads to the overestimate of surface land temperature and underestimate of precipitation over the continents apparent in the control case (due to loss of soil moisture).

Two interactive smoke cases were calculated. In the first, 150 million tonne of smoke was injected into the atmosphere assuming an equal mixing ratio from the surface to 11 km altitude. This was roughly equivalent in magnitude to case 14 of Turco et al. (1983a), although the initial smoke vertical distribution was different in this case due to the constant mixing ratio assumption and the higher mixing of the smoke. The second case involved the injection of only 15 million tonne of smoke. The smoke was assumed to be injected in four discrete regions over the eastern and western U.S., Europe, and western Asia.

The results for the case with 150 million tonne of injected smoke are shown in Figure 5.13 for day 30 after the injection in a simulated three-dimensional view looking north-west from 60°S latitude, 160°W longitude in the South Pacific. Each dot in the diagram represents about 5,000 tonne of smoke particles in the size bin larger than 1 μm. A sequence of such views at various times shows that the smoke has moved upward and southward, with scavenging and dispersion reducing the concentrations in the lower atmosphere. By day 20 the smoke had, in fact, spread nearly uniformly over the Northern Hemisphere, except in low latitudes. A few regions still had

ALL PARTICLES--ALL HEIGHTS--DAY 30

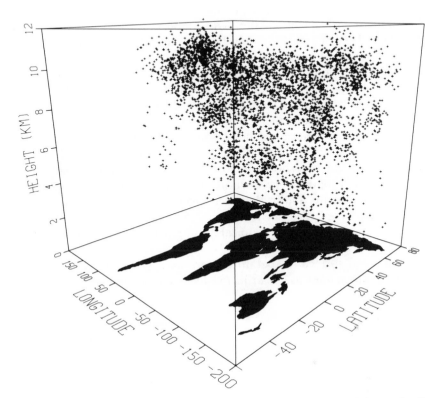

Figure 5.13.   Three-dimensional representation of the large particle smoke distribution 30 days after a summer injection of 150 million tonne of smoke in four regions over the eastern and western U.S., Europe, and western Asia. Results are for an interactive smoke case from MacCracken and Walton (1984). Each dot represents about 5,000 tonne of smoke

extinction optical depths of 10, but most of the hemisphere was covered by an extinction optical depth of 2, which was about the hemispheric average at that time. Smoke had started to spread to equatorial and sub-tropical latitudes of the Southern Hemisphere within the first three weeks.

In order to examine the effects of interactive transport, two other cases were examined. In one case, the smoke was transported by the winds and scavenged by the precipitation taken from the control simulation, i.e., in an atmosphere unaffected by the presence of the smoke. In this passive smoke case, there was virtually no transport of smoke to the Southern Hemisphere during the full 30 days of the simulation; most of the smoke moved toward the pole and spread around the hemisphere. In the second case, the winds and precipitation generated by a fixed, uniform-smoke simulation were used (as explained in Cess et al., 1985). The perturbed circulation used was similar to those developed by Covey et al. (1984) and Aleksandrov and Stenchikov (1983) simulations. This circulation led to modest transport of smoke into the Southern Hemisphere (less than in the fully interactive case), and less transport toward high northern latitudes.

Removal rates due to scavenging by precipitation generated in the fixed-smoke OSU simulation were found to be much slower than were determined based on precipitation rates from the unperturbed atmosphere simulation. In the fully interactive case, the scavenging was not quite so slow, but further analysis is needed to determine the reason for the difference. Precipitation rates in the control case and for three successive 10-day periods in the interactive smoke case are shown in Figure 5.14, for land areas only. Note the marked and progressive reduction of precipitation over the northern mid-latitude continents and in the Inter-Tropical Convergence Zone. A local increase occurred at about latitude 30° N, which could be related to a low-level return flow of moist tropical air compensating for the southward flow at upper levels.

Resulting land surface temperatures were on average only a little colder in the fully interactive case than in the fixed smoke simulation. However,

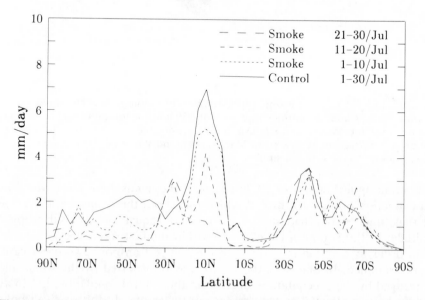

Figure 5.14.   Zonally-averaged precipitation rates over land in the control case (no smoke) and the interative smoke case, as in Figure 5.13, for three successive 10-day periods (from Ghan et al., (1985)

in the interactive case, which started with discrete source regions, there was necessarily patchiness in the smoke distribution, with more extreme cold under the denser smoke clouds, and less cooling where there was less smoke. This was particularly evident in the first week or two. In the subtropics of the Southern Hemisphere, there was evidence of warming by a few degrees, both in the fixed smoke case and in the fully interactive smoke case, a result of increased subsidence induced by the presence of the smoke in the Northern Hemisphere and of surface drying due to reduced precipitation. Presumably, if the model were run for a longer time, allowing sufficient smoke to pass into the Southern Hemisphere at high altitudes, this slight warming would change to a cooling as solar radiation was absorbed aloft by the smoke.

MacCracken and Walton (1984) also show 30-day time series of surface temperatures for a number of typical locations. Two such examples are shown in Figure 5.15, one for a mid-continental site in western Asia and one for a site near the east coast of Asia. Temperatures are compared for the control case and for the interactive case having initially 150 million tonne of smoke. The site in western Asia is far removed from oceanic influences and gives the most severe cooling. In the interactive smoke case this site exhibited a dramatic cooling to some 30–40°C below the control case by the end of the first week, followed by some amelioration at the end of the fourth week as the smoke was dispersed. The coastal site showed little significant difference between the control and interactive smoke cases until the fourth day when smoke moved overhead. Cooling remained around 10°C in the following week, and then almost doubled to a 15–20°C cooling as more smoke moved overhead and winds became more off-shore.

Several factors should be borne in mind in relation to effects in coastal zones. Firstly, in these simulations the sea surface temperatures were held fixed when, in reality, they would slowly cool by a few degrees, or perhaps be affected more dramatically by perturbed ocean motions. Secondly, the vertical resolution of the general circulation model was limited to only two layers and therefore did not adequately resolve the boundary layer. Thus, the model had difficulty simulating low level temperature inversions, which in some synoptic situations would isolate the coastal land areas from maritime air. Thirdly, a surface gravitational outflow of cold air from the continental interior could occur in coastal valleys, similar to nocturnal valley winds, or to the katabatic winds which are a common feature of coastal climates around Antarctica (see Fitzjarrald, 1984; Parish, 1984). Under these conditions, coastal zones may experience extreme cold episodes, including damaging frosts, even though such zones may experience on average much milder conditions. Higher resolution regional and mesoscale models are necessary to predict the occurrence of such cold episodes, which in any case would occur more or less as random events.

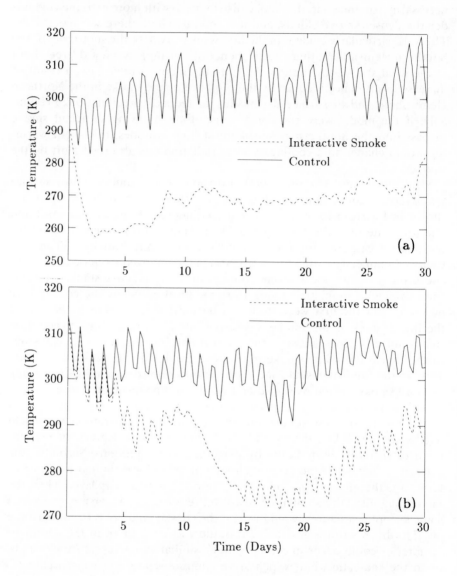

Figure 5.15.   Time series of surface air temperature for control (no smoke) and interactive smoke case (as in Figure 5.13) for (a) mid-continental Asia, and (b) east coast Asia. These curves show diurnal cycles, suppressed when smoke is over-head, and fluctuations due to passing weather systems (in the control case) and moving smoke clouds. Actual variations at a particular location would depend on the chance occurrence of particular weather situations at the time of smoke injec-tion

When only 15 million tonne of smoke were injected into the atmosphere, MacCracken and Walton (1984) found that, if the smoke was assumed to be uniformly distributed, virtually no significant cooling occurred. However, if the smoke was injected from the four discrete source regions, cooling of up to 10°C occurred under the smoke clouds before they dispersed.

Aleksandrov (1984) and Stenchikov (1985) have run the GCM of the Computing Centre of the U.S.S.R. Academy of Sciences, discussed in Section 5.3 above (see Aleksandrov and Stenchikov, 1983; Thompson et al., 1984) in an interactive mode that treated the global circulation of the atmosphere, the heating of the upper atmosphere by absorption of sunlight by the smoke (assuming a fixed size-averaged absorption cross-section), and transport of smoke by atmospheric motions that are modified by the smoke heating itself. As in their earlier non-interactive smoke case, the smoke was injected uniformly between latitude 12° and 90° N.

For a case with an initial absorption optical depth for smoke of 3, and with scavenging rates as in the noninteractive case rather than evolving with the changing climate, the interactive simulation of Stenchikov (1985) produced a surface temperature cooling as shown in Figure 5.16. Comparison with the corresponding non-interactive result (not shown), showed that the surface temperature drop in middle to high northern latitudes was less than in the

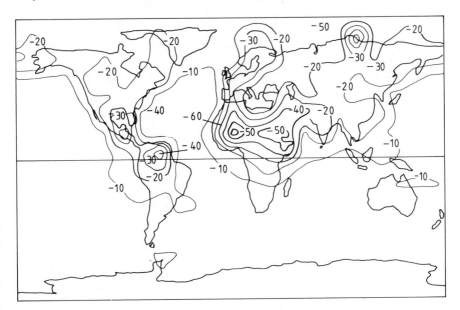

Figure 5.16.   The change in surface air temperature (°C) from the normal annual mean 40 days after the injection of nuclear smoke and dust between latitude 12° and 90° N with an initial mean absorption optical depth for smoke of 3, using the model of Aleksandrov and Stenchikov (1983) (from Stenchikov, 1985)

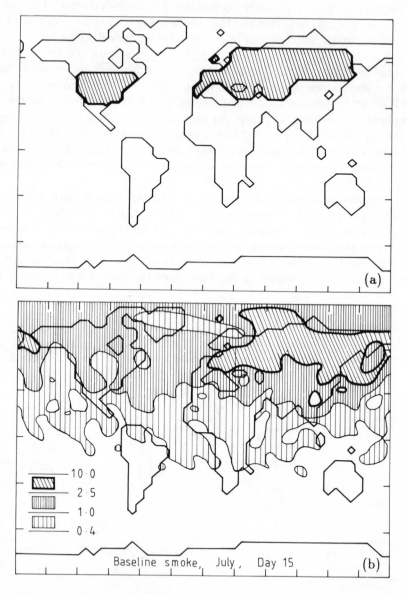

Figure 5.17. (a) Smoke injection regions for the interactive baseline scenario of Thompson (1985) in which 180 million tonne of smoke are injected between 0 and 7 km altitude. (b) The resulting distribution of smoke absorption optical depth after 15 days, for a July injection

non-interactive case, although still as much as 20 to 30°C over parts of North America, Europe, and the U.S.S.R. There was, in addition, considerable cooling over the Middle East by as much as 30°C or more. As could be expected from the circulation changes in the noninteractive simulations, in which the northern arm of the Hadley circulation was reversed, this cooling at more southerly latitudes was due to southward transport of smoke and dust at high altitudes in the model. The cooling was relatively large at lower latitudes because of the high intensity of the solar radiation being intercepted at these latitudes, especially in the annual mean case considered here, and the normally warm temperatures.

The NCAR Community Climate Model has also been run for an interactive smoke case (Thompson, 1985). Smoke was transported by the explicitly calculated large-scale motions, but sub-grid-scale convective transport was ignored. The model has no smoke removal process, so the resulting estimated average cooling is probably too great. It also has no surface heat capacity to slow the rate of temperature change, and no diurnal cycle, which may be important in the calculation of surface temperature change at low optical depths.

In his interactive baseline scenario, Thompson (1985) injected 180 million tonne of smoke distributed uniformly over portions of the NATO and Warsaw Pact countries and between 0 and 7 km altitude in July. The injection regions are shown in Figure 5.17. The smoke was assumed to be purely absorbing with a specific absorption of 2 m²/g. By day five, a smoke layer with an absorption optical depth greater than 1 bridged the North Atlantic, but there remained a gap across the North Pacific Ocean and Alaska. Smoke had, however, already spread to the north, and to the south as far as Mexico, tropical East Africa, and northern India. By day ten, small patches of smoke reached 20° S, and a layer with an absorption optical depth of at least 1 existed over most of northern Africa and parts of the Indian subcontinent. Over North America there were isolated clearer patches and only a small area with optical absorption depths in excess of 2.5. Smoke completely covered the Arctic basin all the way to the North Pole. The distribution of absorption optical depth on day 15 is shown in Figure 5.17b.

The rapid dispersion of the smoke in the NCAR CCM should be viewed with some caution. The particular spectral advection scheme used in the model to compute the horizontal transport suffers from artificial numerical diffusion, especially in areas with sharp gradients in smoke concentration and it has been necessary to develop a somewhat *ad hoc* correction algorithm. Thus, the spreading in the early days of the simulation, where the gradients can be quite high, may be altered somewhat by this scheme.

North-south, zonally-averaged vertical cross-sections show that by day five smoke had risen around the southern edge of the initial injections to altitudes in excess of 12 km. By day 10, a strong southward movement of high altitude smoke had occurred, with a tongue reaching as far as 30° S at 15 km altitude. Northward movement occurred between 5 and 10 km altitude. The vertical cross-section on day 20 is shown in Figure 5.18. The largest smoke mixing ratios on this day were found over the Arctic, but there was a continuing southward movement between 10 and 20 km altitude, with appreciable smoke as far as latitude 40° S.

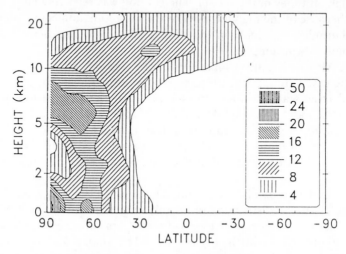

Figure 5.18.   Vertical cross-section of smoke mixing ratio, in units of 10^{-8} g/g, after 20 days, for the July baseline case of Thompson (1985)

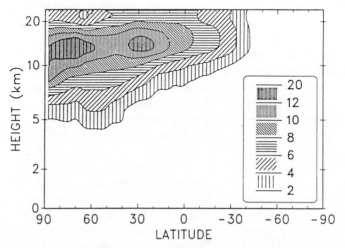

Figure 5.19.   Vertical cross-section of atmospheric heating rates after 20 days, for the July baseline case of Thompson (1985). Units are 10^{-5} °C/s (which is roughly equivalent to °C/day)

The reason for this dramatic change is evident from the vertical cross-section of heating rates (Figure 5.19). Note that around 15 km altitude, the heating rates were as high as 15–20°C/day. The resulting vertical cross-section of temperature on day 20 is shown in Figure 5.20, in which it is evident that the tropopause has been effectively lowered to around 4 km altitude in northern mid-latitudes, and to about 9–10 km in the tropics (i.e., about 7 km below normal). It is probable that this sharp increase in static stability in what was the troposphere would suppress deep convection, precipitation, and baroclinic activity in the Northern Hemisphere.

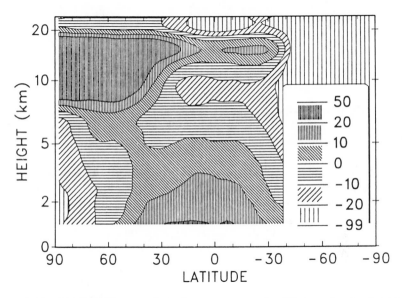

Figure 5.20. Vertical cross-section of temperature after 20 days, for the July base-line case of Thompson (1985). Temperatures in the Northern Hemisphere 10–20 km altitude range have increased by 80°C or more

A major change in precipitation may be inferred from Figure 5.21, which indicates severe suppression of condensation in the former Inter-Tropical Convergence Zone, where release of latent heat normally plays a major role in driving the atmospheric circulation. Condensation in the former mid-troposphere has been strongly reduced over most of the Northern Hemisphere. There have, however, been some local increases in condensation in the bottom 1–2 km of the atmosphere, especially around 30° N. These support the findings of an increase in precipitation at 30° N found by Ghan et al. (1985) (see Fig. 5.14).

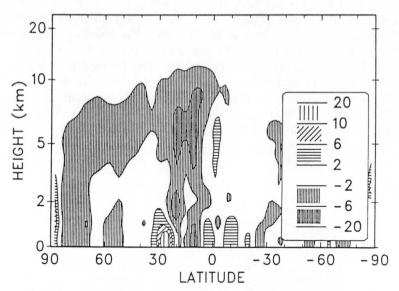

Figure 5.21.   Vertical cross-section of changes in condensation rate, averaged over days 5–20 after smoke injection, for the July baseline case of Thompson (1985). Units are 10^{-6} kg/m^2·s

Changes in surface temperature were consistent with the southward move-ment of the smoke cloud. By day 5 after the smoke injections, very patchy areas of below freezing temperatures appeared in north temperate latitudes, including most of the western U.S., parts of eastern Europe and the Mid-dle East, central Asia, and the Tibetan Plateau. There were suggestions of a slight warming at southern mid-latitudes, which could be due to induced subsidence in the intensified southern arm of the Hadley circulation, as well as reductions in cloudiness and precipitation. Day 10 showed some consol-idation of the below freezing areas over North America and Eurasia, but there were still some comparatively warm patches. Day 15 showed a further consolidation of the cold areas over North America and Eurasia, with a more extensive area below freezing over southeastern Asia including Tibet. Cool-ing was also apparent over portions of South America, southern Africa and inland Australia. The situation on day 20 (Figure 5.22) was quite similar.

The same initial smoke input, but for January conditions, showed only a slight tendency for the smoke to rise above its initial height of injection, and then only to move polewards. Some movement toward the equator was found in the bottom 2 km, however, where by day 30 the smoke had reached about latitude 10° N (Figure 5.23). This very different behavior was due to lack of strong solar heating in the winter hemisphere. The low level movement towards the equator could be largely counteracted by washout, since limited

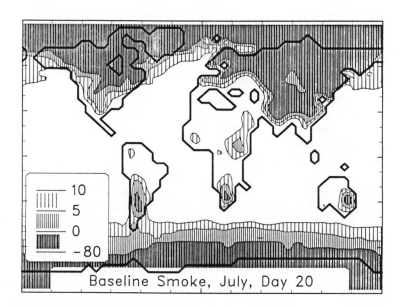

Figure 5.22. Surface temperatures, in °C on day 20 after smoke injection, for the July baseline case of Thompson (1985)

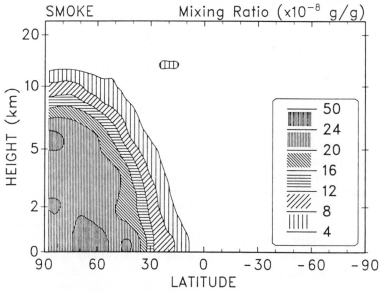

Figure 5.23. Vertical cross-section of smoke mixing ratio in units of 10^{-8} g/g after 30 days for the January baseline case of Thompson (1985)

solar heating will mean much less tendency to set up a more stable thermal structure. On the other hand, it is conceivable that, if enough smoke did reach the northern subtropics, solar heating would begin to drive a thermal circulation that might bring more smoke southward. It would be necessary to include particle scavenging and to run the model for a longer period in order to investigate this possibility.

Surface temperature changes in the winter case were much smaller than in summer, with the principal effect being cooling along the southern edge of the smoke around latitude 20-40° N. The cooling averages about 5°C, but was occasionally as large as 15-20°C.

An important question raised by the winter simulation is the rate of removal of smoke from the Arctic winter atmosphere, where precipitation is normally very small. Infrared cooling normally leads to descending air over the winter pole, which would tend to bring the smoke layer to lower levels where it could be efficiently scavenged in late spring and summer by Arctic stratus clouds. However, direct heating of the smoke layer in the spring could affect the transport and scavenging processes. Further model studies concentrating on the Arctic Basin are needed to resolve this issue.

Thompson (1985) also examined a summer case with smoke injection between the surface and 4 km altitude. In this case, within 15 days lofting due to solar heating raised the smoke to 15 km and a tongue of smoke even moved as far as latitude 20° S.

A case with three times the baseline smoke levels, i.e., with an injection of 540 million tonne of smoke, was also run. This amount of smoke seems unlikely with present nuclear arsenals unless much more smoke is generated in mass fires than in smaller-scale fires (NRC (1985) gave an upper range value of 650 million tonne). Alternatively, an absorption optical depth equivalent to this case might be possible if sources of elemental carbon other than smoke from fires, such as soil and surface carbon lofted into the stratosphere by the nuclear fireballs themselves, were to be much larger than currently believed likely (Galbally et al., 1985; see Chapter 3).

In the case of a 540 million tonne injection of smoke, severe surface cooling was indicated by day 15 over much of the tropics, including below freezing temperatures in parts of Africa and South-east Asia. Large areas were also calculated to drop below freezing in the Southern Hemisphere subtropics, which would normally be experiencing a mild winter.

A case with 60 million tonne of smoke, representing the potential injections from a more limited exchange, was also run for the Northern Hemisphere summer. This simulation led to significant cooling by day 15, with below freezing temperatures at the surface only over northern Canada, northern Europe, and Siberia. The location of these sub-freezing patches would depend on the variable wind patterns, and must therefore be considered to be more or less random within the continental interiors of this latitude zone.

Malone et al. (1986) have used a 20-level version of the NCAR Community Climate Model in an interactive mode with aerosol scavenging by the model-predicted precipitation. In the unperturbed atmosphere, the simulated latitudinal distribution of precipitation is fairly realistic, but the total amount is too large, especially in the tropics, and the precipitation is generated too low in the atmosphere.

The basic case considered by Malone et al. (1986) had an injection of 170 million tonne of smoke with an assumed visible specific absorption of 2 m^2/g. Removal rates were based on a simple empirical relationship between the height-dependent precipitation rate (within the cloud) and the fraction of particles scavenged. Below-cloud scavenging, which is generally much less efficient, was neglected and gravitational settling was included but proved to be unimportant. Two passive and two interactive simulations were carried out. The passive tracer cases (i.e., cases in which the aerosols have no effect on radiation or circulation) consisted of a "low" injection between 2 and 5 km altitude and "middle" injection between 5 and 9 km. The interactive cases were a "low" injection (2–5 km), to compare with the passive case, and a "NAS" case based on the U.S. National Academy of Sciences (NRC, 1985) "baseline" case, i.e., a constant density injection between 0 and 9 km. Smoke was injected over the U.S. and over west and east Europe, with maximum injection rates on day 0, declining linearly to zero on day 7.

Aerosol residence times were calculated from the model for the passive "low" and "middle" cases and compared with observed residence times for natural and anthropogenic aerosols. The "low" case gave a residence time of 5–6 days, and the "middle" case 9–10 days, which are both within the range estimated from observations of the fate of other tracers (Pruppacher and Klett, 1978). Figure 5.24 shows the calculated aerosol remaining as a

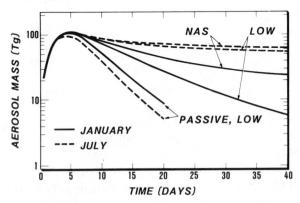

Figure 5.24. Smoke aerosol mass (1 Tg = 1 million tonne) remaining in the atmosphere as a function of time after injections in January (full lines) and July (dashed lines) for the passive and interactive smoke simulations, and for "low" and "NAS" injection profiles (from Malone et al., 1986)

function of time after the start of injection for January (full lines) and July (dashed lines) simulations, for (a) the passive "low" case, (b) the interactive "low" case, and (c) the interactive "NAS" case. Note that in the passive "low" case, the lifetime of the smoke was longer for a winter injection, but in both of the interactive cases the lifetime was greater for a summer injection. This is because lofting of the smoke due to solar heating in summer would quickly take it above the precipitation level. The "NAS" case, with its higher mean altitude of injection, gave longer lifetimes than the "low" injection in both seasons, but the difference was much less in summer because in summer both "low" and higher altitude smoke inputs would be lofted to much the same levels within a matter of a week or so. In the summer cases, lofting increased residence times from normal values of about a week up to 5 or 6 months by day 40. Estimates of the tropospheric residence times are probably not very accurate due to the less-than-perfect simulation of precipitation and the crude parameterization of the smoke removal process. Nevertheless, the relative changes seem reasonable.

The July passive and interactive cases are compared in Figure 5.25 for the "low" injection after 20 days. Not only is the center of mass of the smoke higher in the interactive case, but the total amount of smoke remaining is greater. In the interactive case, the smoke has already reached 25 km altitude and latitude 30° S after 20 days.

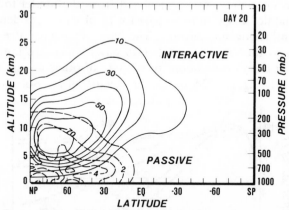

Figure 5.25.   Comparison of the vertical cross-sections of smoke mixing ratios for the passive (dashed lines) and interactive (full lines) smoke cases of Malone et al. (1985), on day 20. Units are 10^{-9} g/g

The separation of the smoke from the precipitation in the summer interactive case, "NAS" injection, is illustrated in Figure 5.26. By this time, most of the remaining smoke (about one third of that originally injected into the atmosphere) had risen to above the tropopause, and the precipitation was confined to the troposphere, most occurring in the lowest 2 km, although it did occur up to about 5 km near the equator.

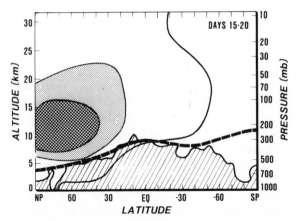

Figure 5.26. Vertical cross-section of the atmosphere, showing the modified position of the tropopause (heavy dashed line) and the precipitation distribution (cross-hatched region below the tropopause), both averaged over days 15–20, and the smoke distribution at day 20 (stippled area mainly above the tropopause). Results are for the July interactive-smoke case with 170 million tonne of smoke injected between 0 and 9 km altitude (from Malone et al. 1986)

Malone et al. (1986) concluded from their study that the "NAS" smoke injection led to substantial reductions in surface air temperature over the continents, relative to the simulated smoke-free climate, for both northern summer and winter injections. Minimum temperatures occurred within one or two weeks and, in summer, the low temperatures would continue for many weeks afterwards. Large temperature reductions did not persist in winter due to the shorter residence time of the smoke, which resulted mainly from the lack of lofting in winter. Lofting in summer carried the smoke to high altitudes, increasing its residence time and horizontal spread, and intensified zonal winds led to rapid longitudinal homogenization in the stratosphere. The spatial distribution in the Northern Hemisphere remained non-uniform after 20 days, but the non-uniformity was essentially in the smoke remaining in the troposphere. By day 40, the distribution was fairly uniform zonally. In January, the more intense zonal circulation into which the smoke was injected produced a rapid longitudinal homogenization of the smoke; the distribution then becoming quite zonally symmetric after 20 days.

The model results also indicated that, for the summer case, transport into the Southern Hemisphere occurred almost entirely at altitudes above 10 km. Malone et al. (1986) suggested that in the NAS case the smoke remaining in the stratosphere would eventually spread more or less uniformly over the globe, leading to an absorption optical depth of about 0.2. This optical depth would correspond to a 20% or greater reduction in sunlight reaching the surface, depending on solar zenith angle. The lifetime of this stratospheric smoke is still an open question. The estimated lifetime of the smoke after 40 days of simulation was 180 days. This figure obviously has considerable uncertainty attached to it due to the short length of the simulation and the limitations of the model. In the current stratosphere, background particle lifetimes are observed to range from 6 months to 2 years (see Chapters 3 and 4 for further discussion). The climatic effect of this amount of stratospheric absorption (i.e., an absorption optical depth of 0.2) and reduction in surface insolation on time scales of months to years has not yet been calculated, but could be considerable.

Malone et al. (1986) indicate concern about important weaknesses in the model, especially in the boundary layer and surface physics, which affect continental surface temperatures and the amount and distribution of precipitation. Their model gives weaker than observed circulation in the unperturbed summer hemisphere, and neglects scattering and infrared effects of the aerosol particles, as well as the evolution of the particles with time.

To one degree or another, all of the multi-dimensional models that have been used to study the climatic impact of nuclear exchanges have both strengths and weaknesses. While research aimed at improving the models and the climate simulations continues, the results already obtained, however, offer considerable insights into the possible climatic impact.

5.5 NONLINEARITIES AND THRESHOLD EFFECTS

As was pointed out in Chapter 4, surface land temperatures are determined by a balance of incoming solar energy and downward infrared energy from the atmosphere with outgoing infrared energy from the surface and heat transferred from the surface to the atmosphere either directly through small-scale convection or indirectly through evaporation of water. If the amount of solar radiation reaching the surface is reduced by small amounts, the balance of the terms is maintained by complementary reductions in the loss of heat through convection and evaporation, with the result that the surface temperature remains relatively constant. However, if a sufficient reduction in solar radiation occurs, convection and evaporation cease entirely, and the surface temperature must begin to decrease more rapidly. If the reduction in solar radiation is caused by the injection of an absorbing aerosol such as smoke, then it is possible, in theory, to specify the amount of smoke necessary to reduce the solar energy reaching the surface to the point where convection is suppressed. Once this point is reached, surface temperatures will decrease rapidly with increasing absorption optical depth until the smoke thickness is sufficient to reduce the solar radiation reaching the surface to nearly zero. This reduction in solar radiation obeys an exponential law: thus, an absorption optical depth of one, for overhead Sun, would allow some 37% of the incident sunlight to reach the surface; an additional absorption optical depth of one would allow 37% of the first 37% (that is, 14%) to get through; and, an absorption optical depth of 3 would allow only 5% to reach the surface. Beyond this point, increasing amounts of smoke in the vertical column will have little effect on the resulting surface temperature, since there is little incoming solar radiation left to be absorbed. This effect is illustrated in Figure 4.8 and discussed in Section 4.5.3.

These concepts (i.e., of a compensation point, beyond which the surface temperature begins to fall rapidly with increasing absorption optical depth, and the exponential nature of light attenuation by absorption) have led to the notion of a "threshold" optical depth, above which a full-fledged climatic cooling would be expected. In principle, such a "threshold" could be defined in terms of the optical depth of a smoke layer or, alternatively, the mass of smoke, if the optical properties are known. In reality, however, the "threshold" concept is imprecise, and liable to be misleading if it is taken too literally, even though it is based on a correct understanding that the effects of smoke injections are highly nonlinear.

The precise absorption optical depth at which convection at the surface would essentially cease might well be taken as the critical absorption depth. However, this quantity is a function of a number of other variables, including at least the vertical distribution of the absorbing aerosol, the ratio of scattering to absorption optical depth (e.g., see Cess et al., 1985), the season of the year, the latitude, and the state of the atmosphere at the time of injection. The vertical distribution and fraction of absorbing material, in turn, would vary greatly with such variables as the mixture of the sources of smoke and the size of the weapons used (see Chapters 2 and 3), while the state of the atmosphere at the time of injection would be virtually unpredictable until shortly before the actual exchange took place (and even then, inadequately known). Thus, it is essentially impossible to relate a zonal or hemispheric-mean threshold optical depth to particular numbers of warheads exploded or total amount of megatonnage detonated.

Moreover, when discrete smoke and dust source regions and the gradual dispersion of the resulting clouds of absorbing material are considered, (see, for example, MacCracken and Walton, 1984; Thompson, 1985; and Malone et al., 1986), it is apparent that optical depths in excess of the critical value would occur locally under these patchy clouds, even though the total amount of absorber would not be sufficient to create large-scale climatic effects if spread uniformly over the hemisphere or globe. Such a phenomenon was particularly apparent in the case study of Thompson (1985) in which 60 million tonne of smoke was injected and in the case of a 15 million tonne injection considered by MacCracken and Walton (1984) (see discussion in Section 5.4.2).

Therefore, as Schneider (1985) has argued, there is no sharp cutoff in smoke amount below which there will be no adverse weather and climate effects. Even if there were, the present level of sophistication of climate models is such that a precise estimate of where that cutoff would be could not be given. Thus, the nonlinearity of the surface temperature dependency on the amount of absorber is not useful in setting an upper limit to the environmentally "acceptable" amount of smoke and dust that could be generated in a nuclear war without causing weather or climatic disasters.

Another potentially important nonlinearity arises as a result of the "lofting" and the reduction in precipitation scavenging rate induced by atmospheric stabilization (Malone et al., 1986). In their studies, both the fraction and amount of smoke lofted increased with smoke amount up to absorption optical depths around 2 or 3 (due to increasing local warming of the smoke layer). For even larger initial smoke amounts, however, the fraction remaining in the atmosphere began to decrease, so that the amount of smoke which was lofted increased less rapidly than the initial smoke amount. The reason the fraction did not continue to increase was because, once thick upper layers of smoke were established, these upper layers shaded the underlying smoke, thereby preventing it from being lofted upward. As a result, the large amount of smoke in the lower layers continued to be exposed to removal processes in the troposphere.

5.6 SUMMARY OF MODELING RESULTS

The results discussed in the preceeding sections are indicative of the recent advances in model sophistication and in the realism of the simulations of the climatic consequences of a nuclear exchange. Three particular model developments stand out as having improved the understanding of the climatic impact: (1) the inclusion of the interaction between the absorption of radiation by the smoke particles and atmospheric motions; (2) the simulation of smoke dispersion from regional sources, as opposed to the assumption of uniformly distributed smoke as an initial condition; and (3) the use of scavenging rates determined by the model-generated precipitation (MacCracken and Walton, 1984; Malone et al., 1986), even though the parameterization of scavenging processes is still fairly simplistic.

The simulations run with these improved models have produced two very significant results. All of the groups which have performed interactive smoke simulations find that, in the northern summer, the feedback causes smoke to be lofted and then to move southwards. In the models with greater vertical resolution (Thompson, 1985; Malone et al., 1986), the height to which the layer rises is around 15 to 25 km and the major transport southward occurs at these levels. For initial smoke injections of around 100–200 million tonne, the lofting and southward transport results, during the first month, in only a slight amelioration of the cold surface temperatures calculated for northern midlatitude continental areas for the fixed smoke case. However, when smoke transport and feedback effects are included, appreciable cooling occurs in the northern subtropics, with some spill-over of smoke into and consequent cooling of the tropics and southern midlatitudes. Since subtropical and tropical plants are more sensitive than temperate climate plants to lowered temperatures, this is a particularly important result (see discussion in Volume II of this report).

The second important result emerging from these new calculations is that the lofting of the smoke and the changed vertical temperature profile could quickly lead to a lowering of the tropopause to levels of around 5 km in the Northern Hemisphere. In this modified atmosphere, most of the precipitation would occur over the oceans and in the lowest 1 or 2 km of the atmosphere. The bulk of the smoke would be above the new tropopause where it would not be efficiently scavenged. This would lead to much longer atmospheric lifetimes for the smoke and, consequently, to prolonged surface cooling and greater effects in the Southern Hemisphere as the smoke is transported southward (Malone et al., 1986).

Results for smoke injections in the northern winter, when there is far less solar energy available to loft the smoke, show little southward movement of the smoke, and removal rates more typical of those in the ambient atmosphere due to less induced atmospheric stability from particle heating. In this case, potential effects in the tropics and Southern Hemisphere would be more critically dependent on the initial height of injection of the smoke and its subsequent lifetime. If sufficient smoke remained aloft into the spring, significant southward transport of smoke might still occur, probably causing surface cooling when it reached more southerly latitudes. However, model simulations have not yet been run for the extended time periods needed to evaluate these effects.

Most of the major uncertainties in the simulations now are related to processes that happen on time and space scales which are smaller than those resolved in the models. These include the efficiency of prompt scavenging in the initial smoke plumes, the effects of coagulation on the optical properties and size of the smoke particles, longer timescale scavenging rates, induced coastal and mesoscale effects, and the effect of sub-grid scale mixing on modifying vertical stability and enhancing lofting of the smoke. These processes will have to be evaluated with higher resolution models and the results incorporated into the GCMs through parameterizations. There is a second group of processes, such as albedo feedbacks and the effect of smoke particles on stratospheric chemistry, that are only beginning to be addressed at this point.

In attempting to summarize the current status of the issue of the climatic consequences of a nuclear war or, to put it more directly, to answer the question "would a major nuclear exchange have severe climatic consequences?", the following four points must be made.

1. No new and substantial work (as opposed to some qualitative expressions of skepticism) has lessened the probability that a major nuclear exchange would cause severe environmental effects (although some of the effects would probably be less extreme than was sometimes suggested in discussions of the early results). Consideration of the lofting of smoke due to solar heating increases the probability and probable duration of significant effects, as a result of the increased lifetimes of the particles and the rapid separation of the smoke layer from regions of precipitation and scavenging.

    Small and Bush (1985; see discussion in Chapter 3) have produced estimates of reduced smoke emissions from wildland fires. Even if these estimates are correct, they make little difference in the overall problem since the total smoke amount is dominated by urban and industrial fire emissions. It is clear, however, that significant collateral damage to urban and industrial sites (including fuel storage facilities) is almost certainly necessary to produce severe climatic consequences.

    Current model results, primarily those from one-dimensional models (Ramaswamy and Kiehl, 1985; Ackerman et al., 1985) and two-dimensional models (Haberle et al., 1985), indicate that inclusion of the infrared absorption and emission properties of the smoke and dust aerosol is unlikely to moderate substantially the predicted surface cooling unless the smoke is very near the surface. This, however, must be verified more rigorously, both by better measurements of representative smoke and dust optical properties, and by the inclusion of infrared effects in the three-dimensional models. Inclusion of aerosol scattering properties is also essential to obtaining firmer conclusions, especially for low extinction optical depths.

2. The results of the interactive models support the possibility of significant environmental effects in the tropics and in the Southern Hemisphere. If large quantities of smoke were injected into the northern mid-latitudes in the northern spring or summer, it now seems likely that considerable smoke would be carried southwards at high altitude within a matter of weeks, which would result in at least some cooling in the tropics and Southern Hemisphere. Models will have to be run out for several months, and include removal processes (including stratospheric chemistry and particulate coagulation), before the extent of these effects in the tropics and Southern Hemisphere can be described with confidence.

    The effects of a similar large injection of smoke and dust during the northern autumn or winter, when lofting due to solar radiation could be almost negligible, is a different problem. The crucial questions here concern the heights of injection and the lifetime of the smoke and dust in the winter atmosphere. In the perhaps unlikely event that large quantities of smoke and dust were to survive in the atmosphere until northern spring and summer, there might be significant effects, even at more southerly latitudes.

3. There are still large uncertainties associated with the problem, some due to poorly understood physical processes and some due to more intangible issues. Some of these uncertainties, when resolved, could increase the severity of the climatic consequences (e.g., determining the timescale of the climate perturbation if the smoke is lofted into the stratosphere and the lifetime of the particles once they are in the stable upper atmosphere). Other uncertainties, such as the possibility of rapid coagulation in dense plumes, could reduce the potential impacts. Perhaps the greatest uncertainties are associated with processes and questions whose effects are completely undetermined and may result in either enhanced or moderated severity. These include items as diverse as the effects of mesoscale and synoptic-scale circulations, revised estimates of smoke production from urban fires and dust production from surface bursts, the fraction of the smoke involved in prompt scavenging, and various aspects of atmospheric chemistry.

One subject that may have a considerable impact on the entire problem and which has not yet been addressed is atmosphere-ocean interactions. The very substantial global-scale changes in the temperature structure of the atmosphere that are predicted by the model simulations imply very substantial changes in atmospheric circulation and stability, which would in turn, greatly affect precipitation patterns and surface wind fields. Thus, it is probable that ocean currents, regions of oceanic upwelling, and land-sea circulations would also be altered. Even in the normal climate system, these phenomena, such as the Southern Oscillation-El Niño and monsoon circulations, account for major year-to-year climatic fluctuations (e.g., see Wyrtki, 1975; Rasmusson and Wallace, 1983). It is likely that there would also be complex effects in coastal zones where land-sea thermal contrasts would change sign in summer, and strong horizontal temperature gradients might set up abnormal mesoscale circulations. Delineation of the effects on the oceans requires fully interactive, coupled atmosphere-ocean models. Research versions of this type of model are just beginning to become available for climate research. Mesoscale ocean-atmosphere effects require investigation with much higher spatial resolution than is used in the present generation of general circulation models. This may be possible by the use of compatible nested models in which a general circulation model is used to set the initial and boundary conditions for one or more limited area mesoscale models.

4. To a large extent, the discussion of climatic consequences has focused on regional and hemispheric responses. Results have been formulated in the context of departures from the model predicted "normal" climatic

means. To understand the actual consequences, these results have to be put into the context of their effect on biological, ecological, and agricultural systems. For instance, the anticipated cooling effects at northern high and mid-latitudes, while very substantial if the smoke injection occurs in the northern summer, are no greater, and indeed may be less, than the cooling which occurs every winter. Nevertheless, if such cooling occurred suddenly during a normal growing season, its biological consequences would far exceed those usually associated with winter, because the normal onset of winter is gradual, anticipated, and prepared for by humans, animals, and plants alike. It is also possible that the effects would be more serious in subtropical latitudes which could experience a smaller, but completely unprecedented, cooling, perhaps accompanied by major decreases in precipitation. The problem of extrapolating from climate predictions to weather variability on the synoptic and mesoscale further complicates the estimation of biological impacts. While these issues are addressed in detail in Volume II of this report, they should remain clearly in focus as additional climatological research is done on this problem.

5.7 EXTRAPOLATIONS FROM THE MODEL RESULTS

Models are tools for correcting, refining, and quantifying the results of analytical thinking. However, as was noted in the preceeding sections of this chapter, climate models have definite limits and cannot provide answers to all the questions that are raised with regard to the consequences of a nuclear exchange. In the absence of detailed modelling results that could be used to answer these questions, qualitative reasoning can be helpful in many cases in attempting to provide first-order estimates of trends and probabilities and to suggest more detailed consequences. In this section, an attempt is made to provide answers to some of the questions that have been posed based on knowledge of the current climate system and the way in which it operates and on reasonable inferences drawn from the studies that have been carried out to date. Obviously, these answers are not "final" in any sense; they represent a preliminary evaluation and should be interpreted within that context.

5.7.1 Effects on the Oceans

Changes in sea surface temperature as a result of the presence of optically-thick aerosol layers in the atmosphere would arise from a number of effects in addition to the reduction in solar radiation reaching the oceans. These include changes in the downward infrared flux due to altered cloud cover, the presence of the aerosol layer itself, cooler atmospheric temperatures; and changes in latent heat fluxes due to altered air temperatures, humidities and wind speeds. There could also be changes in coastal upwelling and ocean currents due to altered wind stress. Changes in upwelling and ocean currents would be highly location-specific and could well be locally dominant over other changes in the surface radiation and heat fluxes.

Most of the general circulation models that have been used to study the effects of smoke and dust following a nuclear war have not been coupled to models of the ocean and have, in fact, assumed either a fixed or seasonally-varying sea surface temperature based on climatological averages. The exceptions are the two-layer, three-dimensional model of Aleksandrov and Stenchikov (1983), the two-dimensional climate model of Mac-Cracken (1983), and the one-dimensional energy-balance model of Robock (1984). However, none of these models includes the actual dynamics of the oceans. In each of these simulations, the essential result is a cooling of the ocean surface by only a few degrees Celsius in the first few months. As an illustration, the temperature changes calculated by MacCracken (1983) are shown in Figure 5.11. The only exception to these small coolings is in the simulation of Robock (1984), where, due to the inclusion of ice-albedo feedback, the cooling at high northern latitudes in the first autumn following a summer war was of the order of 5–10°C. However, the lack of energy conservation in this model simulation makes the results somewhat suspect.

Even if the possibility of changes in the dynamics of the oceans are ignored, realistic estimates of sea surface temperature changes must await the coupling of an ocean model to an atmospheric model incorporating fully interactive smoke and dust. In this case, the simulated changes in atmospheric wind speeds and relative humidities could be used to obtain a more accurate estimation of the latent heat fluxes at the ocean surface.

As already suggested, however, the largest changes in sea surface temperatures would be likely to occur regionally due to changes in ocean dynamics induced by changes in wind stress. Outflow of cold surface air from the continents, as in a winter monsoon or katabatic flow, would lead to coastal upwelling of cold water, as occurs naturally in many parts of the world. This could reduce the moderating influence of the oceans in coastal zones that are not subject to strong on-shore winds.

The most dramatic changes could occur in systems such as the El Niño-Southern Oscillation (ENSO) system, which operates in the equatorial Pacific and is apparently modulated by the strength of the trade wind systems (Wyrtki, 1975). ENSO is a complex phenomenon in which surface winds influence the oceanic state and sea surface temperatures completely across the tropical Pacific Ocean. These sea surface temperature anomalies, in turn, influence the atmospheric circulation, not only in the tropics but in higher latitudes (see, for example, Chervin and Druyan, 1984; Stone and Chervin, 1984). This complex interacting system is the subject of very active research (see NRC, 1983, Kerr, 1984;) both because of its intrinsic interest and its observed influence on mid-latitude weather.

Clearly, it is premature to predict how surface wind changes associated with the effects of optically-thick smoke and dust layers would affect ENSO, but it is clear that such effects would occur. Surface wind patterns would have to be derived from a fully interactive GCM, and these might be expected to change as the situation develops due to the spreading of the smoke and dust and to seasonal variations. Feedbacks between the atmosphere and the ocean could change the situation further. Such changes could lead to significant disturbances to climate both in the tropics and at higher latitudes, even in the absence of smoke and dust layers in these regions. Regions which could be impacted include South America, Australia, New Zealand, and southern Africa (Pittock, 1984).

5.7.2 Effects on the Monsoons

The monsoons are seasonal, continental-scale circulations driven by the contrasts between land and sea-surface temperatures, which change sign between summer and winter (Ramage, 1971; Webster, 1981). They lead to marked dry and wet seasons across most of tropical Africa, southern Asia, and northern Australia, as shown in Figure 5.27.

In the northern summer, solar heating of the north African and the Asian mainlands, and especially of the high Tibetan Plateau, generates low surface pressure and flow of moist air towards the northeast from the tropical Atlantic Ocean, the Arabian Sea, the Indian Ocean, and the South China Sea to produce the southwest monsoon, which provides most of the annual rainfall to the Sahel, the Indian subcontinent, Southeast Asia, China and parts of Japan. At the same time southern sub-tropical Africa and northern Australia experience their dry season.

Six months later the situation is reversed, with high pressure centered over the cold land masses and subsiding dry air flowing out from the northern continents, which are in the middle of their dry season. Now the air is flowing from the north and brings rain to the east side of the Indian peninsula, but nowhere else in Asia. As it crosses the equator it turns to the south-east and brings rain to northern Australia. Similar flows bring rain to the southern subtropics of east and west Africa and to Madagascar.

If the northern midlatitudes and subtropics were covered with a smoke pall during northern summer as a result of a nuclear exchange, it is probable that the southwest monsoon circulation would be switched off (Oboukhov and Golitsyn, 1983). Due to the attenuation of solar radiation by the smoke, the land-sea temperature difference and, in particular, the solar heating of the Tibetan Plateau, would be reduced and then reversed probably within a matter of a few days. The southwest monsoon might be replaced initially by a

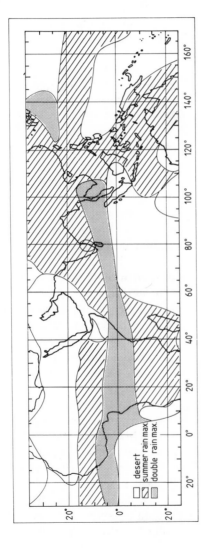

Figure 5.27. Areas normally affected by local summer precipitation regimes are indicated by broad hatching. Stippled areas have double rainfall maxima (summer and winter). These rainfall regimes may be seriously affected by surface land cooling after large injections of smoke into the atmosphere. After Ramage (1971). Reproduced by permission of Academic Press

shallow northeast monsoon, which would bring unseasonally dry conditions to much of south and east Asia and the Sahel zone, and rain to Indonesia and northern Australia. Outflow of cold air from the Asian continent over the warm ocean could, however, lead to increased rainfall in coastal zones. The rain over Indonesia and northern Australia might quickly weaken, with rain becoming confined to the coastal zones of the Asian mainland as upper level smoke moves southward, thereby possibly inducing a return northward flow at low levels in the tropics. There are suggestions of some of these possible effects in Figure 5.14 and 5.21 above.

A more detailed description of these effects must await simulations by higher resolution general circulation models or regional scale models. However, there is little doubt that the normal rainfall pattern over the monsoon regions would be drastically affected.

5.7.3 Coastal Effects

Normal coastal weather is strongly affected by local circulations induced by land-sea temperature contrasts, the familiar land and sea breezes. Similarly, in the presence of an optically-thick smoke layer, coastal effects would arise from the large horizontal temperature gradients which could be set up in the coastal zones between the cold land and relatively warm oceans. This gradient, in conjunction with gravity flows, could produce significant surface outflow of cold air, probably in quite shallow layers, from the interiors of continents out over the warmer ocean waters. These outflows could occur spasmodically, depending in part on weather fluctuations, and could lead to sub-freezing conditions in coastal zones lasting for periods of days or weeks. These extreme episodes would be far more important to the survival of crops and plant communities than average cold conditions (see discussion in Volume II), and are not predictable with the present generation of general circulation models because of inadequate spatial resolution.

Cold outflow conditions might set up situations partially analogous to the "cold outbreaks" that occur most notably off the east coast of Asia in normal winters in association with the northeast monsoon (Zhu, 1983; Lav and Lav, 1984; and Chu and Park, 1984). However, the increased vertical stability at mid-tropospheric levels over the oceans would ensure that any associated convective storms over the ocean would be shallow. It is also possible that the increased vertical stability would be such that cold air outflow would be confined to a very shallow boundary layer similar to the katabatic winds that cross the coast of Antarctica (see eg. Fitzjarrald, 1984; Parish, 1984), or to a less severe nocturnal drainage situation. Such circulations are strongly influenced by local topography and tend to be strongest in coastal valleys, where many cities are located.

Onshore winds, such as might occur on mid-latitude west coasts and low latitude east coasts, likewise could tend to bring only rather shallow inflows of warm moist air moving up over the shallow layers of cold surface continental air. If the on-shore westerlies were sufficiently strong (which might not be the case in northern summer), this circulation might produce situations analogous to lake effect storms, such as occur near the Great Lakes in North America. These storms typically produce heavy snowfalls or intense rain.

Such possibilities are, of course, little more than informed speculation until they are investigated by experiments with mesoscale and general circulation models having sufficient spatial resolution to resolve these probably very shallow circulation features. This may be possible with a series of compatible atmospheric models ranging from global scale down to mesoscale, with the initial or boundary conditions for the high resolution models being set by simulations of the perturbed atmosphere with the coarser resolution models.

5.7.4 Island Effects

The special situations applicable to small land masses surrounded by ocean, such as New Zealand, Tasmania, Japan, Indonesia, the Philippines and the West Indies, should also be investigated. Land surface cooling in these situations might be limited by land-sea breeze circulations. If this mechanism operates, cooling of small land masses might be limited to, at most, some 5 to 10°C below the surrounding ocean, with even less cooling under strong wind conditions. However, in the case of islands close to continental land masses such as the British Isles and Japan, cold winds from the continents could lead on occasions to much more severe conditions.

Serious effects might arise from significantly reduced precipitation over tropical islands in those areas where a high proportion of normal rainfall is associated with sea-breeze convergence and the diurnal cycle. In these cases, there might be a reduced or no sea breeze, and rainfall might tend to occur off the coast.

Island coasts subject to strong on-shore orographic rainfall might not be so strongly affected, although wind patterns might change significantly, even as far south as Tasmania and the South Island of New Zealand. Initially, as lofting and southward movement of smoke occurs in the Northern Hemisphere, subsidence in southern mid-latitudes might tend to suppress rainfall in this zone, but a strengthening of the mid-latitude westerlies and a shift in their latitude of maximum strength might lead to increased precipitation on some windward coasts. If and when the smoke spreads more uniformly with latitude, however, this strengthening of the westerlies might cease and orographic rainfall could diminish.

5.7.5 Precipitation Changes

Changes in precipitation patterns have already been discussed both in this chapter and the preceeding chapter, but it is useful to bring the various discussions together. Above all, it must be stressed that precipitation is one of the most difficult quantities to treat in general circulation models and the models are, in general, only moderately successful at simulating observed precipitation patterns. It follows that the precipitation changes are among the least certain estimates of effects. Nevertheless, it is possible to speculate intelligently on the general trend of precipitation changes from a consideration of the basic physics involved.

All the models predict a general heating of the upper layers of smoke, and cooling at the surface, especially over land. This must lead to increased vertical stability below the smoke, which on average must tend to reduce the precipitation, especially over land. Moreover, since land surfaces would cool more rapidly than ocean surfaces, there would be a tendency for subsidence of air over land, at least in the lower levels. This should lead, in general, to a suppression of precipitation over land. Heating of the upper air would also lead to reductions in relative humidity and upper level cloudiness (Oboukhov and Golitsyn, 1983), except perhaps for thin cirrus that might be generated by convective motion above the heated smoke layer.

Some have argued (e.g., Katz, 1984; Singer, 1984; Teller, 1984) that the initial cooling of lower tropospheric air could lead to considerable rain, snowfall, and fog. Water from the combustion of wood and fuels, and water from entrained boundary layer air would provide additional sources of moisture. The mass of water from the fires actually exceeds the mass of smoke produced, but is not significant compared to the total amount of water normally in the troposphere. The significance of these quantities of water vapor for precipitation and particle scavenging is easily exaggerated because of a common misconception (not necessarily held by the above authors) that large amounts of precipitation derive from the water contained in the volume of air over any given area at that time. Precipitation is usually the result of a dynamic process in which air is lifted in a continuing stream as a result of orographic flow, convection, or wave motion (fronts). Thus, water is removed not just from a single, limited air mass, but from a continuous stream of air flowing through the cloud. The amount of water in a cloud at any one time is not very large compared with the total precipitation often experienced over time at a given point on the ground. Thus the cooling of a particular parcel of air, such as would happen under a smoke pall, would not by itself result in large amounts of precipitation.

However inadequate the parameterization of the precipitation process may be, the models with internally-generated precipitation (i.e., which do not have prescribed rainfall) that have been applied to this problem show reduced precipitation in line with the above general arguments. For example, MacCracken (1983) found, with noninteractive smoke, that the changed thermal structure and circulation led to 25% less precipitation over land and 20% less over the oceans some three months after the smoke was injected. Thompson (1985), in his fully interactive smoke run with the NCAR CCM, found large reductions in condensation in the upper troposphere and in the ITCZ, as shown in Figure 5.21. Similarly, Ghan et al. (1985) found large overall reductions in precipitation (see Figure 5.15).

As discussed in Section 5.7.2 concerning the monsoons, it is also likely that the southwest monsoon, which provides most of the rainfall to southern Asia, and the similar system which waters the Sahel in Africa, would fail if major cooling of land surfaces were to occur in the northern summer. Increased rainfall might occur, however, in coastal zones in these areas (see Section 5.7.3).

Island rainfall was discussed in Section 5.7.4 above. Those tropical islands which get most of their rainfall from sea breeze convergence during the late afternoon or evening would lose much of their rainfall if the diurnal temperature cycle over land were suppressed, since the sea breeze would be reduced. Orographic rainfall would also be affected by changing wind patterns and strengths.

Finally, the direct thermal circulation that would be set up between the Northern and Southern Hemispheres until the smoke became more evenly distributed between hemispheres, could cause a relatively greater subsidence in the southern mid-latitudes, which could tend to decrease rainfall in these latitudes. A return flow at low altitudes could lead to a local increase in rainfall around 30° N latitude (see Figures 5.14 and 5.21).

If sufficient smoke remains aloft after it has spread globally to provide an absorption optical thickness of about one or more, one might expect a gradual weakening of synoptic disturbances and of the hydrologic cycle in general. In this case, globally-averaged precipitation could be significantly below normal for an appreciable period after the initial smoke injection. It has also been suggested that changes in the electrical properties of the atmosphere due to increased background radioactivity might be possible (Izrael, 1983), and that these might have some effect on the hydrologic cycle. No quantitative calculations have been made, however, and the effect is not likely to be noticeable compared to smoke-induced changes.

5.7.6 Effects on the Southern Hemisphere

In the Ambio scenario (Ambio Advisors, 1982), it was assumed that only about 3% of the megatonnage in a 5,700 Mt nuclear war (about 170 Mt) would be detonated in the Southern Hemisphere. According to Galbally et al. (1984), this might generate some 10 million tonne of smoke. By itself, this would almost certainly not be enough to produce widespread and significant surface cooling (although short-term local cooling could occur under thick patches of smoke before they dispersed). Therefore, should a large nuclear exchange occur, the major environmental consequences in the Southern Hemisphere would be the result of both transport of smoke from the Northern Hemisphere and modification of the circulation as a result of perturbations caused in the Northern Hemisphere.

Transport of the smoke to the Southern Hemisphere would most likely be due to changes in the general circulation of the atmosphere induced by the smoke and dust injected into the Northern Hemisphere. The model results discussed in the preceeding sections show that, for smoke injections in the northern spring and summer, solar heating of the smoke layer would tend to produce a direct circulation that would transport smoke and dust southwards at altitudes around 10–20 km. The smoke and dust could reach southern midlatitudes within a matter of a few weeks. This induced circulation would be in marked contrast to the unperturbed circulation, which leads to a very slow exchange of air between the Northern and Southern Hemispheres, with characteristic times of a year or more. The normal rate of exchange is illustrated by the observed lag of approximately one year in the concentration of carbon dioxide in the Southern Hemisphere behind that in the Northern Hemisphere, where most of the anthropogenic carbon dioxide emissions occur (Pearman et al., 1983).

The more recent three-dimensional simulations, in which the heating of the smoke layer is allowed to force dynamical motions, all show this transport into the Southern Hemisphere for smoke injections occurring in the northern summer and, to a lesser degree, spring (see, for example, Figures 5.18 and 5.25). For smoke injections in the 150 million tonne or greater range, appreciable visible absorption optical depths could be reached in the southern subtropics within a matter of two to three weeks. Since the interactive simulations have not been run for time intervals longer than about six weeks, the ultimate extent of the smoke coverage in the Southern Hemisphere has not been determined. However, the model results all indicate that the lofted smoke which would be moving into the Southern Hemisphere would be well above the level of significant precipitation (see especially Figure 5.26), and that lower level precipitation in the Northern Hemisphere would be greatly reduced by the increased lower level static stability (Figures 5.21 and 5.14). This implies that the smoke could have a long lifetime, on the order of months to years. It would then have time to mix much more evenly between hemispheres and even to higher southern latitudes. Thus, in the Southern Hemisphere, optical depths might reasonably be expected to continue increasing beyond the first few weeks and, possibly, to produce significant land surface cooling in continental midlatitudes. This, of course, would not be true if other processes such as coagulation, accretion of sulfuric acid and subsequent gravitational settling or oxidation, were to operate to remove the smoke from the stratosphere.

The magnitude of the effects in the Southern Hemisphere likely would be highly dependent on the season in which the smoke injection occurs. For major injections occurring in the Northern Hemisphere during spring, summer, or early autumn, the transfer of large quantities of smoke to the Southern Hemisphere seems likely. However, smoke injected during the northern spring or early summer would arrive at southern latitudes in the southern winter, when it would have a minimal effect on surface temperatures. In this case the operative question is how much of this smoke would still be in the atmosphere when the following southern spring and summer arrive. In view of the relatively high altitude at which the transport would occur, it seems probable that a large fraction of the smoke would still be in the upper atmosphere unless some chemical process destroys it *in situ* (see discussion in Chapter 6). It is possible that the most serious effects for the Southern Hemisphere would occur in the event of a late summer or early autumn injection in the Northern Hemisphere. In this case, the southward drifting smoke would be continually heated and lofted by the maximum solar intensity during the change of seasons. This heating and the associated circulation probably would produce the greatest transport of smoke into the Southern Hemisphere, and the smoke would be present in the Southern Hemisphere summer, when its effect on surface temperature would be maximized.

If injection occurred in the northern autumn or winter, the smoke and dust probably would be confined to high northern latitudes until northern spring, as indicated in Figure 5.23. Thus, the potential effects on the Southern Hemisphere depend crucially on the lifetime of this smoke and dust in the northern winter atmosphere. The results of Malone et al. (1986) suggest that the lifetime could be very dependent at that time of year on the initial height of injection (see Figure 5.24). For smoke injected between 0 and 9 km with a uniform concentration (the NRC, 1985, scenario), the particle lifetimes after about 40 days of simulated time are estimated to be of the order of 45 days. However, only 14% of the original injected mass remains at day 40. A projection to the following northern spring based on these figures would lead to a considerable reduction in total smoke levels, leaving only some 3–5% of the particle mass remaining 3 months after the initial injection. If this were correct, significant effects in the Southern Hemisphere could only occur for total initial smoke injections of the order of 1000 million tonne or more, which is about an order of magnitude greater than the estimated injection mass (see Chapter 3). There is, however, great uncertainty in these estimated lifetimes, as they are highly dependent on the initial height of injection and on the simulated precipitation rates under perturbed Arctic conditions.

There are processes other than the direct attenuation of sunlight by transported smoke layers by which Southern Hemisphere weather and climate might be affected, especially if significant amounts of smoke were present in the Northern Hemisphere in the northern spring or summer. Perturbations to atmospheric and oceanic circulations could affect winds, temperatures, and precipitation, even in a Southern Hemisphere not covered by smoke. As discussed in Section 5.7.2, the northern summer monsoons might be curtailed by the lack of solar insolation in the Northern Hemisphere. This could lead initially to a surface outflow of cold air from Asia and to a burst of "monsoon-type" rainfall during the normal dry season in northern Australia and the southern subtropical monsoon regions of Africa.

Changes in the global circulation induced by the presence of smoke and dust in the Northern Hemisphere would probably result in a direct, thermally-driven meridional circulation with air moving southward across the equator at about 10–20 km and descending in middle latitudes of the Southern Hemisphere. This could lead to surface warming and a marked decrease in cloud cover and precipitation until smoke and dust were to move overhead. At this point, the surface warming would turn to cooling, but there would not necessarily be any significant increase in precipitation. Changing wind patterns might, however, alter this picture somewhat. Also, the presence of the direct thermal circulation might strengthen the zonal winds in the mid-latitudes of the Southern Hemisphere. This could cause an increase in orographic rainfall on windward exposures, but the effect might be short-lived if the induced meridional circulation were to achieve a fairly uniform mixing of smoke and dust into the Southern Hemisphere within a matter of a few months. After that, one might expect the vigor of the circulation in the lower atmosphere to decrease somewhat due to the reduced energy input into the surface and troposphere. Malone et al. (1986) suggest that this reduction in solar insolation might be about 20%.

Finally, it must be emphasized that the preceeding discussion is, of necessity, somewhat speculative. It is, however, based on reasonable extrapolation from the modelling results and on scientific judgement. While far from established, the conclusions are the best that can be drawn at this time.

5.7.7 Longer-Term Effects

Since, under certain scenarios, severe atmospheric effects resulting from a nuclear exchange could last for weeks to several months, and since large-scale nuclear war would cause other damage to the environment, it is natural to ask whether climatic changes could be induced on time-scales of years. In all probability, climatic perturbations resulting from a nuclear exchange would be, on climatic time-scales, only a sharp transient disturbance. While that transient could have disastrous human and biological consequences, the

Earth's climate would be expected to return to the current "normal" unless the initial disturbance triggered some climatic feedback process that would alter the energy balance of the Earth, either permanently or on a much longer time-scale. While the possibility of such a drastic climatic change cannot be completely dismissed, it appears to be unlikely.

There are, however, several processes that might have effects on the requisite time-scales to cause long-lasting climate perturbations. The recent results which confirm the possibility of lofting smoke to altitudes exceeding 10–20 km raise serious questions about the ultimate fate of these particles. The lifetime of soot particles at these altitudes becomes a critical question, as is their effect on the chemistry and radiative properties of the upper atmosphere. Death of vegetation from fire, radioactive fallout, abnormally cold conditions, or toxic chemicals in the surface air could result in reduced evapotranspiration, which Mintz (1984) and Shukla and Mintz (1982) regard as a significant determinant of regional and global climates. The death of vegatation could also produce changes in surface albedo (e.g., Eaton and Wendler, 1983; Jurik and Gates, 1983) with effects on climate both at middle and low latitudes, according to Otterman et al. (1984) and Sud and Fennessy (1982), although this is disputed by Henderson-Sellers and Gornitz (1984). The length of time that albedo effects would persist is dependent on the rate at which some sort of vegetation would re-establish itself, thereby reducing the surface albedo and restoring evapotranspiration rates towards normal values. The re-establishment could be nullified if some positive feedback process prevented the recovery of the vegetative cover. Several investigators have argued that this has been the case in areas such as the Sahel, where loss of vegetation due to over-grazing and drought has increased the surface albedo and possibly led to increased aridity (Charney et al., 1975). Whether such a mechanism would operate outside certain already very sensitive areas is, however, very doubtful. (See Volume II for further discussion on this point.)

Settling and deposition of soot on snow and ice fields could decrease surface albedo in areas that normally have permanent snow or ice cover, possibly leading to increased solar heating and melting once the air-borne smoke layers have cleared (Warren and Wiscombe, 1985). Such an effect would be moderated by subsequent falls of clean snow covering the layer of soot, or by melting and subsequent run-off. In some circumstances, soot layers could be exposed repeatedly at the surface following melting of over-lying snow. It seems unlikely, however, that an effect lasting more than a few years would be produced.

Robock (1984) studied the effects of snow- and ice-albedo feedback, which he calculated would enhance the cooling in summer, especially in the second summer when there would be less smoke and more solar radiation to be reflected back into space by the increased snow and ice cover. He also pointed out that the transfer of heat from the oceans to the atmosphere in the presence of an optically-thick smoke layer would cause some cooling of the mixed layer of the ocean. This, in turn, would feed back as a cooler lower boundary for the atmosphere in the following years, which could prolong the atmospheric surface cooling. Unfortunately, the lack of vertical resolution of atmospheric processes in his model makes the interpretation of such results problematic.

The potential effect of the deposition of soot onto sea ice has been examined by Ledley and Thompson (personal communication). They find that the largest perturbations of the sea-ice cycle might occur in spring, leading to an increase in the summer ice-free period of from 2 to 3.5 months at latitude 82.5° N. The predicted disturbance in the annual cycle of sea ice that they calculated continued into following years due to the increased absorption of solar radiation by the ice-free surface waters. Large-scale sea-ice changes could significantly affect climate, probably on the time-scale of a few years.

In view of its potential importance, the question of possible long-term effects should be addressed more rigorously than has been possible here. This will require not only improvements in understanding of the global climate system and in models used to simulate it, but also requires better definition of the short-term physical and biological impacts of a nuclear war.

5.8 PROVISIONAL TEMPERATURE EFFECT SCENARIOS

Most of the discussion concerning climatic consequences has been couched in qualitative terms and broad generalities because of the uncertainties associated, in particular, with predicting detailed effects both on temporal and spatial scales. However, as a result of interactions with biologists and ecologists working on Volume II of this report and with other interested persons, it was felt that an attempt should be made to offer a more quantitative assessment of the potential climatic effects. The provisional temperature effect scenarios given in Tables 5.2 and 5.3 are the results of that attempt.

These temperature scenarios are conditional on a wide range of variables including, but not limited to, the total amount of smoke injected, the fraction of amorphous elemental carbon in the smoke, the height of the initial injection, and the season in which the injection occurs. The suggested temperature changes are given for broad geographical regions rather than being location specific, and have wide ranges of uncertainty. They are, in the absence of more definitive knowledge, interim and somewhat speculative conclusions. *These numbers should not be quoted without full acknowledgement of their qualified and tentative nature. They are not "predictions" in the sense of weather forecasts. They are estimates of the plausible ranges of effects, based on a combination of model results and scientific judgement.*

The values presented in Tables 5.2 and 5.3 are estimates of the departure from normal for surface air temperature for the NRC (1985) "baseline" case of 180 million tonne of smoke injected with uniform density between 0 and 9 km altitude. Estimates are made for three time intervals: "acute", meaning the first few weeks after smoke injection, with emphasis on the most extreme effects in space and time (many areas could have lesser or less prolonged effects); "intermediate", meaning the first one to approximately six months; and "chronic", meaning one to several years after the nuclear war. The ranges shown are fairly subjective estimates of confidence limits. They were purposely chosen to be large so that there would be a high probability that the temperature changes would fall within the range given.

The definitions used in the tables for "continental interiors" and "coastal areas" have been left intentionally vague because of inherent uncertainties. As a general guide, a place is effectively "continental" and free from oceanic influences if it normally has a large diurnal range in temperature. This is clearly not just a function of distance from the coast, but also of topography and prevailing wind strengths and direction. As such, the appropriate designation for a particular site can change with the seasons, and could be very different in a perturbed atmosphere. "Small islands" may be loosely defined as islands small enough for land-sea breeze systems to penetrate effectively to the interior. Very small islands have essentially an oceanic climate.

Temporal and spatial variability about these average changes is also of interest. There is little evidence on this issue, but some broad principles can be suggested. Firstly, in the acute phase, initial patchiness of the smoke clouds would induce a large variability in the temperature and precipitation changes. Qualitative estimates could be derived for specific scenarios from a daily series of maps such as that shown in Figure 5.22. Secondly, in the intermediate and chronic phases, when there is greatly reduced patchiness of the smoke and dust cover, the day-to-day variability in temperatures would probably become less than in the natural atmosphere, except perhaps in coastal zones. The reduced variability would be the result of a reduced diurnal cycle, generally less synoptic variability under a more stable thermal stratification, and a smoke veil that would be far more uniform than the normal patchiness of natural cloud cover. Coastal zones are a possible exception since on-shore winds could bring much warmer air from over the oceans, while off-shore winds could bring cold air from the continental interiors. More confident predictions of likely variability in coastal areas must await further modelling.

The tables are loosely based on injections that would be approximately uniform with height. Injections at lower altitudes would tend to decrease the magnitude of the changes given; higher injections would tend to increase them. For late spring or summer injections, the effects of differences in injection height would be fairly small because lofting of the smoke would

TABLE 5.2.
TEMPERATURE ANOMALIES IN °C FOR SMOKE INJECTIONS AS DEFINED
IN THE NRC (1985) BASELINE CASE AND OCCURRING IN THE
NORTHERN HEMISPHERE DURING LATE SPRING OR SUMMER. THE
VALUES OF THE ANOMALIES MUST BE INTERPRETED IN THE
CONTEXT OF THE DISCUSSION IN THE TEXT

Region	Acute (first few weeks)	Intermediate (1–6 months)	Chronic[b] (first few years)
Northern midlatitude continental interiors	− 15 to − 35 when under dense smoke[a]	− 5 to − 30	0 to − 10
Northern Hemisphere sea surface[b] (ice free)	0 to − 1	− 1 to − 3	0 to − 4
Northern Hemisphere coastal areas[b]	very variable. 0 to − 5 unless off-shore wind when − 15 to − 35	very variable, − 1 to − 5 unless off-shore wind when − 5 to − 30	variable. 0 to − 5
Northern Hemisphere and tropical small islands[b]	0 to − 5	0 to − 5	0 to − 5
Tropical continental interiors	0 to − 15	0 to − 15	0 to − 5
Southern midlatitude continental interiors	initial 0 to + 5, then 0 to − 10 in patches	0 to − 15	0 to − 5
Southern Hemisphere sea surface[b] (ice free)	0	0 to − 2	0 to − 4
Southern midlatitude coastal areas	0	0 to − 15 in off-shore winds	0 to − 5
Southern Hemisphere small islands	0	0 to − 5	0 to − 5

Footnotes:
[a] "Dense smoke" refers to smoke clouds of absorption optical depth of the order of 2 or greater, staying overhead for several days.

[b] These values are climatological average estimates. Local anomalies may exceed these limits, especially due to changes in oceanic behaviour such as upwelling or El Niño-type anomalous situations.

TABLE 5.3.

TEMPERATURE ANOMALIES IN °C FOR SMOKE INJECTIONS AS DEFINED
IN THE NRC (1985) BASELINE CASE AND OCCURING DURING
NORTHERN HEMISPHERE WINTER. THE VALUES OF THE
ANOMALIES MUST BE INTERPRETED IN THE CONTEXT OF THE
DISCUSSION IN THE TEXT[c]

Region	Acute (first few weeks)	Intermediate (1–6 months)	Chronic[b] (first few years)
Northern midlatitude continental interiors	0 to − 20 when under dense smoke[a]	0 to − 15	0 to − 5
Northern Hemisphere sea surface[b] (ice free)	0	0 to − 2	0 to − 3
Northern Hemisphere coastal areas[b]	very variable, 0 to − 5 unless off-shore wind when 0 to − 20	very variable, 0 to − 5 unless off-shore wind when 0 to − 15	0 to − 3
Northern Hemisphere and tropical small islands[b]	0 to − 5	0 to − 5	0 to − 5
Tropical continental interiors	0 to − 15	0 to − 5	0 to − 3
Southern midlatitude continental interiors	0	0 to − 10	0 to − 5
Southern Hemisphere sea surface[b] (ice free)	0	0 to − 1	0 to − 1
Southern midlatitude coastal areas	0	0 to − 10 in off-shore winds	0 to − 5
Southern Hemisphere small islands	0	0 to − 5	0 to − 5

Footnotes:

[a] "Dense smoke" refers to smoke clouds of absorption optical depth of the order of 2 or greater, staying overhead for several days.

[b] These values are climatological average estimates. Local anomalies may exceed these limits, especially due to changes in oceanic behaviour such as upwelling or El Niño-type anomalies.

[c] These values allow for a considerable range of variation in smoke removal rates in the Northern Hemisphere winter atmosphere. More rapid removal rates would lead to negligible effects on the intermediate and chronic time scales, less rapid removal to upper limits for effects as indicated here.

tend to wipe out any differences in injection height. For winter injections, however, lower injection heights might significantly reduce the estimated changes, while higher injections might increase the longevity of the effects.

Injections of different amounts of smoke than used in developing the estimates in Table 5.2 and 5.3 would change the estimated responses. However, as yet there has not been as detailed a study of the possible range of effects as for the case given by the tables, but, in recognizing that uncertainties are large, some inferences can be made. For example, injection of about one-third as much smoke (i.e., assuming an injection of smoke containing about 10 million tonne of elemental carbon) in the spring, summer, or autumn would produce shorter and more patchy effects in the acute stage and intermediate effects more like the chronic effects given in Table 5.2; chronic effects would tend to zero. Winter injections of such amounts would probably have only rather small effects in the acute stage that would disappear relatively quickly. Injection of about three times as much smoke (i.e., smoke containing about 100 million tonne of elemental carbon) would, on the other hand, induce effects that would be more extensive and longer lasting. The acute stage in the Northern Hemisphere would not be much worse in mid-continental regions, but the area affected would be larger and the effects would last months rather than weeks. The intermediate effects in the Southern Hemisphere would be similar to those in the Northern Hemisphere and the chronic phase would be more severe and longer-lasting in both hemispheres. As these examples indicate, the effects would not be linear in smoke levels, and thus caution must be exercised in interpolating or extrapolating the data.

It should also be re-iterated that the information given in Tables 5.2 and 5.3 should not be construed as "predictive". The values are offered as guidelines based on the currently available model results and extrapolation from current knowledge of how the atmosphere works. They should be used and quoted only within these constraints.

Environmental Consequences of Nuclear War Volume I:
Physical and Atmospheric Effects
A. B. Pittock, T. P. Ackerman, P. J. Crutzen,
M. C. MacCracken, C. S. Shapiro and R. P. Turco
© 1986 SCOPE. Published by John Wiley & Sons Ltd

CHAPTER 6

Nuclear and Post-Nuclear Chemical Pollutants and Perturbations

6.1 INTRODUCTION

At the time of the 1975 NAS report on the long-term global effects of nuclear war, the major issue was the depletion of stratospheric ozone resulting
from nitrogen oxides formed in fireballs and transported by them into the
stratosphere (Foley and Ruderman, 1973; Johnston et al., 1973; Chang and
Duewer, 1973; Hampson, 1974). The basis for consideration of this effect
was growing theoretical understanding of the importance of nitrogen oxides in determining the stratospheric ozone abundance and concern for the
environmental effects of the nitrogen oxides formed in the engines of aircraft flying in the lower stratosphere (Crutzen, 1970, 1971; Johnston, 1971).
Recently, Crutzen and Birks (1982) suggested other potential impacts of
nuclear war on large-scale atmospheric chemistry and estimated the quantities of smoke and gaseous emissions that could arise from fires ignited by
nuclear explosions. They also suggested that the resulting fires would supply nitrogen oxides, hydrocarbons, and carbon monoxide to the lower 10
or 12 km of the atmosphere that could, under sunlit conditions, result in
widespread ozone and oxidant production in the lower troposphere by the
processes that are known to generate urban photochemical smog.

   In this chapter, some estimates are presented of the quantities of gaseous
and particulate effluents that could be emitted into the atmosphere by the
fires ignited during a nuclear war, and the potential for developing harmful
levels of these materials. The emission estimates are based on the nuclear
fire properties described in Chapter 3. The changes in the concentrations
of atmospheric species that may occur as a result of subsequent photochemical reactions are also considered. Perturbations to stratospheric and
tropospheric chemistry are discussed separately, because of the independent nature of the chemical effects of stratospheric NO_x (produced by fireballs) and gaseous species emitted by fires into the troposphere. Chemical
models of the unperturbed atmosphere are used to estimate these changes.
The assessment could be carried out with greater confidence if the numbers and yields of weapon's detonations were precisely known and the

potential effects of dust and smoke on atmospheric chemistry, dynamics and solar flux could be accurately calculated. As indicated in Chapter 5, however, such projections are extremely difficult to develop. Accordingly, for both the stratosphere and troposphere, there is a wide range of uncertainty in the estimates of photochemical effects. While some limited insight can be provided into what may occur in a perturbed atmosphere, better analyses will be needed in order to take into account the potentially significant changes in dynamics, temperature, and composition that could occur as a result of multiple nuclear explosions, fires, and smoke plumes.

Some of the major consequences of the predicted chemical changes arise from perturbations of the atmospheric ozone concentration. Stratospheric ozone depletion could, in the absence of thick smoke layers, lead to an increase in solar ultraviolet radiation at the Earth's surface sufficiently large to be noticeably harmful to man and the biosphere (NRC, 1984). High concentrations of ozone and other pollutants in the troposphere could be directly harmful to plants, and maybe humans as well. Both of these issues, and others related to the impact of more exotic atmospheric contaminants, are discussed below, and taken up again in Volume II.

6.2 EMISSIONS AND SHORT-TERM POLLUTANT CONCENTRATIONS FROM POST-NUCLEAR FIRES

During a nuclear war many chemical pollutants would be injected into the atmosphere. In Chapter 3, estimates were made of the potential areas of urban fires and quantities of combustible material that could burn as a result of several hundred megatons of nuclear explosions over urban and industrial centers. Although there are considerable uncertainties in estimating the quantities of materials that could burn under such circumstances, the studies by Turco et al. (1983a,b), Crutzen et al. (1984), and NRC (1985) indicate that they may amount to about 2000 to 5000 million tonne of cellulosic material and nearly 1000 million tonne of fossil fuels and fossil fuel-derived products. The flaming combustion of these materials could produce about 100 million tonne of sooty, absorbing smoke particles, which would cause substantial optical and meteorological perturbations in the global atmosphere, as described in Chapters 3, 4 and 5. Extensive smoldering of plastics and cellulosic materials could produce similar or even larger quantities of oily smoke particles that do not absorb sunlight as effectively. In addition, numerous gaseous pollutants could be created and dispersed by the fires. Nuclear explosions, in addition to forming nitrogen oxides as already noted, could disperse industrial chemicals directly into the environment from storage facilities. This section provides estimates of the potential releases of some of the gases that can play a role in the photochemistry of the atmosphere or that reach levels high enough to constitute a health hazard. For a discussion of potential health problems associated with the release of asbestos fibers into the atmosphere, the reader is referred to the NRC (1985) report and the more thorough discussion of Stephens and Birks (1985).

6.2.1 Smoke from Smoldering Combustion

The smoldering combustion of cellulosic materials can produce much more smoke than flaming combustion (McMahon and Tsoukalas, 1978; Bankston et al., 1981). Measured emission factors range between 3% and 20%. Various plastic materials likewise produce large concentrations of smoke by smoldering (Bankston et al., 1981). It is, therefore, clear that several hundred million tonne of smoke particles may be produced by smoldering combustion following nuclear attacks. This smoke would tend to stay at low altitudes, especially because it would be emitted well after the initial intense flaming phase, so that surface cooling and a stable near-surface temperature inversion might have been established. Some of the smoke particles would be in the supermicron range and would thus be deposited in the respiratory tracts of people. Gaseous byproducts and very fine aerosols would likewise be inhaled and absorbed by the lungs. The gaseous and condensed pollutants are likely to include potentially hazardous organic matter; for example, up to 100 ppm of polycyclic organic compounds (Hall and DeAngelis, 1980; McMahon and Tsoukalas, 1978).

6.2.2 Carbon Monoxide

A variety of measurements of the carbon monoxide yield from large fires in cellulosic materials, such as forest and other wildland fires (Crutzen et al., 1985; Greenberg et al., 1984), and also in real building fire situations (Treitman et al., 1980) indicate CO-to-CO_2 molar emission ratios of 12–15%. In the case of building fires, these ratios can be derived from the reported statistical distributions of elevated CO and CO_2 concentrations. Emission rates of about 100 g of CO per kg fuel have been determined by Muhlbaier (1981) for small-scale open biomass combustion, and by Quintière et al. (1982) in smoldering fires in closed compartments. Much higher CO emission ratios, however, are also possible for smoldering fires (Ives et al., 1972) and for the burning of damp forest fuels (Sandberg et al., 1975) and plastics (Terrill et al., 1978). Tewarson (1984) reports an average production from flaming burning of cellulosic materials of 6 g of CO per kg fuel for well-ventilated fires, increasing to 97 g of CO per kg fuel for mixed flaming/smoldering combustion. Here a yield of about 100 g of CO per kg fuel is adopted, after assigning the greatest weight to measurements taken in actual fire situations. This leads to a total emission of about 270 to 750 million tonne of CO from fires following a nuclear war, assuming that the total fuel consumed is within the range given by the Crutzen et al. (1984) value of 2700 million tonne and the NRC (1985) value of 7500 million tonne. With background atmospheric volume mixing ratios of CO varying between 50 ppv in the Southern Hemisphere and 150–200 ppbv at middle and high latitudes in the Northern Hemisphere (Seiler and Fishman, 1981), the total atmospheric mass of carbon monoxide is equal to about 500 million tonne, so that a substantial increase in ambient CO concentrations could occur, especially at middle latitudes of the Northern Hemisphere.

6.2.3 Hydrocarbons

Studies of the release of hydrocarbons from wildland fires (Crutzen et al., 1985; Greenberg et al., 1984) indicate methane to carbon dioxide release rate ratios of about 1%, or 5 g CH_4 per kg fuel. For nonmethane hydrocarbons, the average measured ratio was about 1.3%, or 6.5 g carbon per kg fuel. Both ratios are uncertain by about 30%. The composition of nonmethane hydrocarbons was about 45% alkenes (mostly C_2H_4), 25% alkanes (mainly C_2H_6 and C_3H_8), 13% aromatics (especially benzene and toluene), 6% acetylene, and the rest various oxygenated compounds. In a number of field fires in the U.S., total hydrocarbon emissions varied between 1.4% and 5.4% (7 to 27 g hydrocarbons per kg fuel) (McMahon, 1983). According to a compilation by the U.S. Environmental Protection Agency (EPA, 1972), methane production yields for various categories of fuels were as follows: municipal refuse 15, automobile components 15, horticultural refuse 10, and wood 2 g CH_4 per kg fuel. Wood-burning fireplaces produce only 1.5 g hydrocarbons per kg fuel (Muhlbaier, 1981). Fire tests performed with room furnishings produced typically 5–10 g unsaturated and 5–15 g saturated hydrocarbons per kg fuel (Ives et al., 1972). By assigning the greatest weight to those measurements which were made in large fires, emission factors of about 5–10 g CH_4 and 5–15 g nonmethane hydrocarbons per kg fuel may tentatively be adopted. This would lead to the emission of about 14 to 75 million tonne CH_4 and 14 to 110 million tonne nonmethane hydrocarbons, using the Crutzen et al. (1984) and NRC (1985) estimates of total fuel consumed.

The addition of this amount of methane to the atmosphere is negligible compared to the total of 5000 million tonne that is normally present in the atmosphere (Khalil and Rasmussen, 1983). The emissions of nonmethane hydrocarbons would, however, increase their atmospheric abundances by large factors. For instance, in the case of ethane (C_2H_6), which is currently present in the atmosphere at the ppb level, the global increase could be a factor of 2. For other more reactive compounds, the increase could be much larger, in some cases by orders of magnitude.

These emission estimates do not include the potentially large releases of hydrocarbons from explosions and fires in above ground fossil fuel deposits and natural gas distribution systems. It is also known that about 50% of spilled oil may volatilize within a few days (Jernelöv and Lindén, 1981). Such events may be rather common in a nuclear war, so that this volatilization could very well release on the order of 100 million tonne of reactive alkane hydrocarbons to the atmosphere. The effects of deliberate attacks on natural gas production wells, leading to blowouts, could be even more serious (Crutzen and Birks, 1982), but will not be taken into account in the following analyses.

6.2.4 Oxides of Nitrogen

It has been estimated that 10^{32} molecules of NO are formed per megaton explosion yield (Foley and Ruderman, 1973; Johnston et al, 1973). In a 6000 Mt nuclear war (Chapter 2), this mechanism would produce 30 million tonne of NO. Large-scale savanna fires (Crutzen et al., 1984) give NO-to-CO_2 molar emission rate ratios of about 2×10^{-3}. Laboratory experiments with various types of biofuels have given average molar ratios of 2.5×10^{-3} (Clements and McMahon, 1980). Similar or somewhat smaller values were compiled by EPA (1972) for the open burning of municipal refuse, automobile components, and horticultural refuse. Adopting an average NO_x-to-CO_2 molar emission rate of 2×10^{-3} (2 g NO per kg fuel), the production of NO from fires would be about 5 to 14 million tonne. The total amount of NO produced in the fireballs and in the urban and industrial fires would, therefore, add up to about 35 to 45 million tonne, which is roughly equal to the worldwide, annual production of NO from automotive and industrial combustion processes. This emission may be an underestimate, because it does not take into account the potential production of NO in hot mass fires.

6.2.5 Local Concentrations of Toxic Compounds

Emissions of CO, hydrocarbons, nitrogen oxides, and other primary emitted compounds, when distributed through large portions of the atmosphere, would not lead to concentrations that are lethal or hazardous to health. Of course, for survivors near local fire plumes, dangerous toxic levels may exist. Hazardous levels of primary pollutants might also be reached if, as a consequence of the absorption of sunlight high in the atmosphere, strong temperature inversions were to develop over the continents, particularly in river valleys and lowland areas, while smoldering combustion is still taking place. As an example, assume that strong temperature inversions can limit the vertical mixing of smoldering fire effluents to the lowest 200 m of the atmosphere; that smoke from smoldering fires over an area 5 km across mixes with background air flowing across the area at 5 m/s; and that the fires have a smoldering time of 3 days, an average fuel loading of about 10 kg/m^2, and a smoke yield of 50 g/kg fuel. In this case, the average smoke density in the air flowing out of the smoldering city beneath the inversion would be equal to about 10 mg/m^3 more than the smoke density in air flowing into the city. Such a high smoke density would limit visibilities to about 100 m (Middleton, 1952). Of course, close to the ground in the air immediately leaving the smoldering urban centers, the visibility could be appreciably less. Assuming an emission rate of 100 g CO per kg fuel, the air flowing out of the city could contain about 20 ppmv of carbon monoxide more than the air flowing into the city. Such concentrations would be too low to cause acute health effects. Treitman et al. (1980) estimate that much higher smoke densities of 1 g/m^3 are required to cause immediate respiratory distress and about ten times higher concentrations of CO are required to cause acute health effects (Woolley and Fardell, 1982). On the other hand, the pollutant concentrations calculated using this extremely simplified model are high enough to warrant further consideration of this issue. For instance, the effects of multiple city smoldering in densely populated regions was not considered. The duration of exposure to such pollutants could also be important. Furthermore, fuel loadings and emission yields of CO and smoke may be higher than assumed in these calculations, and the simultaneous presence of a variety of other toxins should also be considered. Inside or in the immediate surroundings of the burning areas low to the ground, CO concentrations could also be much higher than those estimated above.

In many fire environments, CO is the most hazardous gas (Terrill et al., 1978). If this is true in the case considered, other gaseous compounds produced by fires should generally constitute lesser health hazards on larger scales, but this has not yet been adequately studied. Besides CO, perhaps the most significant pyrotoxic gases are acrolein and hydrochloric acid (Terrill et al., 1978; Treitman et al., 1980; Woolley and Fardell, 1982). The same studies indicate much less concern for direct human health effects from HCN (hydrogen cyanide) and NO_x, depending on the conditions of exposure.

The simultaneous occurrence of health problems due to heavy air pollution cannot, therefore, become a matter of concern on continental scales. However, regionally and locally, acute health effects could be much more serious, especially in connection with the special meteorological conditions that may develop as a consequence of large scale nuclear war. Synergistic effects due to the presence of many gaseous and particulate air pollutants could also lower the thresholds for severe health effects considerably (Ives et al., 1972). Potential effects on plants are discussed in Volume II. It is clear that more thorough analysis of potential effects is required.

6.2.6 Other Emissions and Effects

If the $HC\ell$-to-CO_2 emission ratio of about 1%, measured by Treitman et al. (1980) in building fires, is extrapolated to global nuclear war conditions, the total emission of $HC\ell$ (hydrochloric acid) from the war itself could amount to about 30 million tonne. The release of about 1% sulfur from fossil fuel burning—a level consistent with current statistics (Bolin and Cook, 1983)—may lead to the production of about 14 million tonne of sulfur as H_2SO_4 (sulfuric acid). Further, if all oxides of nitrogen from the nuclear-induced fires were converted to HNO_3 (nitric acid), an injection of 5–14 million tonne of nitrogen as HNO_3 would result. These acids are removed naturally by precipitation. After a nuclear war, if the removal of the added acids occurred over one month of normal rainfall, the pH of precipitation over the northern mid-latitudes could be lower than 4 (i.e., almost ten times or more acidic than present polluted rain. The possible formation of cold acid fogs in a thermally stable atmosphere and its effects on the biosphere might be another consequence of the outcome of a nuclear war to be considered in future studies.

Chemical releases from the targeting of industries may lead to local pollution of the water, soil, and atmosphere (Turco et al., 1983a,b; NRC, 1985). As an example, consider the case of chlorine storage. In the United Kingdom, there are about 100 storage containers that can each hold between 20 and 50 tonne of chlorine. These containers are located at water treatment plants, large power stations, and various industrial plants. In addition, there are approximately 10 larger installations where greater quantities of chlorine are stored. These facilities hold between 250 and 2000 tons per site in tanks holding up to 350 tons. Any release of this heavy gas into the environment could create locally severe conditions. Chlorine container failures can, however, also trigger intense fires that could carry the chlorine to higher altitudes, thereby limiting toxic effects to the near vicinity of the accident (J.P.H. Shaw, personal communication). Thus, evaluating potential effects is not straightforward.

The release of numerous organic chlorine compounds may likewise be of concern in and downwind of targeted cities. Turco et al. (1983a,b) have reported the storage of more than 30 million tonne of PCBs (polychlorinated biphenyls) in electrical equipment in the U.S. Turco et al. (1983a,b) and Crutzen et al. (1984) point to the possible production of chlorinated dioxins and dibenzofurans from the smoldering combustion of such chlorine-containing substances. These compounds, and many others that are common in industry, are also persistent in the environment, and can be carcinogenic and mutagenic as well as toxic. Accordingly, long-term pollution effects need to be evaluated from the unprecedented chemical releases in a nuclear war.

6.2.7 Summary of Air Pollution Effects

In conclusion, chemical releases from attacks on industries and cities could cause hazardous pollution levels on local and perhaps regional scales, especially in low ventilation areas near smoldering fires, in water, soil and air affected by chemical spills, and in regions exposed to persistent toxins in the form of gases or combined with smoke. Most of the chemical releases would be likely to occur in or near densely populated areas. Obviously, a more detailed analysis of the release and effects of dangerous substances is needed.

6.3 STRATOSPHERIC CHEMISTRY

6.3.1 Impact of Fireball Nitrogen Oxides on Stratospheric Ozone

Although the current ambient integrated ozone column abundance varies substantially with latitude and season, assessments of the response of stratospheric ozone to perturbations have usually been based on models in which altitude is the single spatial dimension (NRC, 1984). The complexity of stratospheric transport is, in this case, reduced to a specification of characteristic times for vertical diffusion as a function of altitude in a hemispheric average sense. The result is a model that approximates mid-latitude or global average conditions based on the assumption that horizontal mixing is much more rapid than vertical mixing (i.e., that it is instantaneous) in the stratosphere. Recognizing this limitation, the one-dimensional models are useful for comparing the importance of chemical processes in the stratosphere and for estimating the general magnitude of stratospheric response to various perturbations. An observed mid-latitude ozone profile (WMO, 1982) and the altitude profile of ozone from a current model are shown in Figure 6.1. The nonuniform vertical distribution arises from the interaction of chemical processes, which are driven chiefly by solar ultraviolet radiation, and transport, which is most significant in the lower stratosphere where the ultraviolet flux has been reduced by absorption at higher altitudes. In the ambient atmosphere, production of ozone by ultraviolet photolysis of molecular oxygen is balanced by several chemical recombination processes (NRC, 1984). The two reactions involving NO_x ($NO + NO_2$):

$$NO + O_3 \rightarrow NO_2 + O_2$$
$$NO_2 + O \rightarrow NO + O_2 \quad ,$$

are the most important (in current one-dimensional models) (e.g., Crutzen, 1970; NRC, 1984; Connell and Wuebbles, 1986). The remaining ozone destruction is distributed among processes involving chlorine radical species,

$HO_x (= OH + HO_2)$ radical species, and, to a much lesser extent, transport to the troposphere.

Most of the ambient stratospheric NO_x is produced by the reaction of excited atomic oxygen with N_2O, which is emitted at the surface by various combustion processes and soil bacteria and transported into the stratosphere. The total abundance of NO_x in the present stratosphere is $0.5-1.5 \times 10^{11}$ moles (compared to 7×10^{13} moles of O_3). In a cooling nuclear fireball, NO_x production results when the equilibrium dissociation reaction, $N_2 + O_2 = 2NO$, is rapidly quenched from high initial temperatures (≥ 2000 K), such that a high non-equilibrium abundance of NO remains in the rising and expanding nuclear fireball. Theoretical estimates of the production of NO_x in this environment have been discussed by Gilmore (1975) and by the NRC committees (NAS, 1975 and NRC, 1985). Approximately 1×10^{32} molecules of NO (about 5000 tonne) are produced per megaton of nuclear energy release, with an uncertainty of perhaps a factor of 2. Hence, 400–900 Mt of nuclear explosions would double the existing stratospheric NO_x abundance. In 10000 Mt global nuclear war scenarios considered in the past (NAS, 1975; Turco et al., 1983b; Chang and Wuebbles, 1983), with the majority of the

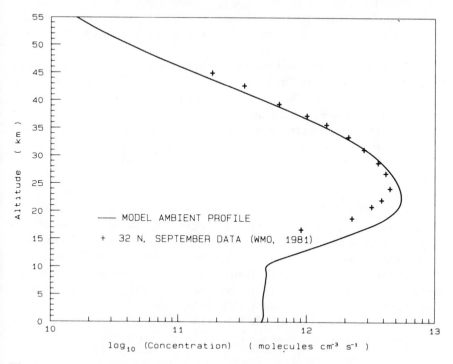

Figure 6.1.   Observed vertical ozone concentration profile compared with that predicted by a one-dimensional model (Connell and Wuebbles, 1986)

explosive energy in weapons with individual yields ≥ 0.5 Mt, stratospheric NO_x would be increased by 15–20 times over the natural background. The more recent scenarios, especially NRC (1985) and Crutzen and Birks (1982) indicate stratospheric NO_x injections that are several times less. The 5000 Mt baseline scenario of Turco et al. (1983a), which included some of the larger weapons that still exist in the arsenals, resulted in an NO_x injection between the high and low cases just mentioned.

Because the various stratospheric chemical species interact with each other as well as with ozone, their net effect on ozone is not a simple sum of the effect of each species calculated independently. At each altitude, the overhead burden of ozone also affects the solar flux in the photolytically active ultraviolet region. Hence, the local change in the ozone abundance depends on the altitude distribution of injected NO_x and the resultant change in the ozone profile.

The ozone depletion depends on the heights of injection of NO_x, and therefore on the top and bottom altitudes of the stabilized nuclear clouds. These heights vary with the explosive yield, as discussed in Chapter 1. At middle latitudes, it is expected that weapons of about 0.4–0.5 Mt would loft substantial amounts of NO_x above 17 km (Peterson, 1970; Foley and Ruderman, 1973; Glasstone and Dolan, 1977; NRC, 1985). The tropopause lies at about 11–13 km, and weapons as small as ~ 100 kt would inject NO_x into the lower stratosphere. However, in the model, NO_x injected below ~ 17 km results in a small net *production* of ozone mainly as a byproduct of the oxidation of methane. The efficiency of this process increases with higher concentrations of NO_x. Thus, the calculated perturbations of stratospheric ozone depend strongly on the assumed nuclear detonation yields, but less directly on the total megatonnage. This is clearly demonstrated in the case of the Ambio baseline scenario (used by Crutzen and Birks, 1982) in which 5740 Mt of low-yield weapons produced essentially no net change in the stratospheric ozone column, although the vertical distribution of ozone was modified.

Since the 1975 NAS report, as a result of improving knowledge of relevant laboratory chemical kinetics, the importance of NO_x in calculating ozone depletion, and the crossover altitude between ozone production and ozone destruction, have varied. For massive injections of NO_x high into the stratosphere (e.g., from 10000 Mt in weapons of greater than 1 Mt each), the calculated ozone change has not been sensitive to the changes in kinetics parameters over the last decade (NAS, 1975; Duewer et al., 1978; Crutzen and Birks, 1982; Turco et al., 1983a,b; NRC, 1985). For smaller injections, and especially for scenarios assuming individual weapon yields smaller than 1 Mt, the calculated ozone depletion is more sensitive to chemical reaction rate coefficients whose measured values have varied with improvements in laboratory techniques. The simulations presented in this chapter were obtained with atmospheric models using current estimates of chemical reaction rates.

Even when an assumed scenario includes weapons of sufficient yield to penetrate the tropopause, the projected ozone depletion depends on the distribution of injection heights. A number of major factors come into play. First, since the ozone-dissociating ultraviolet solar flux intensity increases and air density decreases with altitude, the density of atomic oxygen—in steady state with respect to production by ozone photolysis and loss by combination with molecular oxygen—increases also. This increases the efficiency per NO_x molecule of the ozone-destroying NO_x chemical reactions by increasing the rate of the reaction of NO_2 with O, while NO_2 photolysis (an ozone neutral reaction) is unaffected. Second, about two-thirds of the stratospheric ozone column lies below about 25 km, so that large relative changes in the upper stratosphere can make a smaller contribution to the total change in the ozone column than smaller relative changes near the ozone maximum. Thirdly, methane oxidation reactions come into play that can increase ozone. Finally, if ozone is diminished above its concentration maximum at about 23 km, the subsequent increase in the ultraviolet flux at lower altitudes increases the rate of molecular oxygen photolysis and ozone production. Dissociation of oxygen by solar ultraviolet produces oxygen atoms and subsequently ozone, so the increase in oxygen photolysis can partially compensate for the decrease in the upper stratospheric ozone column.

Figure 6.2 shows the vertical distribution of NO_x injection that would result from three nuclear exchange scenarios. Since the distribution of injected NO_x is calculated from the yields of the individual weapons, widely different distributions can result from different scenarios involving the same total megatonnage. (See Chapter 2 for details of these scenarios.) The NRC 6500 Mt and the Ambio 5740 Mt scenarios are based on estimates of near-term nuclear arsenals assuming continuing trends to smaller warheads (however, see Chapter 2). The potential effect of including higher yield weapons, some of which may still be present in the arsenals, is illustrated by the Knox (1983) scenario, which included several 20 Mt warheads that would loft NO_x into the upper stratosphere. The NRC (1985) report also considered a case in which large yield weapons were assumed to be detonated over very "hard" targets. Although most of the larger warheads have been or are expected to be retired in the near future, some may remain and others could be added in the future.

While calculated ozone perturbations are very sensitive to the vertical distribution of NO_x injections, and therefore the assumed yields of individual weapons, smoke emissions are generally much less sensitive. Accordingly, calculated ozone changes may vary widely among scenarios estimated to produce roughly equivalent quantities of smoke. For the nuclear scenarios

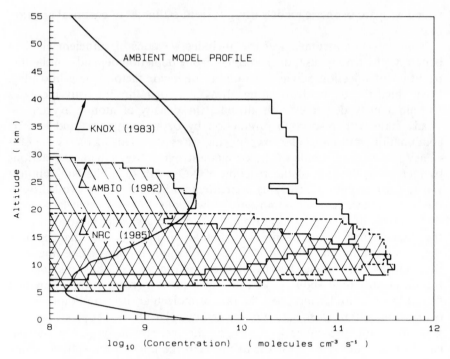

Figure 6.2.   NO$_x$ input from fireballs

described above, the predicted average hemispheric-scale ozone reductions are shown in Figure 6.3. The differences arise mainly from variations in the number of explosions with energy yields between 0.5 and 20 Mt (which deposit NO$_x$ into the model stratosphere). The Knox (1983) scenario, which includes some high yield weapons, produces a maximum ozone column depletion of 44% after 6 months. The NRC 6500 Mt scenario excludes weapons larger than 1.5 Mt, but corresponds to a somewhat higher average injection height than the Ambio scenario, which contained many small weapons. The NRC scenario produced an ozone column decrease of about 17% and the Ambio scenario produced a 4% maximum decrease at a somewhat later time. The greater time delay in the Ambio case occurs because the injected NO$_x$ is slowly transported and mixed upward to the region where it can affect the ozone. In general, maximum ozone depletions (in an atmosphere unperturbed by smoke) are found to range up to perhaps 50% for scenarios of ~5000 Mt including high yield weapons; the peak depletion is reached in 6 to 12 months, and a sustained depletion of 10% or more can persist for 3 to 6 years. On the other hand, with only low yield weapons, the peak ozone depletion may never reach even 10%. The 5000 Mt baseline model of Turco et al. (1983a,b) predicted peak ozone depletions of about 20–30%,

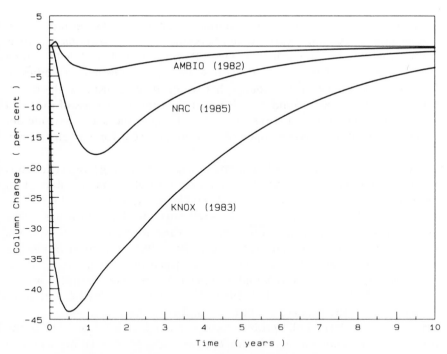

Figure 6.3. Total column ozone change vs time computed with the one-dimensional model, of Connell and Wuebbles (1985)

representing the impact of mixed high and low yield weapons. Izrael et al. (1983) have also considered the effects of nuclear-generated NO_x on stratospheric O_3, and estimated peak depletions of 30 to 50% consistent with the above discussion.

Several factors that can lead to much larger ozone depletions have not been considered in these studies, including the likely induced changes in atmospheric dynamics and temperature, and possible reactions of O_3 with injected aerosols. These will be discussed in more detail in the following section. Moreover, the instantaneous meridional and longitudinal spreading assumed in one-dimensional models very probably underestimates potential ozone reduction for the first few months in the northern mid-latitude zone, where the injected NO_x may tend to remain concentrated.

In addition to the problem of long-term hemispheric-scale ozone depletion, there may also be deep, transient, short-term regional-scale depletions. These could result from detonation of many large weapons within a confined area, such as an ICBM field, leading to a local NO_x concentration hundreds of times greater than the ambient value. Before dispersive processes dilute the injected NO_x, ozone in the affected region of the atmosphere could essentially be completely removed. Luther (1983) projected ozone column decreases of up to 70% after a few hours and persisting for several days—an "ozone hole"—assuming various numbers and sizes of weapons for areas characteristic of an attack by the U.S.S.R. on a U.S. ICBM field. These calculations ignored the effect of varying wind directions and velocities in different atmospheric layers (shear) that would tend to disperse the ozone hole as seen from the ground and also neglected the great mass of dust and water that would accompany the NO_x. Accordingly, the derived ozone depletion is probably an overestimate of what might occur for a given affected location.

The increases in ultraviolet radiation at the ground arising from reductions in total ozone depend on latitude and season (as well as on any absorption and scattering by intervening clouds of smoke, dust, and ice). The biological impacts of UV-B radiation (~ 280–320 nm) also depend on the action spectra (absorption times quantum yield) for various physiological responses to the radiation in individual organisms (NRC, 1984). For most organisms the responses are uncertain, although some effects on certain crops, insects and marine micro-organisms would be expected in light of recent laboratory studies (see Volume II and NRC, 1984). In humans, increased UV-B accumulated over many years can lead to a number of disorders, including skin cancer (NRC, 1984). While probably not a major health factor itself in the aftermath of a major nuclear exchange, enhanced UV-B radiation would be another factor degrading the post-war environmental state.

Nachtwey and Rundel (1982) have discussed the calculation of changes in biologically active ultraviolet radiation given a particular percentage reduction in total ozone. Based on that discussion, UV-B increases can be estimated for the scenarios discussed earlier. For a 40 to 50% ozone decrease at 30° N, a factor of five increase in biologically active, wavelength-integrated UV would result. At 30° with a 10% ozone decrease, biologically active UV would increase by about 25%. With all low yield weapons scenarios, a lower limit for the average change in surface UV-B radiation could essentially be zero (also see below). On the other hand, Luther (1983) calculated that a 70% ozone reduction in ozone holes would increase the surface flux of 300 nm radiation by about a factor of twelve. Accordingly, ultraviolet radiation intensities could be enhanced in some regions at certain times, by up to an order of magnitude, and over several years, on average, by up to a factor of around twice the estimated percentage ozone depletion.

6.3.2 Stratospheric Chemistry in an Atmosphere Perturbed by Smoke

The injection of nitrogen oxides from rising nuclear fireballs is only one of the potential influences of a nuclear war on the chemistry of the stratosphere, particularly on its ozone concentration. If nuclear explosions take place near the surface, substantial amounts of soil dust (and possibly soil carbon and water vapor) could be injected into the stratosphere (see Chapter 3). Reactions with dust could lower ozone concentrations. Smoke and gases generated in fires ignited by the nuclear exchange, particularly from very intense city fires, could also be lofted to stratospheric altitudes (see Chapters 3 and 4). In addition to direct injections into the stratosphere, perturbations of atmospheric temperatures and circulation could loft substantial quantities of smoke and debris from the troposphere to the stratosphere (see Crutzen et al., 1984, and Chapter 5) and could dynamically redistribute stratospheric ozone and other trace constituents. Other potential nuclear war induced perturbations include the introduction of additional compounds (e.g., soot, HCl and H_2O) that affect chemical kinetics, changes in temperature that alter chemical reaction rates, and a lowering of the tropopause and changes in the dynamic coupling between the troposphere and stratosphere (via gravity and planetary waves) that determine stratospheric residence times.

There are two fundamental problems to be dealt with in order to calculate these effects. First, a reasonable estimate must be developed of what materials may be injected, and how the atmosphere could be perturbed. Second, since changes in the ozone concentration would induce temperature changes that in turn could alter the atmospheric response, the capability must be developed for interactively and simultaneously calculating the effects of all of these processes and perturbations as the atmosphere evolves following a nuclear conflict. Because neither of these problems has been completely solved, it is only possible to speculate on some of the possibilities. In doing so, it is assumed here that the smoke injection is relatively large and occurs in summer, thereby inducing a large perturbation to the atmospheric circulation (see Chapter 5). The effects of a winter war would likely be less dramatic, but could still be significant.

Smoke injected into the middle and upper troposphere could dramatically increase the normally slow vertical mixing in the stratospheric layers above the smoke and could lift the lower stratospheric air mass upward and towards the equator ahead of the warming smoke. The enhanced vertical mixing could bring ozone-destructive gases higher into the ozone layer than would otherwise occur, probably leading to deeper ozone reductions. The displacement of the ozone reservoir in the lower stratosphere to higher altitudes and toward the equator would also probably lead to further ozone depletions.

The strong upward air movement induced in regions of the Northern Hemisphere during the first weeks to months would be balanced by a large scale, slow downward motion in the Southern Hemisphere, which might well allow transport of stratospheric air having relatively high ozone concentrations to the surface (for example, under some current situations ozone concentration levels can occur briefly as a result of thunderstorm-induced downdrafts or other intense vertical mixing). Once the smoke spreads to the Southern Hemisphere, however, the Hadley circulation may become less intense (based on current GCM results, see Chapter 5), the upper troposphere may be stabilized and deepen the stratosphere, and stratospheric contributions to tropospheric ozone concentrations might decrease. The induced movement of air to the Southern Hemisphere may, however, also carry nitrogen oxides injected in the Northern Hemisphere, thereby leading to greater ozone reduction in equatorial and southern latitudes than would occur if the circulation were not perturbed. Although mixing of the NO_x into the Southern Hemisphere might somewhat reduce the ozone reduction in the Northern Hemisphere, this enhanced horizontal spreading of the NO_x is likely to lead to a greater average ozone reduction worldwide. Whether the surface flux of ultraviolet radiation would increase, however, would also depend on the concentrations of other radiative absorbers such as smoke particles.

Solar heating of the smoke lofted into the stratosphere by induced circulation changes could also produce substantial increases in stratospheric temperatures, in some cases by 20 to 50 °C or more, depending on the season. Such temperature changes alone would cause substantial reductions in the ozone concentration. For example, when the NRC (1985) estimates of middle latitude temperature changes following a nuclear exchange are assumed (which basically create an isothermal lower stratosphere at a temperature of about 240 K) the vertical ozone column is found to be reduced by 18% in the absence of any other effects. However, note the concurrent presence of smoke, which would block a fraction of the enhanced solar UV-B radiation. On the other hand, the smoke would also lead to absorption of shorter wavelength UV radiation, which is active in producing ozone by molecular oxygen dissociation (Crutzen et al., 1984). This may be another important factor leading to ozone depletion. Warming of the stratosphere and upper troposphere by smoke is also projected to extend the atmospheric residence time of stratospheric constituents (see Chapter 5), thereby extending the recovery process from months to perhaps a few years.

In addition to absorbing UV radiation, ozone is an active constituent in determining the visible and infrared radiation balances of the atmosphere. Thus, changes in ozone can affect temperatures and circulation patterns (NRC, 1985). Solar absorption by ozone in the middle and upper stratosphere, for example, may provide a stable temperature inversion at 35–40 km, preventing smoke particles from rising beyond this level (Malone et al., 1986). Ozone infrared cooling of upper stratospheric air normally contributes to the wintertime descent of that air in polar regions; this sinking motion comprises one of the cleansing processes for the stratosphere. The infrared emission of smoke aerosols during the polar night may augment this cooling and thereby induce downward movement of the smoke to levels where it may be more subject to removal by precipitation scavenging.

Oxidation of smoke by ozone (and other reactive species, perhaps especially OH) may provide an important long-term removal mechanism for injected smoke particles. Oxidation of the nonabsorbing hydrocarbons comprising smoke at ozone levels typical of the unperturbed stratosphere could be relatively rapid, although at the temperatures expected in the perturbed stratosphere, and with depleted ozone, this possibility requires a more thorough review. The oxidation of light-absorbing graphitic soot should be even slower (R. Fristrom, private communication). At present, there is no quantitative evidence to suggest that the physical and optical properties of the injected smoke would be significantly altered over short periods as a result of chemical attack, but such an effect cannot be discounted.

This analysis of possible effects is certainly not complete and must be acknowledged as uncertain. It will require considerable research to answer the most important questions. Quite clearly, however, the changes in stratospheric chemistry that have been proposed could have important global influences; at this time, the effects cannot be accurately quantified. It should be recognized, however, that previous calculations of ozone reductions based on a smoke free atmosphere (as described in Section 6.3.1) probably do not represent the conditions likely to prevail after a nuclear war. Larger, longer lasting and more widespread reductions in stratospheric ozone would now seem to be a possibility.

6.4 TROPOSPHERIC EFFECTS

The chemistry of the troposphere is qualitatively different from that of the stratosphere. Most gaseous species emitted at the Earth's surface are removed from the air in a relatively short time by photochemical reactions and by a number of dry and wet physical scavenging and removal processes. The photochemistry of the troposphere is driven by chemical radicals, of which OH is most generally reactive. Ozone is produced as a byproduct of hydrocarbon decomposition via chain reaction mechanisms involving peroxy radicals and NO. Species that are relatively inert chemically, such as N_2O, CH_4, H_2O, COS and many fluorocarbons are transported upward through the tropopause in substantial quantities.

Crutzen and Birks (1982) and Birks and Staehelin (1985) have suggested a number of chemical changes that might occur in the troposphere as a result of nuclear war. Incorporating species emitted by fires into existing models of the atmosphere has been used to provide initial estimates of the expected perturbations. Major changes in atmospheric structure and climate caused by the smoke emissions could, of course alter current tropospheric processes and characteristics such as vertical mixing, temperature profile, and wet and dry deposition processes, and should eventually be factored into these studies. In such a perturbed state, atmospheric processes currently important, and therefore reasonably treated by models, might lose importance, while new processes not properly treated could be dominant. For example, the interaction of gaseous species with aerosols in a smoky atmosphere could control the overall composition of the troposphere, unlike present conditions (Birks and Staehelin, 1985).

The photochemical oxidation of hydrocarbons released to the troposphere by fires, in the presence of sunlight and sufficient quantities of NO would lead to the production of ozone. Crutzen and Birks (1982) have shown that the oxidation of one molecule of CO can yield one molecule of ozone, while the yield from ethane can be as much as six ozone molecules. The production of NO from nuclear explosions and fires is projected to be so large that the NO concentration should not be a limiting factor for ozone formation. The hydrocarbons and NO_x would also be mixed together in polluted air masses. By implication, the formation of a few hundred million tonne of tropospheric ozone from the oxidation of the hydrocarbon emissions discussed earlier could occur within a week or so. The average ambient mixing ratio of surface ozone is about 50 ppbv and the total tropospheric ozone burden is roughly four hundred million tonne. Therefore, in principle, the potential exists for noticeable ozone enhancements over large regions of the northern mid-latitudes. However, such an outcome requires sunlight sufficient to drive the necessary photochemical processes. The light intensity, in turn, depends on the optical properties, distribution, and residence time of the smoke injected by the fires. If a substantial quantity of smoke is not promptly removed, it is likely that NO and NO_2 would be transformed by reactions not requiring sunlight and deposited on particles or on the surface as follows:

$$NO + O_3 \rightarrow NO_2 + O_2$$
$$NO_2 + O_3 \rightarrow NO_3 + O_2$$
$$NO_3 + NO_2 + M \rightarrow N_2O_5 + M$$
$$N_2O_5 + H_2O \ (aq) \rightarrow 2HNO_3$$
$$HNO_3 \rightarrow cloud \ droplets, \ aerosol$$
$$aerosol \rightarrow deposition$$

Hence, photochemical ozone formation would be less likely to occur after the eventual removal of the smoke from the atmosphere.

A multi-dimensional model including coupled treatments of dynamics, radiation, and homogeneous and heterogeneous chemistry would be necessary to properly investigate this problem. Without such a model, and lacking the detailed experimental information needed to construct it, the potential tropospheric effects can only be sketched. An analysis of the problem has been carried out by Penner (1983), who concluded that large tropospheric ozone increases covering wide areas would be an unlikely outcome of a nuclear war. The surface ozone effects expected based on present calculations would be negligible against the direct effects of nuclear warfare, except perhaps in localized areas where enough pollutants and sunlight were present to generate large ozone concentrations.

Many natural and anthropogenic gases are removed from the atmosphere by reactions with hydroxyl radicals (OH), formed by photochemical processes. With large amounts of smoke in the atmosphere, the necessary sunlight to drive these reactions might not be available; the smoke itself could be a strong sink for hydroxyl radicals. Under these circumstances, there would likely be a buildup of undesirable gases that are now present in the atmosphere only at very low concentrations (Birks and Staehelin, 1985). No quantitative evaluation has, however, been presented and this problem may also turn out to be of secondary importance. For instance, the emission of H_2S (hydrogen sulfide) over the continents is about equal to 50 million tonne S per year under normal conditions. If this much H_2S were spread over the Northern Hemisphere mid-latitudes ($10^{14} m^2$) and mixed through a depth of a few kilometers, there could be a build-up of H_2S to about 10 ppbv over one month, assuming no removal. Although this level is quite high compared to normal, it is still probably not high enough to cause major health problems compared to the direct effects of a nuclear war.

6.5 SUMMARY

The potential impacts of a nuclear war on atmospheric chemistry have been investigated for more than a decade. During this period, a better recognition of the effects that may be most important has developed, although we have no assurances that all of the crucial issues have been investigated.

The potentially significant chemical consequences discussed in this chapter are summarized in Table 6.1. The impact of these changes on biological systems is discussed in Volume II. In the absence of fires and smoke emissions, the reduction in stratospheric ozone and consequent increases in surface ultraviolet radiation would likely be the most important effect of nitrogen oxides generated by nuclear explosions. With smoke in the stratosphere, changes in circulation and temperatures, as well as interactions between soot particles, solar UV radiation, and ozone, could lead to strongly enhanced ozone destruction. In the presence of large quantities of smoke particles, the variety of gaseous chemical emissions could also result in climatic changes through alterations of stratospheric composition (such changes have not yet been considered in the model studies reported in Chapter 5). These effects could delay the recovery of the atmosphere significantly. Further detailed analysis of these interactions is required. On the other hand, the formation of high ozone concentrations by photochemical reactions in the troposphere is not now considered to be an important problem.

On local and regional scales, the most important potential chemical effects

TABLE 6.1.
SUMMARY OF POTENTIALLY SIGNIFICANT ATMOSPHERIC
CHEMISTRY EFFECTS

Time period after explosions	Spatial coverage	Effect
Hours to Days	Close to fires	High levels of CO and pyrotoxins Local releases of various hazardous pollutants from chemical factories and storage facilities
Days to Weeks	In unventilated areas such as river valleys and other low areas near smoldering fires; high values require strong temperature inversions	High particle and gas concentrations: Particles: 0.01–0.1 g/m^3 CO: 15–150 ppmv $HC\ell$: ≈ 1–10 ppmv Aldehydes: ≈ 0.1–1 ppmv
	In limited regions surrounding areas of many high yield explosions	Large increases in UV-B radiation in "ozone holes" if not shielded by smoke
Weeks	Northern Hemisphere mid-latitudes	Precipitation acidities with pH on the order of 4, assuming no change in other emissions Tropospheric ozone concentrations could increase at edges of smoke clouds, but unlikely to be very significant
Months	Northern Hemisphere	Several times increase in UV-B in smoke-free parts of atmosphere due to reductions in stratospheric ozone by up to 30% (or more as a result of changes in dynamics and presence of smoke in stratosphere)
Years		Unresolved (see text)

would include: the build-up of pollutants from smoldering fires, particularly in poorly ventilated, cooled air masses trapped in valleys and lowlands; spills and dispersal of highly toxic industrial chemicals and pyrototoxins; and, possibly, in very limited regions and for only a few days, severe stratospheric ozone depletion.

A nuclear war such as considered in this study could also lead to changes in atmospheric chemistry over months to decades after the initial releases of pollutants by the nuclear explosions and fires. Such alterations might be coupled to potential long-term changes in the biosphere that are described in Volume II. At the present level of knowledge, only a few of the possibilities deserving further study can be suggested. Although direct emissions of CO_2 from post-nuclear fires are roughly equivalent to only one year's emissions from current fossil fuel combustion (and are, therefore, climatically insignificant), the subsequent death of extensive plant communities, as suggested in Volume II, and release of CO_2 through decay and fires could, over a number of years, raise CO_2 levels by a few tens of percent if not balanced by regrowth of vegetation. Similarly, alteration of land and marine ecosystems over large areas could modify the production and release of trace gases such as methane, could alter air-sea gaseous exchange rates, and could affect the hydrological cycle. Uncertainties related to these and other longer term perturbations remain to be addressed.

Environmental Consequences of Nuclear War Volume 1:
Physical and Atmospheric Effects
A. B. Pittock, T. P. Ackerman, P. J. Crutzen,
M. C. MacCracken. C. S. Shapiro and R. P. Turco
© 1986 SCOPE. Published by John Wiley & Sons Ltd

CHAPTER 7
Radiological Dose Assessments

7.1 INTRODUCTION

Nuclear explosions create highly radioactive fission products; the emitted neutrons may also induce radioactivity in initially inert material near the explosion. In this chapter the potential doses associated with these radionuclides are assessed. Our focus is on the consequences outside the zone of the initial blast and fires. Prompt initial ionizing radiation within the first minute after the explosion is not considered here, because the physical range for biological damage from this source is generally smaller than the ranges for blast and thermal effects (see Chapter 1).

In this assessment of the potential radiological dose from a major nuclear conflict, the contributions from "local" (first 24 hours) and more widely distributed, or "global" fallout will be considered separately. Global fallout will be further subdivided into an intermediate time scale, sometimes called tropospheric, of 1 to 30 days; and a long-term (beyond 30 days) stratospheric component. Mainly the dose from gamma-ray emitters external to the body is considered. Contributions from external beta emitters are not estimated because of the limited penetration ability of beta radiation, but there is the possibility that in areas of local fallout, beta radiation can have a significant impact on certain biota directly exposed to the emitters by surface deposition (Svirezhev, 1985; see also Volume II). Potential internal doses from ingestion and inhalation of gamma and beta emitters are estimated in only an approximate manner as these are much more difficult to quantify (see also Volume II, Chapter 3).

The total amount of gamma-ray radioactivity dispersed in a nuclear exchange is dominated by the weapon fission products, whose production is proportional to the total fission yield of the exchange. Exposure to local fallout, which has the greatest potential for producing casualties, is very sensitive to assumptions about height of burst, winds, time of exposure, protection factor, and other variables. For global fallout, the dose commitments are sensitive to how these fission products are injected into various regions of the atmosphere, which depends on individual warhead yield as well as burst location. The distribution of fallout in time and space from

237

the atmospheric weapons testing programs of the 1950s and early 1960s has been studied extensively as a basis for developing a methodology for treating these many dependencies (see, for example, Glasstone and Dolan, 1977; UNSCEAR, 1982).

Despite this dependence of potential radiological dose on the details of an exchange, a scenario-independent methodology is presented—if you will, "user's guides"—to allow interested researchers to estimate doses for the scenarios of their choice. In this chapter, these methods are applied to scenarios typical of those that have been reported in the literature. For local fallout, aspects of the baseline scenario outlined in Chapter 2 are considered. For global fallout, both the 5300 megaton baseline scenario reported by Knox (1983), and the TTAPS 5000 megaton reference nuclear war scenario (Turco et al., 1983a) are considered.

Some previous assessments of radiological fallout have relied on assumptions that are no longer valid. For example, the 1975 study by the U.S. National Academy of Sciences (NAS, 1975) predicted global dose levels significantly lower than those reported here. The NAS study was devoted to the assessment of long range effects and specifically excluded local and short term effects derived from the deposition of radioactive fallout, even though they were acknowledged to be of significance. The total yield of the NAS scenario was 10,000 Mt consisting principally of weapons having a 1 or 5 Mt yield. This contrasts markedly with current scenarios that generally assume use of weapons having yields of 0.5 Mt or less. These lower yield weapons inject most of their radioactivity into the troposphere, where it is more rapidly deposited at the surface. Such injections can, therefore, deliver higher radiation doses than stratospheric injections. In the study by Shapiro (1974), lower doses were also found because its dose assessments were based on scaling from past atmospheric tests. Again, the mix of yield, burst locations, and meteorology in these tests were very different from present weapons arsenals and scenarios.

Previous studies have not considered the potential effects on radiological estimates of the possible climatic perturbations described in Chapter 5. By considering the possible effects of perturbed conditions here, earlier assessments have therefore been extended. These efforts have only begun; thus, the present results must be viewed only as indicative of what may happen, given current understanding and relatively simple assumptions.

7.2 LOCAL FALLOUT

Local fallout is the early deposition of relatively large radioactive particles that are lofted by a nuclear explosion occurring near the surface in which large quantities of debris are drawn into the fireball. For nuclear weapons, the primary early danger from local fallout is due to gamma radiation.

Fresh fission products are highly radioactive and most decay by simultaneous emission of electrons and gamma-rays. The most intense radiation occurs immediately after a nuclear explosion. Elements that are less radioactive, however, linger for long periods of time. An approximate and conservative rule-of-thumb for the first six months following a weapon detonation is that the gamma radiation will decay by an order-of-magnitude for every factor of seven in time (Glasstone and Dolan, 1977). Thus, if gamma activity at 1-hour after detonation produces a radiation level of 1000 rad/h, then at 7-hours the dose rate would be 100 rad/h. In two weeks it would be ~1 rad/h. For the sake of comparison, a lethal whole-body radiation dose would be about 450 rads delivered within 48 hours or 600 rads received over several weeks. The lethal dose level also depends on the presence of other trauma as well as on the amount of medical attention available (i.e., a lower dose could prove fatal if untreated).

If the implausible assumption is made that all of the radioactivity in the fresh nuclear debris from a 1 Mt, all-fission weapon arrives on the ground 1-hour after detonation and is uniformly spread over grassy ground such that it would just give a 48-hour unshielded lethal dose (i.e., 450 rad) then approximately 50,000 km^2 could be covered. Given such a "uniform deposition" model, it would require only about 100 such weapons to completely cover Europe with lethal radiation. In reality, because of a variety of physical processes, the actual areas affected are much smaller. Most of the radioactivity is airborne for much longer than an hour, thus allowing substantial decay to occur before reaching the ground. Also, the deposition pattern of the radioactivity is uneven, with the heaviest fallout near the detonation point where extremely high radiation levels occur. When realistic depositional processes are considered, the approximate area covered by a 48-hour unshielded lethal dose is about 1300 km^2, i.e., nearly a factor of 40 smaller than the area predicted using the simplistic model above. This large factor is partially explained because only about one-half of the radioactivity from ground bursts is on fallout-sized particles (DCPA, 1973). The other portion of the radioactivity is found on smaller particles that have very low settling velocities and therefore contribute to global fallout over longer times. Portions of this radioactivity can remain airborne for years. For airbursts of strategic-sized weapons, virtually no fallout-sized particles are created, and all of the radioactivity contributes to global fallout.

Calculating the physical processes governing the amount, time, and location of the deposition of the radioactive particles is an exceedingly complex and difficult task requiring computer simulation, but it is extremely important to do this properly because of the large variations that can occur (as indicated above). If less accurate information is sufficient, then semi-quantitative approaches which have been derived from sophisticated models are available. These models are based largely on nuclear test observations.

A semi-quantitative model that has been widely used for impact analysis and planning purposes has been presented by Glasstone and Dolan (1977). Semi-quantitative models are useful where scenarios are neither too complex nor wind shears too different from those used to derive the model. However, if solutions are needed that require consideration of complex wind systems, time-of-arrival of radioactivity, or overlay of doses from many fallout patterns, then more sophisticated models should be used. In this study, to facilitate analysis of yet undefined scenarios, a simple graphical method is presented that can be used to generate rough estimates of gamma radiation patterns from multiple nuclear weapon scenarios. Based on a complex computer simulation, this graphical model was chosen for its ease of usage. Time-of-arrival of radioactivity has been accounted for in its development. Overlap from multiple bursts is considered in Section 7.2.3.

7.2.1 Phenomenology

Lofted radioactive fallout particles that have radii exceeding 5 to 10 μm have sufficient fall velocities to contribute to local fallout. Most of these local fallout particles can be seen by the unaided eye. Particles can be as large as several millimeters in radius. These paticles have settling velocities that range from a few centimeters per second to many tens of meters per second. They are lofted by the rising nuclear debris cloud and are detrained anywhere from ground level to the top of the stabilized cloud.

Horizontal wind speeds usually increase with height up to the tropopause and, frequently, wind directions have large angular shears. Nuclear clouds disperse due to atmospheric shears and turbulence. The larger the debris cloud, the faster its radius grows since the rate of eddy mixing increases as the size of the cloud increases (for a discussion of scale-dependent eddy mixing see Walton, 1973). The arrival of radioactivity at a given location can occur over many hours, with large particles from high in the cloud usually arriving first at a downwind location.

Rainout effects have been suggested as being potentially significant contributors to local fallout effects from strategic nuclear war (Glasstone and Dolan, 1977). However, the inclusion of rainout processes would probably not significantly affect the answers to generic questions pertaining to large-scale nuclear war phenomena (for example, "What percent of Western Europe would suffer lethal levels of gamma radiation from local fallout in a large-scale nuclear exchange?"), especially if a substantial portion of the weapons are surface-burst. This is particularly true for strategic weapon yields of greater than 30 kt, because the radioactivity on the small particles most affected by rainout rises above all but the largest convective rain cells. Thus lethal doses from rainout should occur only from large convective rain cells, and this should occur only over relatively small areas (i.e., beneath moving convective cells). However, for any given radioactive air parcel, the overall probability of rainout the first day from a convective cell is quite low for yields greater than 30 kt. Rainout may also occur over large areas associated with frontal systems, but in the case of strategic yields, the radioactivity on small particles must diffuse downward from levels that are often above the top of the precipitation system in order to produce rainout. As a result, radiological doses from debris in precipitation would be substantially lower than early-time doses associated with local fallout. In either case (frontal or convective rainout), for a large-scale multi-burst exchange, the size of the expected lethal-dose rainout areas should typically be small (i.e., well within the range of modeling uncertainty) compared to the size of the fallout areas created by particles with large settling velocities. Thus, first order rainout areas can be ignored in calculating the radiological hazard from a large-scale nuclear war scenario. However, for lower yield (≤ 30 kt) tactical war scenarios, or at specific locations, rainout could lead to important and dominant radiological effects.

Fallout of radionuclides from commercial nuclear reactors has been suggested as a potentially significant contributor to local fallout doses. Calculations indicate that fallout from a reactor and nearby stored nuclear waste facilities can exceed fallout from a single nuclear explosion, if the reactor core can be fragmented and its stored nuclear waste is lofted in the same manner as occurs for weapon radioactivity. However, in a large scale nuclear exchange, if the most dangerous early time impact (that is, gamma radiation deposited by local fallout) is considered, then, in the critical time period during the first week after detonation, the gamma radiation from the reactors will contribute a relatively small portion of the gamma radiation generated by the weapons used in the attack, even if all the radioactivity from all reactors is lofted. In the longer term (i.e., one year or longer), the reactors' radioactivity has the potential to be more important than the weapons' radioactivity. However, the dose rates would be orders of magnitude lower than during the first 48 hours from weapons radiation. (See Appendix 7A for further discussion of the potential radiological dose from an attack on nuclear fuel cycle facilities).

7.2.2 Single-Weapon Fallout Model

To calculate the time of arrival of radioactivity at a location with reasonable accuracy, all significant processes must be taken into consideration. Once the duration and amount of radioactivity arriving at a particular point have been calculated, the dose is obtained by an integration over time, taking into account the decay of the radionuclides. For this work the KDFOC2 computer model (Harvey and Serduke, 1979) was used to calculate fallout fields for single bursts, which in turn were used to develop a semi-quantitative model for preparing rough estimates of fallout areas for typical strategic weapons. A wind profile (including shear) characteristic of mid-continental Northern Hemisphere summer conditions was selected from observations and baseline fallout calculations were performed for several explosion yields assuming all-fission weapons. (A procedure is given below to scale from all-fission to various fusion-fission weapon configurations.) As an example of the results, a one-megaton fallout pattern is shown in Figure 7.1. Figure 7.2 gives the area versus minimum dose relationship for several different yields. Fallout areas are shown rather than maximum downwind extents for various doses since areas are less sensitive to variations in wind direction and speed shears, and should be more useful for analysis. For example, numbers of people or hectares of land can more easily be determined from estimates of area covered than of downwind extent. These areas correspond to doses associated with external gamma-ray emissions. All of the local fallout estimates given below are based on the KDFOC2 model and the wind pattern leading to Fig. 7.1.

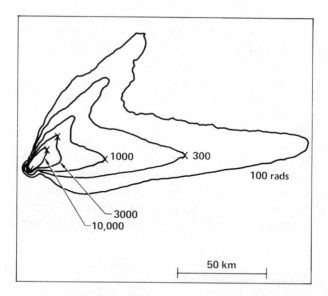

Figure 7.1.   48-hour dose predictions for a 1-Mt all-fission weapon detonated at the surface. A mid-continental Northern Hemisphere summer wind profile was used. The double-lobed pattern is due to a strong directional wind shear that is typical during this season. For a 1-Mt weapon, the lofting of radioactivity is so high that topographic features are not expected to play a large role in pattern development; thus, a flat surface has been used. The protection factor is 1. The local terrain is assumed to be a rolling grassy plain

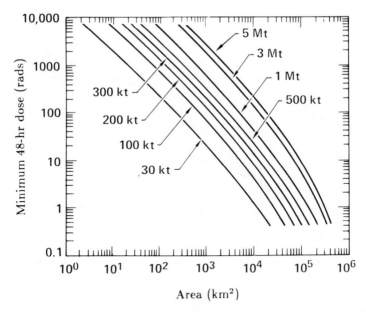

Figure 7.2. Fallout areas versus minimum 48-hour doses for selected yields from 30 kt to 5 Mt. The weapons were surface-burst and all-fission. The wind was that used in the calculation to produce Figure 7.1. These curves include an instrument shielding factor of 25% (Glasstone and Dolan, 1977). Doses within the area defined would exceed the minimum dose

To convert from areas for the 48-hour curves shown in Figure 7.2 to areas for minimum doses over longer times, an "area multiplication factor", AMF, is given in Figure 7.3. For example, if the 2-week, 300-rad area is needed, first the 48-hour, 300-rad area is found from Figure 7.2, then the appropriate AMF is read from Figure 7.3. The 2-week, 300-rad area is the product of the 300-rad, 48-hour area and the 2-week, 300-rad AMF. For example, a 1-Mt, all fission weapon, has a 2-week, 300-rad area of

$$\sim 2000 \ km^2 \times 1.30 \approx 2600 \ km^2.$$

There are two scaling laws that allow weapons design and various sheltering to be factored into dose calculations. The first scaling law permits consideration of weapons that are not all fission. Most large yield weapons (> 100 kt) are combined fission-fusion explosives with approximately equal amounts of fusion and fission (Fetter and Tsipis, 1981). The fission fraction (ρ) is the ratio

$$\rho = \frac{\text{fission yield}}{\text{total yield}} \ .$$

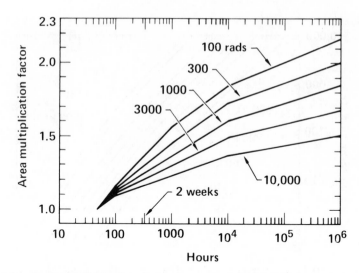

Figure 7.3.   Area multiplication factors to extend the dose integration time from 48 hours to longer times. These factors must be used in conjunction with the areas given in Figure 7.2

To find a 48-hour minimum dose area for a particular fission fraction using Figures 7.2 and 7.3, the dose of interest, D, should be multiplied by $1/\rho$ before reading the values of the area and the area multiplication factor. For example, to obtain the 450 rad, 48-hour dose area for a 50% fission weapon, the area for the scaled dose of 900 rad would be obtained from Figure 7.2. For a 1-Mt, 50% fission weapon, the estimated 450-rad dose area is found to be 720 km^2. The rationale for this scaling law is that the thermodynamics and hydrodynamics of fallout development are insensitive to fission fraction because particle characteristics and lofting altitudes are determined predominantly by total energy yield. For yields that are only part fission, each particle has a fraction of the gamma radioactivity that it would otherwise have if the weapon were an all-fission weapon. This scaling law is appropriate for fission fraction ratios above ~0.3; smaller ratios can lead to situations where neutron induced radioactivity becomes a significant factor. For such cases, careful consideration of surrounding materials may be necessary to produce accurate fallout estimates.

The second scaling law accounts for "protection factors" (K) against ionizing radiation that would be provided by sheltering. The 48-hour minimum dose areas given in Figure 7.2 are appropriate for a person or other organism located on a rolling grassy plain. In other configurations, radiation exposure varies according to how much shielding is obtained while remaining in the area. For example, a person leading a normal lifestyle is likely to achieve an average K of 2 to 3 for gamma radiation from time spent inside buildings and other structures. Basements can provide K's of 10 to 20. Specially constructed shelters can provide K's of 10 to 10,000 (Glasstone and Dolan, 1977).

To determine the radiation area for a dose of D when shielding with a protection factor K is available, the scaled dose KD from Figure 7.2 should be used. For example, for those in an undamaged basement with $K = 10$ for the first 48-hours, Figure 7.2 indicates that the 450 or more rad effective dose area from a 1-Mt, all-fission weapon is about 130 km^2. This is obtained by using a scaled dose of 4500 rads. For comparison, the 450-rad minimum dose area is about 1300 km^2 for people with no shelter, greater by a factor of 10 than the area for those with a K of 10.

Other factors that could reduce the effects of fallout on the population over long time periods (≥ 1 month) include weathering (runoff and soil penetration), cleanup measures, relocation, and the ability of the body to repair itself when dose is spread over time or occurs at lower rates. These consideratons can be taken into account with existing computer models, but are not treated here. Several factors that could enhance the effects of fallout are mentioned below.

7.2.3 Dose Estimation From Multiple Explosions

In a major nuclear exchange, there could be thousands of nuclear warheads detonated. For such an exchange, realistic wind patterns and targeting scenarios could cause individual weapon fallout patterns to overlap in complicated ways that are difficult to predict and calculate. Even though acute doses are additive, a *single* dose pattern calculated for a weapon cannot be used directly to sum up doses in a multi-weapon scenario, except under limited conditions. For example, if the wind speed and direction are not approximately the same for the detonation of each weapon, then different patterns should be used. Thus, only under limited conditions may a single dose pattern be moved around a dose accumulation grid to sum total doses from many weapons.

The number of possible fallout scenarios far exceeds the number of targeting scenarios. This is because, for each targeting scenario that exists, the possible meteorological situations are numerous, complex, and varying. Probabilistic analysis, however, may be used to obtain probability distribution functions which could be analyzed to answer questions of planning and impact analysis.

Two relatively simple multi-burst models can be developed for use in conjunction with the semi-quantitative model presented here. These cases can provide rough estimates of fallout areas from multiple weapons scenarios; however, their results have an uncertainty of no better than a factor of several, for reasons explained below, and are neither upper nor lower case limits. The no-overlap (NO) case is considered first; this could occur when targets are dispersed, there is one warhead per target and the fallout areas essentially do not overlap. Second, the total-overlap (TO) case is examined; this approximation would arise when targets are densely packed and the same size warhead is used against each. A large number of warheads used against, say, a hardened missile field site would be more closely modeled by the TO model than the NO model. Possible incoming warhead fratricide should also be considered in developing any credible scenario for closely packed targets.

As an example of the use of the NO and TO approximations, a case with 100 1-Mt, 50% fission, surface-detonated explosions is considered and estimates are developed for the 450-rad, 48-hour dose areas for both cases. For the NO case the fallout area can be obtained by determining the area for a single 1-Mt weapon (900-rad scaled dose from Figure 7.2) and multiplying by 100. This gives 7.2×10^4 km^2 for the 450-rad, 48-hour dose contour. For the TO model, the area is obtained for a single 1-Mt weapon, 9-rad scaled dose from Figure 7.2. One hundred of these, laid on top of each other, would give 450 rads for 50% fission weapons. The area in this case is 3.3×10^4 km^2. These results differ by about a factor of two, with the NO case giving a larger area.

Although these models are extremes in terms of fallout pattern *overlap,* neither can be taken as a bounding calculation of the extremes in fallout *areas* for specified doses. It is very possible that a more realistic calculation of overlap would produce a greater area for 100 weapons than either of these models. Such a result is demonstrated by a more sophisticated model prediction that explicitly takes overlap into account (Harvey, 1982). In this study, a scenario was developed for a severe case of fallout in a countervalue attack on the U.S. where population centers were targeted with surface bursts. Figure 7.4 shows the contours of a 500-rad minimum 1-week dose where overlap was considered. The 500-rad area is about three times greater than that predicted by the NO model, and six times that of the TO model. Note also that the distribution of radioactivity is extremely uneven. About 20% of the U.S. is covered with 500-rad contours, including nearly 100% of the northeast, approximately 50% of the area east of the Mississippi, 10% of the area west of the Mississippi, and only a small percentage of the area in the Great Plains.

Results of this scenario, as well as those postulated by others, clearly show that such estimates are very scenario-dependent and that detailed estimates should be made with care. For example, the regional results shown in Figure 7.4 could be significantly different if military targets (e.g. , ICBM silos) were included as well. Although the NO and TO cases presented in this chapter are simple to apply, they must be used only to develop rough estimates of total area coverage within regions with relatively uniformly dispersed targets.

When the density of targets of one area is as large as in the northeastern U.S. and another is as dispersed as in the western U.S., regional models should be used to develop specific regional estimates. Even then, multiple weapon fallout estimates should be considered to have uncertainties no smaller than a factor of several, with the uncertainty factor increasing as the model sophistication decreases.

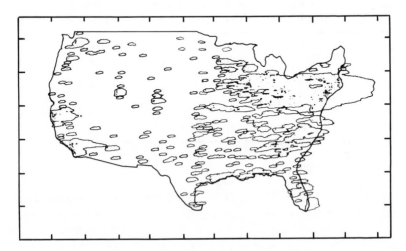

Figure 7.4. A fallout assessment that explicitly takes fallout pattern overlap into account. Shown are 500-rad, 1-week minimum isodose contours. This scenario was intended to emphasize population dose. Approximately 1000 population centers in the U.S. were targeted, each with a 1-Mt, 50%-fission weapon. The assumed winds were westerly with small vertical shear and were nearly constant over the continent (taken from Harvey, 1982)

7.2.4 Sample Calculation of Multiple-Weapon Fallout

To illustrate the fallout prediction method presented here, an escalating nuclear exchange scenario, which is consistent with that developed in Chapter 2, is used to estimate fallout areas. In this scenario there are four sequential phases of attack against five different regions. The five regions are: Europe (both east and west), western U.S.S.R. (west of the Urals), eastern U.S.S.R., western U.S. (west of 96° west longitude), and eastern U.S. The four phases of attack are: initial counterforce, extended counterforce, industrial countervalue, and a final phase of mixed military and countervalue targeting. The weapon yields and the number of warheads that are employed for just the surface bursts during each phase are shown in Table 7.1. Airbursts are omitted since they do not produce appreciable local fallout.

TABLE 7.1.
SURFACE-BURST WARHEADS IN A PHASED NUCLEAR EXCHANGE.
ALL WEAPONS ARE ASSUMED TO HAVE A 50% FISSION YIELD

	Number of warheads				
Weapon yield (Mt)	Initial counter- force phase	Extended counter- force phase	Industrial counter- value phase	Final phase	Full baseline exchange
0.05	0	300	0	250	550
0.1	975	150	50	8	1183
0.2	0	250	50	121	421
0.3	500	250	0	125	875
0.5	1000	200	0	25	1225
1.0	250	495	160	125	1030
5.0	0	50	15	8	73
Total surface- burst yield	~1000	~1000	~250	~250	~2500

In the first phase, land-based ICBM's are the primary targets. These are assumed to be located in the western U.S. and the U.S.S.R. at sites containing 125 to 275 missiles. The geographical distribution of missile silos in the U.S.S.R. is assumed to be fifty percent east and fifty percent west of the Urals. Each missile silo is attacked with a surface-burst and an air-burst weapon. For a given site, the TO model is used to calculate the fallout pattern. All U.S. ICBM sites are attacked with 0.5 Mt weapons. Each of five U.S. ICBM complexes are presumed to have 200 missile silos, while each of 6 U.S.S.R. complexes are presumed to have between 125 and 275 missile silos, with a total of 1300. The Soviet sites are attacked with 1, 0.3, and 0.1 Mt weapons. During this phase, each side employs a total of about 1000 Mt. Besides the attack on Soviet missile silos, 425 0.1-Mt weapons are assumed to be surface-burst against other Soviet military targets, with approximately 28 Mt west of the Urals and 14 Mt to the east. The 425 fallout patterns from these weapons have been modeled with the NO model.

In the second phase of the attack, there are an additional 1000 Mt of surface-burst weapons employed. These are employed against each region with 20, 40, and 40% of the weapons being used against targets in Europe, the U.S. and the U.S.S.R., respectively. Here, Europe includes both the NATO and Warsaw Pact countries. To roughly account for population distribution, the weapons employed against the U.S. are divided up as two-thirds in the eastern U.S. and one-third in the western U.S.; for Soviet targets it is assumed that two-thirds are detonated west and one-third east of the Urals.

TABLE 7.2.
PERCENT OF LAND MASS COVERED BY A MINIMUM 450 RAD,
48-HOUR DOSE

	Initial counter-force phase	Extended counter-force phase	Industrial counter-value phase	Final phase	Full baseline exchange
Europe	0	2.9	0.6	0.8	4.3
Eastern U.S.S.R.	0.5	0.5	0.1	0.2	1.3
Western U.S.S.R.	1.6	2.3	0.7	1.7	6.3
Eastern U.S.	0	4.7	1.0	1.4	7.1
Western U.S.	4.4	2.3	0.7	6.6	8.0

For all the weapons employed in the second, third and fourth phases, the fallout pattern is calculated using the NO model. The results, in terms of percent of land covered by at least a 450 rad, 48-hour dose, are shown in Table 7.2. No shielding has been assumed in calculating these percentages. Similar areas were found for 600 rad over two weeks.

Care must be taken in interpreting these results. To begin with, there is an uncertainty factor of several in the NO and TO modeling schemes, as discussed earlier. Another substantial bias is introduced by neglecting the radioactivity that is blown into or out of a region. For example, the western U.S.S.R. would likely receive substantial amounts of radiation from weapons detonated in eastern Europe because the wind usually blows from Europe toward the Soviet Union. Thus, the area percentages shown in Table 7.2 for Europe would be expected to decrease since some of the area credited to Europe would actually be in the Soviet sector. Similarly, the percentage of the western U.S. is probably overestimated, assuming typical wind conditions. For the eastern U.S., the area covered would be increased by radioactivity originating in the central U.S. and decreased as a result of radioactivity blowing out over the Atlantic Ocean.

There are a number of factors that could change these local fallout assessments.

• Shielding is probably the most sensitive parameter in reducing the effective dose to a population. This effect has been ignored in these calculations. Protective measures could substantially reduce the human impact of fallout.
• Choosing a scenario that exacerbates local fallout (e.g., surface bursting of cities) could increase lethal areas by factors of several.

- Large differences in doses could arise because of irregularities in fallout patterns in the local fallout zones that could range over orders of magnitude. Relocation could substantially reduce a population's dose.
- Debilitating, but not lethal, radiation doses ($\sim$200 rad or more) would be received over much larger areas than areas receiving lethal doses.
- Fission fractions of smaller modern weapons could be twice the baseline assumption of 0.5. Adding these to the scenario mix could increase lethal fallout areas by up to 20% of the baseline calculation.
- Tactical weapons, ignored in the baseline scenario, could increase lethal local fallout areas in certain geographical regions; particularly within Europe, by about 20% of the baseline calculations.
- Internal radiation exposure could increase the average total doses to humans by about 20% of the external dose.
- External beta exposure, not treated here, could add significantly to plant and animal exposures in local fallout areas.
- Targeting of nuclear fuel cycle facilities could contribute to radiation doses (see Appendix 7A).

7.3 GLOBAL FALLOUT

Global fallout consists of the radioactivity carried by fine particulate matter and gaseous compounds that are lofted into the atmosphere by nuclear explosions. One may distinguish two components to global fallout—intermediate time scale and long-term. Intermediate time scale fallout consists of material that is initially injected into the troposphere and is removed principally by precipitation within the first month. The fractional contribution to intermediate time scale fallout decreases as the total weapon yield increases above 100 kt. The importance of intermediate time scale fallout has grown with reductions in warhead yields. Long-term fallout occurs as a result of deposition of very fine particles that are initially injected into the stratosphere. Because the stratosphere is so stable against vertical mixing and the fine particulate matter has negligible fall velocities, the primary deposition mechanism involves transport of the radioactivity to the troposphere through seasonal changes in stratospheric circulation. Once within the troposphere, these particles would normally be removed within a month by precipitation scavenging.

7.3.1 Methodology

Given a specific nuclear war scenario, it is possible to use experience gained from atmospheric nuclear tests to estimate the fate of both intermediate time scale and long-term fallout particles if the atmosphere is not perturbed by smoke. GLODEP2 (Edwards et al., 1984), an empirical code that was designed to match measurements from atmospheric testing has been used. The model contains two tropospheric and six stratospheric injection compartments. By following unique tracer material from several atmospheric nuclear tests in the late 1950s, combined with subsequent balloon and aircraft measurements in the stratosphere and upper troposphere and many surface air and precipitation observations, it was possible to estimate the residence time of radioactivity in the various stratospheric compartments and the interhemispheric exchange rate in the stratosphere. Radioactive material that is placed initially into the troposphere is also handled by the GLODEP2 model (Edwards et al., 1984). From this information, surface deposition tables were prepared. The GLODEP2 model has never been tested against atmospheric nuclear tests in middle latitudes since no extensive series of explosions have occurred in this region. As a result, there is some uncertainty in the results of explosions centered around the Northern Hemisphere middle latitudes, but little uncertainty in the Northern Hemisphere sub-polar latitude calculations since the stratospheric fallout there would deposit much the same as the global fallout from the polar bursts used to generate the polar deposition tables in the model.

In this section, a simple table, based on GLODEP2 calculations is prepared that enables readers to obtain dose estimates for their own scenarios. Table 7.3 presents the 50-year external gamma-ray dose commitment, in rads, for single nuclear explosions of 0.1 to 20 Mt yield. All bursts are assumed to occur at the surface, and to be all fission. For an airburst (where the fireball does not touch the ground), the tabular values must be doubled since about twice as much radioactivity is available for global fallout for an airburst as compared to a surface burst. Recall that about half the radioactivity dispersed in a surface burst is deposited within 24 hours as local fallout. Two burst latitudes, 40° N and 55° N, were selected as median latitudes for strikes against the U.S., Europe, and the U.S.S.R., respectively.

Table 7.3 should be used only (a) for surface bursts or (b) for airbursts whose height is below 3 km but above the height where the fireball touches the surface. The height of an airburst may be defined by the relation $H \geq 870Y^{0.4}$, where Y is the total yield of the explosion in megatons and H is in meters (Glasstone and Dolan, 1977).

As an example of how Table 7.3 can be used, average dose estimates are derived at 30–50° N latitude for an arbitrary, illustrative, simplified nuclear exchange during the Northern Hemisphere winter season. Table 7.4 presents the results of this example. The doses per weapon in column 7 were obtained from Table 7.3, interpolating between yield columns where necessary.

Using the Table 7.3 on this illustrative scenario gives a total 30–50° N dose of 8.8 rads, while the computer version of GLODEP2 gives 8.1 rads. The small difference is due principally to interpolation between total yield categories and the fact that tabular values are given to only one significant

TABLE 7.3.

GLOBAL EXTERNAL GAMMA-RAY DOSE (IN RADS) FROM A SINGLE NUCLEAR WEAPON EXPLODED AT THE SURFACE AS CALCULATED BY GLODEP2. DOSES ARE DUE TO THE RADIOACTIVITY DEPOSITED AT THE SURFACE AND ARE INTEGRATED OVER 50 YEARS, ASSUMING NO WEATHERING. ALL WEAPONS ARE ASSUMED TO BE 100% FISSION. FOR AIRBURSTS MULTIPLY TABULAR VALUE BY TWO

NORTHERN HEMISPHERE WINTER

Latitude	Bursts at ~40° N Total Yield (Mt)						Bursts at ~55° N Total Yield (Mt)					
	0.1	0.3	1	3	10	20	0.1	0.3	1	3	10	20
70–90N	4×10^{-5}	2×10^{-4}	1×10^{-3}	2×10^{-3}	4×10^{-3}	8×10^{-3}	2×10^{-3}	4×10^{-3}	1×10^{-3}	2×10^{-3}	4×10^{-3}	8×10^{-3}
50–70N	1×10^{-3}	3×10^{-3}	4×10^{-3}	8×10^{-3}	2×10^{-2}	3×10^{-2}	5×10^{-3}	1×10^{-2}	4×10^{-3}	8×10^{-3}	2×10^{-2}	3×10^{-2}
30–50N	4×10^{-3}	9×10^{-3}	6×10^{-3}	1×10^{-2}	2×10^{-2}	4×10^{-2}	2×10^{-3}	4×10^{-3}	5×10^{-3}	1×10^{-2}	2×10^{-2}	4×10^{-2}
10–30N	8×10^{-4}	2×10^{-3}	2×10^{-3}	3×10^{-3}	6×10^{-3}	1×10^{-2}	5×10^{-5}	3×10^{-4}	2×10^{-3}	3×10^{-3}	6×10^{-3}	1×10^{-2}
10S–10N	1×10^{-5}	5×10^{-5}	3×10^{-4}	5×10^{-4}	6×10^{-4}	1×10^{-3}	6×10^{-6}	4×10^{-5}	3×10^{-4}	5×10^{-4}	6×10^{-4}	1×10^{-3}
10–30S	3×10^{-6}	2×10^{-5}	2×10^{-4}	4×10^{-4}	1×10^{-3}	3×10^{-3}	3×10^{-6}	2×10^{-5}	2×10^{-4}	4×10^{-4}	1×10^{-3}	3×10^{-3}
30–50S	3×10^{-6}	2×10^{-5}	1×10^{-4}	8×10^{-4}	4×10^{-3}	8×10^{-3}	2×10^{-6}	2×10^{-5}	1×10^{-4}	8×10^{-4}	4×10^{-3}	8×10^{-3}
50–70S	1×10^{-6}	8×10^{-6}	6×10^{-5}	5×10^{-4}	3×10^{-3}	6×10^{-3}	1×10^{-6}	8×10^{-6}	6×10^{-5}	5×10^{-4}	3×10^{-3}	6×10^{-3}
70–90S	7×10^{-8}	4×10^{-7}	3×10^{-6}	1×10^{-4}	7×10^{-4}	1×10^{-3}	7×10^{-8}	4×10^{-7}	3×10^{-6}	1×10^{-4}	7×10^{-4}	1×10^{-3}

NORTHERN HEMISPHERE SUMMER

Latitude	Bursts at ~40° N Total Yield (Mt)						Bursts at ~55° N Total Yield (Mt)					
	0.1	0.3	1	3	10	20	0.1	0.3	1	3	10	20
70–90N	3×10^{-5}	1×10^{-4}	5×10^{-4}	1×10^{-3}	3×10^{-3}	6×10^{-3}	1×10^{-4}	3×10^{-3}	7×10^{-4}	1×10^{-3}	3×10^{-3}	6×10^{-3}
50–70N	1×10^{-3}	3×10^{-3}	3×10^{-3}	6×10^{-3}	2×10^{-2}	3×10^{-2}	4×10^{-3}	9×10^{-3}	3×10^{-3}	6×10^{-3}	2×10^{-2}	3×10^{-2}
30–50N	3×10^{-3}	7×10^{-3}	4×10^{-3}	1×10^{-2}	2×10^{-2}	5×10^{-2}	1×10^{-3}	3×10^{-3}	4×10^{-3}	1×10^{-2}	2×10^{-2}	5×10^{-2}
10–30N	6×10^{-4}	1×10^{-3}	1×10^{-3}	3×10^{-3}	7×10^{-3}	1×10^{-2}	4×10^{-5}	2×10^{-4}	1×10^{-3}	3×10^{-3}	7×10^{-3}	1×10^{-2}
10S–10N	7×10^{-6}	3×10^{-5}	2×10^{-4}	3×10^{-4}	5×10^{-4}	9×10^{-4}	3×10^{-6}	3×10^{-5}	2×10^{-4}	3×10^{-4}	5×10^{-4}	9×10^{-4}
10–30S	2×10^{-6}	1×10^{-5}	1×10^{-4}	3×10^{-4}	1×10^{-3}	2×10^{-3}	2×10^{-6}	1×10^{-5}	1×10^{-4}	3×10^{-4}	1×10^{-3}	2×10^{-3}
30–50S	2×10^{-6}	1×10^{-5}	9×10^{-5}	7×10^{-4}	4×10^{-3}	7×10^{-3}	2×10^{-6}	1×10^{-5}	9×10^{-5}	7×10^{-4}	4×10^{-3}	7×10^{-3}
50–70S	9×10^{-7}	6×10^{-6}	4×10^{-5}	6×10^{-4}	3×10^{-3}	7×10^{-3}	9×10^{-7}	6×10^{-6}	4×10^{-5}	6×10^{-4}	3×10^{-3}	7×10^{-3}
70–90S	1×10^{-7}	7×10^{-7}	6×10^{-6}	2×10^{-4}	1×10^{-3}	2×10^{-3}	1×10^{-7}	7×10^{-7}	6×10^{-6}	2×10^{-4}	1×10^{-3}	2×10^{-3}

figure. This close comparison suggests that increasing the number of yield columns the number of significant figures in the body of the table is not warranted.

TABLE 7.4.
DOSES (IN RADS) AT 30–50°N FOR AN ILLUSTRATIVE NUCLEAR WAR SCENARIO

No. of weapons	Yield (Mt)	Total fission fraction	Burst height (m)	Burst height factor[a]	Burst latitude	Doses from Table 7.3 (rads)	Total dose[b] (rads)
1000	1.0	0.5	1500	2	40°N	6×10^{-3}	6.0
55	20.0	0.5	0	1	40°N	4×10^{-2}	1.1
135	1.5	0.5	0	1	55°N	7×10^{-3}	0.7
52	9.0	0.5	2500	2	55°N	2×10^{-2}	1.0
						Total	8.8 rads

[a] Factor = 1 for surface bursts, 2 for airbursts.
[b] Total dose is the product of columns 1, 3, 5 and 7.

7.3.2 Global Dose in an Unperturbed Atmosphere Using Specific Scenarios

A variety of scenario studies have been performed using GLODEP2 (Knox, 1983; Edwards et al., 1984) Dose calculations for scenarios (A) and (B), which are described in Table 7.5, are presented in detail in Table 7.6. The atmospheric compartments in Table 7.5 refer to those used in the GLODEP2 model. The Ambio reference nuclear war containing 5700 Mt and 14,700 warheads has not been considered here. Its preponderance of low-yield warheads would produce even higher dose estimates than scenarios (A) or (B).

As indicated in the illustrative example, dose assessment is sensitive to yield, and so a somewhat larger dose is expected from (B) than from (A) because of its lower average yield per warhead. From a comparison of GLODEP2 results for the (A) and (B) scenarios for a Northern Hemisphere winter injection (Table 7.6, columns A_1 and B_1), it is seen that the Northern Hemisphere averages for (A) and (B) are about 16 and 19 rads respectively, while Southern Hemisphere averages are more than a factor of 20 smaller. The maximum appears in the 30–50°N latitude band, where scenarios (A) and (B) yield 33 and 42 rads, respectively. All the doses reported here for global fallout are integrated external gamma-ray exposure over 50 years and assume no sheltering, no weathering, and a smooth plane surface.

TABLE 7.5.
NUCLEAR WAR SCENARIO

Scenario A Knox (1983) 5300 Mt baseline nuclear war		Scenario B TTAPS (Turco et al., 1983a) 5000 Mt reference nuclear war			
Total yield/warhead (Mt)	Total fission yield injected (Mt)	Total yield/warhead (Mt)		Total fission yield injected (Mt)	
20.0	305	10.0		125	
9.0	235	5.0		125	
1.0–2.0	355	1.0		213	
0.9	675	1.0		319	
0.75	15	1.0	2	5	
0.55	220	0.5		187	
0.3–0.4	115	0.5		125	
0.1–0.2	110	0.3		113	
<0.1	1	0.3	7	5	
		0.2	5	0	
		0.2	7	5	
		0.1	7	5	
		0.1	1	2	

Mt of fission products injected into atmosphere

	Scenario A	Scenario B
Polar troposphere	226	369
Lower polar stratosphere	1234	898
Upper polar stratosphere	571	226
High polar atmosphere	0	25
TOTAL	2031	1520
Fraction of yield in surface bursts	0.47	0.57
Fission fraction	0.5	0.5
Total number of explosions	6235	10400

For scenario (A), 55% of the dose emanates from the tropospheric injections. The corresponding value for (B) is 75%. This emphasizes the sensitivity of dose to the yield mix of the scenario. As individual warhead yields decrease, the fractional injections into the troposphere increase, resulting in much larger doses on the ground due to more rapid deposition. Tropospheric radioactivity injections per megaton of fission can produce doses on the ground about a factor of 10 greater than those resulting from lower stratospheric injections, which in turn contribute about 3 to 5 times higher dose compared to upper stratospheric injections (Shapiro, 1984). Injections of radioactivity above the stratosphere as a gas or as extremely fine particles would produce relatively negligible doses at the ground.

TABLE 7.6.
GLOBAL FALLOUT DOSE ASSESSMENTS (RADS) FOR AN
UNPERTURBED ATMOSPHERE WITH NO SMOKE

A = 5300 Mt baseline nuclear war (Knox, 1983)
B = 5000 Mt reference nuclear war (Turco et al., 1983a)

Latitude band	A_1	B_1	A_2	B_2	A_3	B_3
70–90N	4.5	3.7	2.9	2.5	7.8	8.2
50–70N	27.3	28.8	21.7	22.7	21.3	24.6
30–50N	32.9	41.7	27.4	33.7	22.3	23.9
10–30N	6.9	8.3	5.6	6.6	7.6	7.2
10S–10N	0.8	0.6	0.5	0.3	1.3	1.0
10–30S	0.6	0.4	0.4	0.2	0.6	0.4
30–50S	0.8	0.4	0.6	0.4	0.7	0.4
50–70S	0.5	0.3	0.5	0.3	0.5	0.3
70–90S	0.1	0.0	0.2	0.1	0.2	0.1
Area averaged—N.H.	16.2	19.1	13.1	15.2	12.8	13.7
Area averaged—S.H.	0.6	0.4	0.5	0.3	0.7	0.4
Area averaged—Global	8.4	9.8	6.8	7.8	6.8	7.1
Global population dose ($\times 10^{10}$) person-rads	6.7	8.2	5.5	6.6	5.3	5.5

A_1 = Winter injection using GLODEP2
B_1 = Winter injection using GLODEP2
A_2 = Summer injection using GLODEP2
B_2 = Summer injection using GLODEP2
A_3 = Summer injection using GRANTOUR with stratospheric contributions from GLODEP2
B_3 = Summer injection using GRANTOUR with stratospheric contributions from GLODEP2

Table 7.6 includes calculated values for the global population dose. This quantity is calculated by multiplying the dose in each 20° wide latitude band by the population of that latitude band, and then summing over all latitudes. For a given scenario, this number is one measure of the potential global biological impact. The global population dose as calculated by GLODEP2 for (A) and (B) are 7 and 8×10^{10} person-rads, respectively. Essentially all of this dose occurs in the Northern Hemisphere because 90% of the world's population and higher doses prevail there.

Figure 7.5 illustrates the time behavior of the buildup of the dose to the 50-year lifetime value as a function of latitude for scenario (A). The bulk of the dose is caused by deposition (mainly from the troposphere) and exposure during the first season after the war, followed by a gradual rise to the 50-year value.

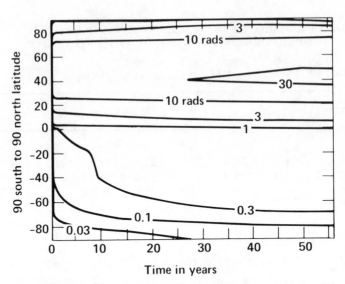

Figure 7.5. Global fallout: accumulated whole body gamma dose (rads) from 6235 explosions totaling 2031 Mt of fission products (scenario A). An 8 day tropospheric deposition decay constant, characteristic of a winter injection, is assumed

A comparison of the GLODEP2 results for the TTAPS scenario (B) and Turco et al. (1983a) results (using an entirely different methodology) reveals that GLODEP2 doses are 19 rads for the Northern Hemisphere average and 42 rads for the 30–50° N latitude band, while Turco et al.'s estimates give corresponding doses of 20 rads and about 40 to 60 rads.

Other studies that have been undertaken using GLODEP2 and the 5300 Mt scenario (A) have led to the conclusions:

Winter vs Summer Injection: GLODEP2 contains an exponential tropospheric deposition model with a variable time constant τ that depends on the season. Values used for τ are 8.2 days for the Northern Hemisphere winter and 18.2 days for summer. For the 30–50° N latitude band, comparison of two runs for scenario (A) yields 27 rads for summer injections compared to 33 rads for winter injections. The corresponding figures for the global population dose are about 6×10^{10} person-rads (summer) and 7×10^{10} person-rads (winter). The population averaged dose per person is 12 rads (summer) and 15 rads (winter). Because of a decrease in the frequency and intensity of large scale precipitation systems in summer, the doses from the troposphere and lower polar stratosphere are reduced somewhat in comparison to winter, while the upper stratospheric contribution is increased. These results indicate that the predicted differences between summer and winter are not large, the dose commitments are not very sensitive to τ, and that other sources of uncertainty would predominate.

Scenarios with Smaller-Yield Devices. The long-term consequences of the shift in the nuclear arsenals from larger to smaller yield devices has been assessed. This shift in average yield has been going on for about the past two decades as targeting accuracy improved, although the trend appears to have halted (see Chapter 2). Table 7.7 presents results comparing the 5300 Mt baseline scenario with two variations. In scenario (Aa), the number of devices in the baseline scenario (A) is increased from 6235 to 13250 while the total yield is held at 5300 Mt. In scenario (Ab), smaller yields have been used, but the number of devices is constant at 6235 (the total yield consequently is reduced by 25% from 5300 to 4000 Mt). The figures presented are for the 50 year gamma-ray dose. For the same total yield, it is seen that a shift to smaller weapons in the baseline scenario has approximately doubled the dose (scenario Aa). For case (Ab), the dose remains about the same even with a 25% drop in the total yield.

TABLE 7.7.
GLOBAL FALLOUT: SENSITIVITY OF DOSE TO WARHEAD YIELD.
THE SAME FISSION FRACTION AND GROUND BURST FRACTION AS
ASSUMED AS IN SCENARIO A

Scenario	Total yield (Mt)	Number of explosions	Avg. Yield per warhead	30°–50° N dose (rads)	Global avg. dose per person (rads)	Global pop. dose (10^{10} person-rads)
A	5300	6235	0.85	33	15	6.7
Aa	5300	13250	0.40	64	27	12.5
Ab	4000	6235	0.64	33	14	6.5

7.3.3 Global Fallout in a Perturbed Atmosphere

Following a large scale nuclear exchange, the large quantities of smoke and soot lofted to high altitudes could decrease the incoming solar radiation, resulting in tropospheric and stratospheric circulation changes (see Chapter 5). Over land in the Northern Hemisphere, the presence of smoke and soot would probably result in less precipitation and a lowering of the tropopause; these changes could decrease the intermediate time scale (tropospheric) fallout and, depending on changes in stratospheric circulation, could alter the stratospheric contribution to fallout in the Northern Hemisphere. However, before the stratospheric burden is carried into the troposphere, a sizeable fraction would be transported to the Southern Hemisphere by the accelerated interhemispheric transport, resulting in doses there that are likely to be increased over those calculated for an unperturbed atmosphere.

Both the GLODEP2 and the Turco et al. (1983a) models assumed fission product depositions from a normal atmosphere in calculating global fallout. Preliminary studies have been conducted with radionuclides in a perturbed atmosphere using a three-dimensional version of the GRANTOUR model (see MacCracken and Walton, 1984). GRANTOUR is a three-dimensional transport model driven by meteorological data generated by the Oregon State University (OSU) general circulation model (Schlesinger and Gates, 1980). Particulate matter appearing as an initial distribution or generated by sources is advected by wind fields, locally diffused in the horizontal and vertical, moved vertically by convective fluxes and the re-evaporation of precipitation, and removed by precipitation scavenging and dry deposition. Information, in the form of mixing ratios of curies per kg of air, is carried by Lagrangian parcels that move with the prescribed winds. It is assumed that the fission products are in the form of particulate material in two size ranges, greater than and less than one micrometer in diameter. The significance of the two size ranges lies in the assumption that the large particles are scavenged by precipitation with greater efficiency than the small ones. Thus, the surface dose will depend upon the assumed division of the radioactivity between the two size ranges. Coagulation from small to large particles is not treated in the version of the model used here. All meteorological information is specified on a fixed spatial grid and is interpolated to the parcel locations. In turn, when mixing ratios are needed on the fixed grid, they are obtained from weighted averages of the parcel values. The removal processes cause material to be accumulated on the ground and this information is saved in a history file that can be used for post-processing. The radioactive decay of the fission products is not calculated in GRANTOUR, but rather in a post-processor. Knowing the time of injection and the amount and time of arrival at a grid point, it is possible to compute the dose for any time interval.

Studies focused on comparisons of radiation dose assessments with smoke in the atmosphere (interactive atmosphere) and without smoke (noninteractive); other relevant parameters were also explored, including consideration of particle size distribution, source location, different initial meteorology, and averaging doses over land areas only. All of the GRANTOUR simulations reported here are for the Northern Hemisphere summer season and use five radioactivity and smoke source locations of equal strength. The locations include two in the U.S., two in the U.S.S.R., and one in western Europe. This division of sources is similar to that assumed in our earlier discussion on local fallout. Sources were initially injected with a Gaussian distribution whose amplitude was 10% of the maximum at a radius of 15° along a great circle. The total amount of smoke injected was 150 teragrams (equivalent to the urban smoke contributions used by Turco et al. (1983a) and NRC (1985)). MacCracken and Walton (1984) describe the induced climatic perturbations (also see Chapter 5). The vertical distribution of the radioactivity injections were distributed, as was the smoke, with the same vertical distribution as the source term injections calculated using the GLODEP2 injection algorithm. Most of the calculations assumed the radionuclides were attached to particles of two diameter sizes; ($> 1 \mu$m and $< 1 \mu$m), with an initial distribution of 43% of the radioactivity attached to the larger particles and 57% to the smaller particles. Deposition was followed for 30 days in most calculations. A single 60 day run indicated that 30 days is sufficient to account for 90% of the deposition. Results are compared for a 50 year unsheltered, unweathered, external gamma-ray dose.

GRANTOUR treats only the troposphere and splits it into three vertical layers extending from 800–1000, 400–800 and 200–400 mbar. In a normal atmosphere, these layers reach up to 2.0, 7.1 and 11.8 km. In the comparisons, GLODEP2 was used to estimate the dose contributions from the stratospheric injections, which were added to the doses calculated by GRANTOUR assuming altered climatic conditions. The results for GRANTOUR's $10° \times 10°$ (latitude-longitude) grid size were then suitably averaged to obtain results for the nine $20°$ wide latitude bands in order to facilitate comparison with GLODEP2. Average doses were also calculated for only the land masses.

Scenarios A and B were used in the calculations. Columns A_2 and B_2 in Table 7.6 display a comparison of the predictions of GLODEP2 for these two scenarios. Column A_3 and B_3 list the results from GRANTOUR, assuming an unperturbed atmosphere (no smoke; no climatic perturbation) for the same two scenarios. There is reasonable agreement (i.e., generally within about 50%) between the GLODEP2 only and GRANTOUR/GLODEP2 methodologies for an unperturbed atmosphere (cases 1 and 3), providing some confidence that the results of GLODEP2 and GRANTOUR can be combined for simulations with a perturbed atmosphere, although the initial accelerated interhemispheric mixing of radionuclides in the stratosphere has not yet been considered. This may lead to a small underestimate of the long term Southern Hemisphere dose.

Table 7.8 compares calculations for a perturbed atmosphere (interactive smoke) with estimates for normal July conditions. These results are also shown in Figures 7.6 and 7.7. and indicate that the perturbed atmosphere lowers the average dose in the Northern Hemisphere by about 15%. Because the principal mechanism for radionuclide removal from the troposphere is precipitation, the GRANTOUR calculations are roughly consistent with the thesis that precipitation is inhibited when large amounts of smoke are introduced. The transfer of fission product radionuclides to the Southern Hemisphere is somewhat enhanced by the perturbed climate, resulting in higher doses than for the unperturbed case. The increases in Southern Hemisphere dose, however, are not large, and the resulting doses are still about a factor

TABLE 7.8.
GLOBAL FALLOUT DOSE USING THE THREE-DIMENSIONAL
GRANTOUR MODEL (SUMMER SCENARIO) COMPARISON OF
PERTURBED ATMOSPHERE (SMOKE) AND UNPERTURBED ATMOSPHERE
(NO SMOKE) EXTERNAL GAMMA-RAY DOSES ARE IN RADS. BECAUSE
GRANTOUR ONLY CALCULATES THE TROPOSPHERIC CONTRIBUTION,
THE DOSES HERE INCLUDE THE CONTRIBUTIONS FROM THE
STRATOSPHERE AS CALCULATED BY GLODEP2

Latitude band	A_3 (no smoke)	A_4 (smoke)	B_3 (no smoke)	B_4 (smoke)
90–70N	7.8	6.4	8.2	5.8
70–50N	21.3	17.2	24.6	18.0
50–30N	22.3	20.1	23.9	20.4
30–10N	7.6	7.5	7.2	7.2
10N–10S	1.3	1.6	1.0	1.4
10–30S	0.6	0.8	0.4	0.6
30–50S	0.7	0.8	0.4	0.5
50–70S	0.5	0.5	0.3	0.3
70–90S	0.2	0.2	0.1	0.1
Area averaged—N.H.	12.8	11.5	13.7	11.5
Area averaged—S.H.	0.7	0.8	0.4	0.6
Area averaged—Global	6.8	6.1	7.1	6.1
Population average—Global	11.5	10.7	12.0	10.7
Global population dose ($\times 10^{10}$) person-rads	5.3	4.9	5.5	4.9

A_3 = 5300 Mt (Knox, 1983), unperturbed atmosphere (no smoke)
A_4 = 5300 Mt (Knox, 1983), perturbed atmosphere (smoke)
B_3 = 5000 Mt (Turco et al., 1983a), unperturbed atmosphere (no smoke)
B_4 = 5000 Mt (Turco et al., 1983a), perturbed atmosphere (smoke)

of 20 lower than in the Northern Hemisphere. This is because the increased transfer to the Southern Hemisphere is mitigated by the decay in activity during the time before the radionuclides are deposited on the ground.

Preliminary conclusions from other parameter studies include:

Land area averages: For each GRANTOUR calculation reported above, dose calculations were repeated, averaging only over the land areas. Since the source locations are centered over land masses, one would expect land average values to be higher than average values that include both land and ocean areas. Averaging doses over only the Northern Hemisphere land areas increased the calculated tropospheric dose by about 30% above the combined average for land and oceans in all of the cases presented. Considering the total dose, including the contribution from the stratosphere, the percentage increase was smaller, ranging from 10 to 20%.

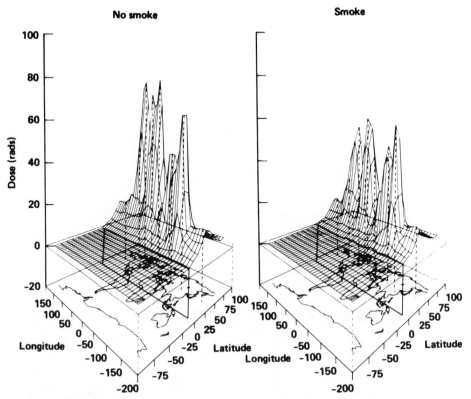

Figure 7.6. Comparison of radionuclide global dose distribution for cases with unperturbed and smoke-perturbed climates (tropospheric contributions only)

Hotspots: Figures 7.6 and 7.7 reveal longitudinal, as well as latitudinal, details that are not apparent in the averages of Table 7.8. Scenario B is illustrated here since the changes due to smoke-induced effects are more apparent. The five original sources have produced four discernible peaks in the tropospheric dose distribution, and the two U.S. sources have merged in the 30 day dose distribution. The tabulated values presented in Table 7.8 are averages over 20° latitude bands. The dose in "hotspots" can be examined by looking at peaks on the 10° × 10° grid. Typically the highest value for a grid square ($\sim 5 \times 10^5$ km^2) is about a factor of 6 to 8 higher than the Northern Hemisphere average dose. There will also be local areas much smaller than the 10° × 10° grid size where the peak doses would be considerably higher.

Particle size. By changing the initial assumed distribution of radioactivity on large and small particles from 43 and 57% to 70 and 30%, respectively, the average dose in the Northern Hemisphere increases about 25%. This is due to more rapid deposition of the larger particles.

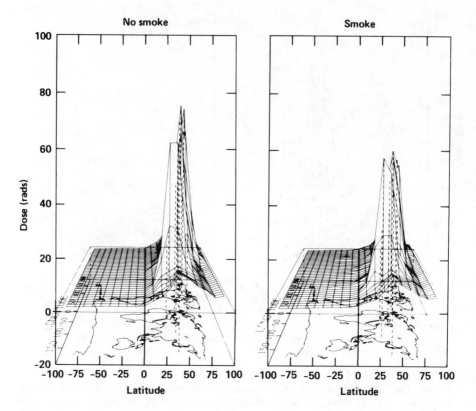

Figure 7.7.   Same as Figure 7.6, but a different viewing angle

Source locations. By shifting the source about 5° on a great circle, zonal changes in dose of 10–20% are observed, but the hemispheric averages do not change significantly. The zonal changes are primarily due to the source strength shifts, but variations in local weather on the first day of the OSU meteorological input to GRANTOUR also play a role.

Initial weather conditions. By starting on day 10 of the Oregon State University July climate (rather than day 1), dose estimates for the northern mid-latitude bands change significantly (about 30%), but the Northern Hemispheric average is unchanged. This indicates that initial weather conditions may produce significant variations in local dose, but that these may average out over hemispheric areas.

As GRANTOUR treats only the troposphere and GLODEP2 has been used for the stratospheric contributions (which assumes an unperturbed stratosphere), additional calculations using a computer model that includes the perturbed stratosphere should be undertaken.

7.4 INTERNAL DOSE DUE TO INHALATION
AND THE FOOD CHAIN

One serious problem following a large-scale nuclear exchange is radioactive contamination of drinking water. Those cities that are damaged would undoubtedly lose their water system due to power loss and ruptured supply pipes. Suburban residents within the local fallout pattern would encounter heavily contaminated water supplies and would have to rely on stored water. Surface water supplies would be directly contaminated by fission products.

During the first few months in areas extending several hundred kilometers downwind of an explosion, the dust, smoke, and radioactivity could cause severe water pollution in surface waters. The dominant fission product during this time would be ^{131}I (iodine-131). Beyond a few months, the dominant fission product in solution would be ^{90}Sr (strontium-90) (Naidu, 1984). Many of the fission products would remain fixed in fallout dust, river and lake sediments and soils. In rural areas, intermediate and long-term fallout would pollute water supplies to a lesser extent than the city and suburban supplies. In the absence of additional contamination from runoff, lakes, reservoirs and rivers would gradually become less contaminated as water flowed through the system.

Initially groundwater supplies would remain unpolluted but they may be difficult to tap. Eventually, however, some groundwater could become contaminated, and remain so for some tens of years after a nuclear war. It would take hundreds or thousands of years for an aquifer to become pure (or nearly so) (van der Heijde, 1985). Doses from drinking this water would be small, but, nonetheless, possibly above current water quality standards. In the long term, ^{90}Sr and ^{137}Cs (cesium-137) would be the major radionuclides affecting fresh water supplies.

The GLODEP2 fractional deposition rates have been used to calculate ^{90}Sr surface concentrations. The results are given in Table 7.9 for the Northern Hemisphere winter and summer seasons. The values are based on the Knox (1983) 5300 Mt baseline scenario A, and are expressed in mCi/km^2 for a 6-year period over 20° latitude bands. The maximum deposition occurs between 30–70° N. The concurrent deposition values for ^{137}Cs can be obtained by multiplying the ^{90}Sr values by 1.6. These values assume an unperturbed atmosphere. As stated earlier, introducing smoke and soot into the troposphere and stratosphere would probably slightly reduce Northern Hemisphere values and slightly increase ^{90}Sr deposition in the Southern Hemisphere.

Significant doses to individual human organs can also arise from specific radionuclides via food pathways. Such doses are caused by consumption of radioactively contaminated milk, meat, fish, vegetables, grains, and other foods. For a normal atmosphere, various researchers (ICRP30, 1979; Kocher,

TABLE 7.9.
AVERAGE ACCUMULATED STRONTIUM-90 DEPOSITION
(MCI/KM^2) AFTER SIX YEARS AS A FUNCTION OF LATITUDE

	Latitude band							
Season	70–90N	50–70N	30–50N	10–30N	10S–10N	10–30S	30–50S	50–70S	70–90S
Winter	271	937	862	234	39	25	47	26	3
Summer	226	946	978	237	26	19	39	30	10

1979; Ng, 1977; Lee and Strope, 1974) have provided means to calculate organ doses for a number of radionuclides and food pathways. However, in a post-nuclear war atmosphere perturbed by large quantities of smoke, the results of the above studies may not be valid since the dose in rads/Ci from soil to animal feed to humans are highly variable geographically and depend upon the degree of perturbation of weather and ecosystems.

However, the internal total body dose (the sum of the dose to each organ weighted by the risk factor due to consumption of various foods) has been very roughly estimated by J. Rotblat (private communication)to be about 20% of the external dose from local fallout, about equivalent for intermediate time scale fallout, and somewhat greater than the external dose from long-term fallout. These estimates are very uncertain. Further consideration of the pathways of fission products into the food chain is given in Volume II.

7.5 SUMMARY

Methods for estimating doses from radionuclides have been studied for more than thirty years. During this period, a better recognition of the effects that may be most important has developed, although there are no assurances that all of the crucial issues have been investigated.

For radionuclides, the most important short-term consequence is the downwind fallout during the first few days of relatively large radioactive particles lofted by surface explosions. The deposition of fresh radioactive material in natural and induced precipitation events could also contribute to enhanced surface dose rates over very limited areas (hotspots) both near to and far away from detonation sites. For both local fallout and distant hotspots, dose rates can be high enough to induce major short- and long-term biological and ecological consequences (see Volume II).

Calculations of local fallout fields were performed using the KDFOC2 model and an escalating nuclear exchange scenario (described in Chapter 2). In this illustrative example where simple assumptions are made about the overlap of fallout plumes, these estimates indicate that about 7% of the land surface in the U.S., Europe, and the U.S.S.R. would be covered by lethal external gamma-ray doses exceeding 450 rads in 48-hrs, assuming a protection factor of 1 (i.e., no protective action is taken). A similar area estimate is obtained for lethal doses exceeding 600 rads in 2-weeks. More realistic overlap calculations would suggest that these areas could be greater (by a factor of 3 in one specific case). For those survivors protected from radiation by structures, these areas would be considerably reduced. Areas of sub-lethal debilitating exposure (≥ 200 rads in 48 hrs) would, however, be larger. A good approximation is that these areas are inversely proportional to the 48 hr dose. In local fallout fields of limited area, the dose from beta rays could be high enough to significantly affect surviving biota. Variations in fallout patterns in the local fallout zones could range over orders of magnitude. If large populations could be mobilized to move from highly radioactive zones or take substantial protective measures, the human impact of fallout could be greatly reduced.

The uncertainties in these calculations of local fallout could be several factors. In addition, using different scenarios (e.g., all surface bursting or little surface bursting of weapons) could modify the calculated lethal areas by several factors. There are a number of other factors that could change these local fallout assessments. Fission fractions of smaller modern weapons could be twice the baseline assumption of 0.5. Adding these to the scenario mix could increase lethal fallout areas by about 20% of the baseline calculation. Tactical weapons, ignored in the baseline scenario, could increase lethal local fallout areas in certain geographical regions, particularly within western Europe, by up to 20% of the baseline scenario. Internal radiation exposure could increase the average total doses to humans by up to 20% of the external dose. Targeting of nuclear fuel cycle facilities could contribute to radiation doses (see Appendix 7A).

For global fallout, different computer models and scenarios have been intercompared. The calculations predict that the 50 year unsheltered, unweathered average external total body gamma-ray dose levels in the Northern Hemisphere would be about 10 to 20 rads, and about 0.5 to 1 rad in the Southern Hemisphere. The peak doses of 20 to 60 rads appear in the 30° to 50° north latitude band. Values predicted for the global population dose using the assumptions made in this study are typically about 6×10^{10} person-rads. The doses in the maxima grid points using a $10^\circ \times 10^\circ$ latitude and longitude mesh size, are a factor of 6 to 8 higher than the Northern Hemisphere averages. Fifty to seventy five percent of the global fallout dose would be due to the tropospheric injection of radionuclides that are deposited in the first month. These results were obtained assuming a normal (unperturbed) atmosphere, and have an estimated confidence level of a factor of 2 for a given scenario. The most sensitive parameter that affects global fallout levels is the scenario (e.g. total yield, yield mix, surface or airburst, burst locations).

Additional calculations involving a perturbed atmosphere indicate that the above dose assessments would be about 15% lower in the Northern Hemisphere, and marginally higher (approximately 1 rad) in the Southern Hemisphere compared to predictions for the unperturbed atmosphere. These results are consistent with the projection that smoke injections can increase vertical stability, inhibit precipitation, and increase interhemispheric transport.

Estimates of dose contributions from food pathways are much more tenuous. Rotblat (private communication) has roughly estimated that internal doses would be about 20% of the external dose from local fallout, about equivalent for intermediate fallout, and somewhat greater than the external dose from long-term global fallout.

APPENDIX 7A

Radioactivity from Nuclear Fuel Cycle Facilities

Three potential effects of radioactivity from nuclear fuel cycle facilities are considered in this report, although there is considerable controversy over the subject of the possible targeting with nuclear warheads of nuclear fuel cycle facilities. There is general agreement that enormous reservoirs of long-lived radionuclides exist in reactor cores, spent fuel rods, fuel reprocessing plants and radioactive waste storage facilities. Disagreement arises when the feasibility and extent of such a targeting strategy are considered. Even if one adopts the view that "what if" questions must be considered, there is still disagreement over the quantitative treatment of the potential dispersal of the radioactivity contained in these sources. In the present treatment, some of the assumptions regarding radioactivity release are considered highly improbable by a number of researchers. The results, therefore, should not be separated from the assumptions and large uncertainties associated with them.

7A.1 INTRODUCTION

A gigawatt nuclear power plant may be a valuable industrial target in a nuclear war. If a targeting rationale is proposed that the largest possible amount of Gross National Product be destroyed in an attack on a nation's industry (one measure of the worth of a target to a nation), then large (~ 1000 MW(e)) nuclear power plants could become priority targets for relatively small (≤ 125 kt) strategic weapons (Chester and Chester, 1976). In the U.S. there are about 100 such targets, and worldwide about 300. There are also military reactors and weapons facilities that could be targeted. Since these facilities may be targeted, reactor-generated radioactivity should be considered as part of the potential post-attack radiological problem.

Whether the radioactivity contained in a reactor vessel can be dispersed in a manner similar to a weapon's radioactivity is debatable. Nuclear reactor cores are typically surrounded by a meter-thick reinforced concrete building

267

that has about a 1 cm thick inner steel lining, many heavy steel structural elements inside the containment building, and an approximately 10 cm thick reactor vessel. Inside the reactor vessel are fuel rods and cladding capable of withstanding high temperatures and pressures. For the core radioactivity to be dispersed in the same way as the weapons radioactivity, all of these barriers must be breached. The core itself must be at least fragmented and possibly vaporized, and then entrained into the rising nuclear cloud column along with possibly hundreds of kilotons of fragmented and vaporized dirt and other materials from the crater and nearby structures, including the thick concrete slab that supports the reactor building. Under certain conditions of damage, there is a possibility of a reactor core meltdown resulting in the release of some of the more volatile radionuclides to the local environment. If this were to occur, however, the area of contamination would be relatively small compared to the contamination by a reactor core if it were to be pulverized and lofted by a nuclear explosion.

Some believe that if the reactor is within the weapon's crater radius that the core could potentially contribute to global and local fallout. Others believe that it cannot be fragmented and lofted in a manner similar to the weapon's residual radioactivity. Considering potential future terminal guidance technology, it is likely that the containment building would be within both a weapon's crater and fireball radius, *if* the containment structure were targeted with a surface-burst weapon.

Even if these barriers were secure, the primary contributor to the long-term dose at a nuclear power plant would not be the core. The most hazardous radioactivity, when assessing long-term effects (≥ 1 yr after attack), is that held in the spent-fuel ponds, if the reactor has been operating at full power for a few years. Since the spent-fuel storage usually has no containment building nor reactor vessel to be breached, it is much more vulnerable to being lofted by a nuclear weapon than the core materials. Unless spent-fuel is located at sufficient distance from a reactor, it could potentially become part of the local fallout problem.

Other nuclear fuel cycle radioactivity may also be significant. Reprocessing plants, although not as immediately important economically as power plants contain a great deal of radioactivity that could significantly contribute to the long-term doses. Also, military reactors developing fissile material and their reprocessing plants might be important wartime targets. They also hold significant amounts of radioactivity in their waste ponds and reactor cores.

Military ships fueled by nuclear power could be prime targets as well. Ships' reactors typically produce less power (~ 60–250 MW(t)) than commercial reactors (Ambio Advisors, 1982). They could, however, have substantially radioactive cores, depending on the megawatt-hours of service a shipboard reactor has produced since refueling. A large nuclear powered ship with more than one reactor, designed for years of service without refueling, can have nearly as much long-lived radioactivity (e.g., ^{90}Sr) on board as an operating commercial reactor (Rickover, 1980). Such shipboard reactors may also be more vulnerable to vaporization than commercial reactors.

Figure 7A.1 shows the gamma radiation dose rate-area integrals from a 1-Mt, all-fission nuclear weapon and from possible commercial fuel cycle facilities. In the first few days, the higher activity of the nuclear weapon debris dominates over the gamma radiation of the reactor. Likewise, gamma radiation levels from a light water reactor (LWR) is greater than that of 10 years worth of stored spent fuel for about one year after the detonation. Subsequently, the spent fuel would be relatively more radioactive. Similarly, the gamma radiation from 10 years of spent fuel is greater than the radioactivity of a 1 Mt fission weapon after about two months because of the greater abundance of long-lived gamma emitters in the spent fuel.

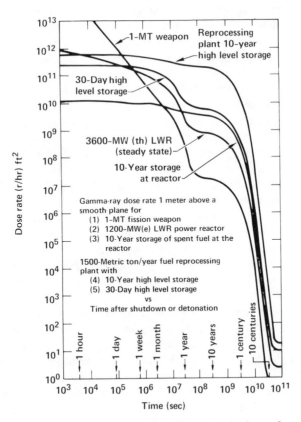

Figure 7A.1. Gamma-ray dose rate area integral versus time after shutdown or detonation (Chester and Chester, 1976)

Thus, for doses from a 1 Mt all-fission weapon detonated on a reactor, the core gamma radiation would be comparable to the weapon's radiation at about five days. By two months the gamma radioactivity from the weapon would have decayed by a factor of over 1000 from its value at 1 hour. Beyond about one year the gamma radiation from the weapon is insignificant compared to a reactor's radiation; however, the dose levels are no longer acutely life threatening.

7A.2 LOCAL FALLOUT

For dose estimates from local fallout, two timeframes are considered—the short-term, where there is acute lethal radiation, and the long term, when chronic doses become important. In the short-term, the gamma radiation is the main hazard. Later, specific radionuclides become important concerns for doses via food pathways.

For doses received within the first 48-hours, the nuclear weapon gamma radiation pathway for a high-yield (~ 1 Mt) warhead dominates the fuel-cycle gamma radioactivity, even if one assumes a worst case assumption in which all the radioactivity from the attacked nuclear fuel cycle facility is lofted with the weapon products. For lower yields and thermonuclear weapons, the core gamma radiation becomes more important, and could potentially dominate the dose, even at very early times. However, since there are now only approximately 100 nuclear power plants available for targeting in the U.S., and possibly a few hundred shipboard reactor targets which are dispersed over the globe (Ambio Advisors, 1982), and because there are typically more than a thousand other U.S. targets in major nuclear exchange scenarios, the impact of fuel cycle radiation to the total U.S. 48-hour external gamma-ray dose would likely be less than 10%.

In the long-term, the radioactivity from the core and spent-fuel ponds could have a dominant effect, both around the reactor and at substantial distances downwind. Because of the long-lived nature of the core radioactivity, civil defense measures (e.g., using expedient shelters) might also require modification when reactor radioactivity is contributing to the local fallout effects.

After about one year, the products from the nuclear fuel cycle could make a substantial contribution to the total gamma-ray dose fallout patterns over the U.S. Certainly, if released, fallout gamma radiation from a large reactor would dominate the dose of a 1 Mt weapon over the long-term (see Figure 7A.2).

In terms of radiological effects, individual radionuclides (e.g., ^{90}Sr) become more important over the longer time-frame than the whole-body gamma radiation. Assuming 50% fission weapons, it is possible to have more ^{90}Sr in a single reactor and its spent fuel pond than that produced in a

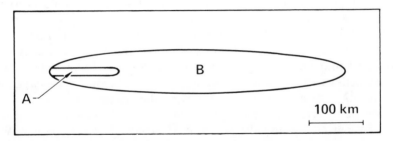

Figure 7A.2.   Contours of 100 rad fallout dose during one year's exposure, starting one month after the detonation of (A) a 1 Mt bomb, and (B) a 1 Mt bomb on a 1 GW(e) nuclear reactor (Rotblat, 1981)

1000 Mt attack. Most of the ^{90}Sr is in the spent fuel pond and thus could be more easily lofted as fallout than the ^{90}Sr in the heavily shielded reactor core. Accordingly, in the long term, the fuel-cycle ^{90}Sr contribution can dominate over the weapon contribution. For example, Chester and Chester (1976) calculated levels of ^{90}Sr much higher than the current maximum permissible concentration (MPC) over much of the U.S. farmland one year after an attack on the projected nuclear power industry of the year 2000. Scaling down their results to an attack on a 100 MW(e) nuclear power industry, they calculated that about 60% of the U.S. grain-growing capacity would be in areas that exceed current ^{90}Sr MPC levels.

The previous discussion emphasizes the effects on U.S. targets since past studies have focused on these. The conclusions, however, are more general.

7A.3 GLOBAL FALLOUT

In calculation of the potential global fallout, assumptions have been made that facilitated calculations and allowed estimation of expected dose. For example, it was assumed that each nuclear facility would be surface targeted by a high yield, accurately delivered warhead that would completely pulverize and vaporize all of the nuclear materials, and that these materials would then follow the same pathways as the weapon materials (a worst case assumption). It was assumed further that the major nuclear facilities in a 100 GW(e) civilian nuclear power industry would also be attacked. The results should be viewed as providing estimates that approach maximum global fallout for an attack on a commercial nuclear power industry of 100 Gw(e). Higher estimates would be obtained, however, using the same assumptions by including military facilities and a larger civilian industry.

This hypothetical reactor attack scenario assumed that, as part of the 5300 Mt exchange of Knox (1983), some of the warheads would be targeted on nuclear power facilities. Specifically 0.9 Mt weapons would be surface burst on 100 light water reactors (LWR's), 100 10-year spent fuel storage (SFS) facilities, and one fuel reprocessing plant (FRP). With a 0.9 Mt surface burst on each facility, 2% of the radioactive fission products would be injected into the troposphere and 48% into the stratosphere. The remaining activity (50%) would contribute to local fallout. Such large yields were assumed because of the hardness of the nuclear reactor. If smaller yield weapons were used to target the nuclear facilities, the relative injections of radioactivity into the troposphere would be much greater. While the weapons radioactivity would result in higher doses on the ground, this would not be true for the nuclear facility radioactivity. This is because of the relatively slow decay of the facilities' radioactivity. Hence, a faster deposition time would not significantly affect the 50 year dose. The patterns and local concentrations of fallout deposition would, however, be affected.

Using GLODEP2 and a Northern Hemisphere winter scenario, the resulting unsheltered, unweathered doses are shown in Table 7A.1. The largest value of 95 rads for the total of weapons plus the nuclear power industry occurred in the 30–50° N latitude band. The doses obtained for the Southern Hemisphere were about a factor of 30 smaller than in the Northern Hemisphere. The majority of the dose contributions came from the spent fuel storage facilities and the high level waste in the reprocessing plant.

TABLE 7A.1.
FIFTY-YEAR EXTERNAL GAMMA-RAY GLOBAL FALLOUT DOSE IN RADS FOR NINE LATITUDE BANDS ASSUMING A FULL NUCLEAR ATTACK, INCLUDING A FULL-SCALE, TOTALLY EFFECTIVE ATTACK ON A 100 GW(E) NUCLEAR POWER INDUSTRY. THESE VALUES DO NOT ACCOUNT FOR WEATHERING, SHELTERING OR RAINOUT

	Latitude bands								
Source	70–90N	50–70N	30–50N	10–30N	10N–10S	10–30S	30–50S	50–70S	70–90S
Weapons	4.5	27.3	32.9	6.9	0.8	0.6	0.8	0.5	0.09
LWR[a]	1.8	6.3	9.1	3.0	0.6	0.3	0.3	0.1	0.01
SFS[b]	6.7	23.8	32.7	11.3	2.3	1.0	1.0	0.4	0.03
FRP[c]	4.1	14.6	20.1	7.0	1.4	0.6	0.6	0.2	0.02
Total	17.1	72.0	94.8	28.2	5.1	2.5	2.7	1.2	0.15

[a] LWR = 100 light water reactors
[b] SFS = 100 spent fuel storage facilities
[c] FRP = fuel reprocessing plant

Figure 7A.3 is a plot of accumulated dose in the 30°–50° N latitude band as a function of time out to 50 years (200 quarter years) for the 5300 Mt scenario (Northern Hemisphere winter injection) with and without the targeting of nuclear power facilities. The bulk of the dose from the weapons alone for this scenario resulted from deposition in the first year. The relative contributions of the nuclear facilities were minimal in the first year, but became larger with time. At 50 years, the contribution of the nuclear facilities would be approximately double that of the weapons alone. In addition, while the weapons-only curve at 50 years is almost flat, the nuclear facilities curve has a positive slope with the radioactivity continuing to directly affect future generations.

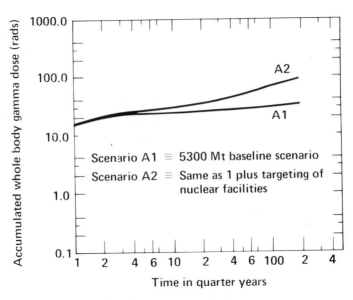

Figure 7A.3. Accumulated dose at 30°–50° N vs time for scenario A, with (A2) and without (A1) an attack on U.S. nuclear facilities

An attack on all of the world's civilian nuclear fuel cycle facilities (approximately 300 GW(e)) would scale the above results up by about a factor of three, although this scenario is even less likely. The potential effect is growing in time; the world's nuclear capacity has been projected to grow to 500 GW(e) by 1995. A significant contribution could also come from targeting military nuclear facilities, with results qualitatively similar to those obtained from attacking power plants.

In summary, using some "worst case" assumptions for a speculative nuclear war scenario wherein 100 GW(e) of the nuclear power industry is included in the target list, the 50 year global fallout dose is estimated to increase by a factor of 3 over similar estimates wherein nuclear power facilities are not attacked.

Accounting for possible moderate to heavy attacks on civilian and military nuclear facilities, for the internal doses necessarily accompanying the external doses (perhaps doubling or tripling these) over generations, the formation of localized hotspots with up to ten times the average radioactivity—in combination with all the other sources of radioactivity—it seems that reactor debris could result in significant long-term radiological problems for humans and ecosystems. Many of these problems involving the radiological assessments associated with nuclear facilities are unresolved and uncertain, but deserve more thorough attention.

Environmental Consequences of Nuclear War Volume I:
Physical and Atmospheric Effects
A. B. Pittock, T. P. Ackerman, P. J. Crutzen,
M. C. MacCracken, C. S. Shapiro and R. P. Turco
© 1986 SCOPE. Published by John Wiley & Sons Ltd

CHAPTER 8

Research Recommendations

Discussions in the preceeding chapters have pointed out numerous areas of uncertainties in our understanding of the chain of events linking a major nuclear exchange to environmental consequences. These uncertainties affect the prediction of the magnitude, longevity and detailed nature of the effects. However, knowledge of the general characteristics of the potential short-term atmospheric response is based on fundamental principles which are less subject to uncertainty. Nevertheless, it is important to reduce the many uncertainties in order to more effectively portray the potential consequences of a nuclear war.

Some of the remaining uncertainties can never be resolved short of a nuclear war itself. Besides questions of how nuclear weapons might be used, there are questions of scale (city-sized fires, hemispheric-scale smoke palls) that cannot be addressed experimentally and, therefore, cannot be confirmed directly. However, there are many other uncertainties that can be substantially reduced through careful research. Such research could also prove useful in areas of atmospheric dynamics, physics, and chemistry that currently are not adequately understood. It is the purpose of this chapter to identify those areas where experimentation and analysis may prove to be illuminating.

The following research recommendations are the result of discussions among the authors of this volume and the participants of the SCOPE-ENUWAR Synthesis Workshop held at the University of Essex in June 1985. The recommendations are divided into several categories, essentially along the lines of the various technical areas discussed in the study. The final section is an exception, offering a few broad recommendations to the science community at large.

8.1 STRATEGIC DOCTRINE

The issue of the potentially severe, global-scale environmental effects of a nuclear war is relatively new. The strategic implications of this issue should be discussed with and factored into the thoughts and concepts of government and military planners. There should be ongoing discussions between the planners and the research scientists so that issues such as weapons inventories, targeting strategies, and plausible expectations about the manner in which nuclear exchanges might develop can be treated using assumptions that are widely agreed upon.

It is recommended that:

- A wider range of plausible scenarios and associated conditions must be developed in order to serve as a basis for considering the sensitivity of atmospheric effects to issues such as weapons types and use, trends in weapons development, the proximity of targets and cities, and other factors.

8.2 ELECTROMAGNETIC PULSE

The International Union of Radio Scientists has begun to study the implications of the predicted electromagnetic pulse (EMP) on the operation of electrical and electronic equipment in a nuclear conflict. This important, but very uncertain, effect must be better understood if more robust international communication systems for crisis control are to be developed. Thus, it is recommended that:

- The impact of EMP generated by high-altitude nuclear detonations during an international crisis should be more accurately characterized. Implications for future systems designed to prevent the breakdown of international communications and safety mechanisms at critical facilities such as nuclear reactors should be investigated.

8.3 DUST

While dust has played only a secondary role in the most recent studies of the climatic effects of nuclear war, it is clear that the contribution of dust to the overall problem—in terms of atmospheric and radiological effects—could be significant. Although it would be quite difficult to reduce all of the uncertainties in estimates of the physical and optical properties of dust, some of them could be narrowed. With this in mind, the potential properties of dust raised by nuclear explosions at continental sites where detonations might occur should be more accurately defined.

In particular, the following efforts should be carried out:

- Studies of soil characteristics, including the size distribution and index of refraction of the soil dust;
- A reanalysis of archived nuclear test data for the amount and characteristics of the soil dust generated; and
- A careful determination of the organic contents of surface and subsurface soils that might contribute to the absorption of sunlight aloft or to the production of soot in the fireball.

8.4 FUEL LOADING

The research carried out thus far clearly indicates that estimates of fuel loadings in and adjacent to target areas are an important aspect of the total problem. While rough estimates of potential fuel loadings have been made from limited fuel surveys in cities, biomass assays in wildlands, and statistics on world production and consumption rates of flammable materials and fuels, the figures obtained so far are only preliminary. Moreover, little information has been developed on the geographical distributions of the fuels or their proximity to likely target zones. It also has become apparent that the storage of fossil fuels, either directly or indirectly in fossil fuel products, may dominate smoke and soot emission estimates, surpassing, for example, the importance of wood used in construction.

Therefore, specific information should be obtained with regard to:

- Inventories of fossil fuel storage;
- Combustible fuel burdens in urban areas (cities, suburbs, and industrial zones);
- Combustible fuel burdens in rural areas (forests, grasslands, and agricultural zones), particularly in areas that are in the vicinity of potential targets of nuclear weapons and;
- Inventories of the storage of dangerous and/or toxic chemicals in potential blast zones that could be spilled or dispersed in a nuclear exchange.

8.5 SMOKE PRODUCTION AND PROPERTIES

The production and properties of smoke from large fires are areas of great uncertainty and great importance. While the issue of scaling from small or moderate size fires to the massive fires expected in the advent of a nuclear war cannot be resolved at this time, there is clearly a great deal of research that can be done to resolve some of the related uncertainties in this area.

Essential information on the quantity of smoke produced, its elemental carbon fraction, and its morphological and optical properties could be obtained from large-scale experimental fires.These might, for instance, be fires involving pools of liquid fuels such as oil or kerosene spread over areas of 10^2 to 10^4 m^2 or structural fires set for experimental purposes. Fires of even larger size in stands of forest could be helpful for understanding smoke production in large fires (as well as aiding studies of the plume dynamics for such fires). Obviously, very careful and critical planning will be required if useful scientific goals are to be achieved in such experimental situations. For example, redundant measurements and cross calibrations of instruments should be employed, as well as consistent *in situ* and ground-based observations.

In field environments of large-scale fires, measurements of the following quantities should receive the highest priority:

- Smoke production efficiencies and yields, especially in large and very hot fires;
- The composition of smoke, with particular emphasis on the fraction of elemental carbon;
- The morphology (shape) of the smoke particles, both initially and as a function of time;
- Size distributions and coagulation rates as a function of time
- The radiative properties of smoke at visible, near infrared, and thermal infrared wavelengths, including variations with time (aging);
- Microphysical interactions of smoke with condensed water in convective plume situations.

A large body of laboratory data already exists on the formation and properties of smoke and soot. Nevertheless, most of the important properties associated with climatic and environmental effects are not well defined. Clearly, further interpretation of laboratory experiments should be made and new measurements should be carried out to refine our knowledge and fill some of the gaps in the data base.

Laboratory studies are needed on a number of properties of smoke, such as:

- Smoke and elemental carbon yields from different types of fuels under varying conditions of heating and ventilation;
- Particle agglomeration rates, size distributions, and morphology;
- Optical properties of fresh and aged smoke; and
- Nucleation properties of smoke particles from different sources.

In addition to the measurement programs on experimental and laboratory fires, several other avenues of research should be explored, including:

- Designing properly-scaled, experimental fires involving house-sized fuel arrays to obtain rough estimates of smoke production on this scale;
- Performing theoretical studies of the optical properties of soot agglomerates to check and to allow for extrapolation from laboratory results;
- Conducting analog scattering and absorption experiments using microwaves on inhomogeneous and agglomerated particles; and
- Considering the feasibility of measuring smoke characteristics at unplanned fires in cities and wildlands using mobile instrument packages.

8.6 FIRE AND FIRE PLUME MODELING

Since experimental studies of massive urban fires will not be possible, model studies must be carried out. Research should continue on:

- The development of large-scale fires in cities following nuclear detonations;

- The sensitivity of fire growth and spread to fuel loadings and meteorological conditions; and
- The effect of adjacent fires on fire development.

Detailed models of cloud dynamics and microphysics are being developed to study natural convective systems in relation to observations. However, only preliminary assessments of large-scale fire processes have been carried out with these models. Modeling studies are needed especially to define the dynamics, microphysics, and smoke lofting by cloud systems evolving from fires as large as those expected to be ignited by nuclear detonations, because such very large fires are unattainable experimentally. Studies of natural convective systems would help provide an understanding of the fundamental processes which control fire plumes. Unfortunately, many of the basic microphysical processes in clouds are not well understood, and progress on detailed plume simulations will be limited accordingly.

Nevertheless, useful information could be obtained from model studies of:

- Fire plumes
  - in different weather environments,
  - with spatially and temporally varying heat sources, and
  - with interactive smoke microphysics;
- The formation of massive convective plumes and cumulonimbus storms over large fires;
- The interaction and merging of multiple plumes in close proximity;
- The microphysical processes of smoke scavenging and precipitation removal in large-scale, fire-induced storms.

As noted above, studies of experimental fires are important in order to determine the quantity of smoke produced and its optical properties. Experimental fire studies are equally important in the context of fire dynamics in order to validate plume models. Measurements that should be made are:

- Ambient meteorological conditions;
- The extent, intensity, and time history of the heat source;
- The height of the smoke injection and vertical extent of the plume;
- The frequency of occurrence and water content of capping clouds.

8.7 PLUME DISPERSION AND MESOSCALE EFFECTS

Mesoscale models that could be usefully applied to the unique meteorological situations that might occur after a nuclear war are currently under development as a result of public and scientific interest concerning improved weather prediction, acid deposition, and warning of severe storms. The science of mesoscale modeling is, however, still in a formative stage. Extensive observations and model validation studies have not yet been carried out, although many interesting features of local and synoptic scale meteorological events could be studied in this manner.

Modeling of smoke cloud dispersion and scavenging on the mesoscale, leading to the global-scale, should be initiated and extended to treat:

• The evolution of extensive smoke-filled ice anvils downwind from large conflagrations;
• The interaction of smoke plumes with synoptic weather systems;
• The response of the boundary layer and lower troposphere to blanketing by thick smoke plumes that may be intermittent or patchy;
• The potential trapping of toxic air pollutants in stable cold layers near the ground.

8.8 CLIMATE MODELING

General circulation models have already been modified to a limited degree to investigate smoke cloud effects following a nuclear war. The resulting studies have revealed the possibilities of accelerated global spreading and stabilization of massive smoke clouds, the potential for self-lofting of the smoke as a result of solar heating and for sudden temperature drops of land surfaces under dense smoke patches, even to temperatures below freezing during the warm season. The GCM simulations should be refined and extended to provide further insight into the problem.

With regard to GCMs, the following research recommendations are offered:

• Improve the treatments and parameterizations of key physical processes, such as
  – solar and infrared radiation transfer in mixed environments, including smoke and water and ice clouds or low level fog,
  – smoke scavenging and removal by clouds and precipitation,
  – diurnal and boundary layer effects, particularly under conditions of strong static stability. Where necessary, more sophisticated parameterizations of heat transport and cloud/fog formation should be developed.
• Perform model intercomparisons to determine the importance of different physical representations and grid resolution in calculating responses.
• Investigate the sensitivity of the results to the season of smoke injections, and extend simulations to cover periods of a year or more.
• Perform simulations using higher resolution and nested mesoscale and cloud scale models.

- Investigate the impact of patchiness of the smoke cloud on the radiative response of global climate models.
- Interpret in more detail the multi-dimensional climate and dynamics simulations that have already been carried out.
- Refine the treatment of stratospheric smoke in the models by including chemical and physical removal processes and interactive ozone photochemistry.

A broad base of information exists in meteorological and climatological records, astronomical data, satellite photographic archives, and the results of measurements made during atmospheric nuclear testing. A number of historical events have been uncovered that are relevant to the problem of climatic change induced by smoke and dust aerosols. Other examples may exist. Reasonable proposals to search archives and records (from World War II, for example) should be considered. Therefore,

- Historical data should continue to be reviewed for pertinent information on climatic perturbations, smoke cloud effects, and related physical phenomena.

Analyses of effects over periods of years to decades following a nuclear war have not yet been undertaken, although a number of speculations have appeared. This is an enormously complex problem, involving the forecasting of both physical and biological feedbacks for which analytical tools are poorly developed. Nevertheless, ideas should be pursued and estimates of effects obtained to see if critical factors have been omitted from existing studies, and to lay the groundwork for future studies when appropriate analytical tools become available.

Studies should be made of the long-term perturbations that are possible after a major nuclear exchange, including:

- Effects of climate feedback mechanisms such as ice-albedo coupling;
- Interactions between the oceans and the atmosphere over month to decadal time periods;
- The implications of massive chemical emissions during a war.

8.9 CHEMISTRY

Some of the earlier concerns about the indirect effects of nuclear weapons centered on the destruction of stratospheric ozone. Numerous studies of the effects on the ozone layer have been carried out, but none have included the possible addition of aerosol particles. Furthermore, the subject of tropospheric chemistry, especially near the surface following soon after the fires, has not been addressed in any detail. Therefore, a variety of studies need to be carried out, including:

- Estimate the emission of toxic gases during the burning of fossil-fuel derived products, such as plastic, and from large fires in general;
- Calculate the possible alterations to tropospheric chemistry (smog and acidic precipitation, for example) resulting from chemicals released by nuclear explosions and fires;
- Determine the response of organic and soot particles to ultraviolet irradiation; and
- Consider the interaction between sooty particles and reactive gases, particularly ozone.
- Simulate the concentrations of air pollutants and toxic materials that will build-up in confined river valleys, lowlands, and other sheltered areas as a result of smoldering fires.

8.10 RADIOACTIVITY

While considerable effort has been made to understand the effects of radioactivity and radioactive fallout, there are still important areas of uncertainty. Further research is recommended to:

- Calculate the local fallout in a manner that realistically treats overlap of radioactivity from adjacent surface bursts and includes the details of population distribution and land use patterns;
- Evaluate internal dose contributions in a post-nuclear war environment;
- Improve understanding of the radiological dose commitments associated with the potential targeting and damaging of civilian and military nuclear fuel cycle facilities;
- Extend the calculations of global fallout to include the stratospheric contribution in a smoke-perturbed atmosphere.

8.11 GENERAL RECOMMENDATIONS

The direct and indirect effects of a nuclear exchange are among the greatest problems facing the human race at this point in history. While the magnitude of the potential direct effects of a major nuclear exchange has been recognized since the first bombs were dropped on Hiroshima and Nagasaki, only recently have the major indirect effects come to be seen as comparably important on a global scale. Clearly, a great deal of research remains to be done. Furthermore, the study of indirect effects involves many scientific disciplines that must contribute and interact to bring together findings from scientists around the world. In addition, the findings must be expressed in a manner convincing to scientists, government, and the public around the world.

To provide continued improvement in the further understanding of this problem, we recommend that:

- An international committee or coordinating body of scientists be established to follow the events, research, and progress on the problems of the global effects of nuclear war, and to report on a regular basis to governments and national and international scientific unions on the status of work and understanding. The committee could also, upon request, provide information to those seeking to pursue research on the problem.

- Interactions between physical scientists and biological scientists should be continued as a means to promoting interdisciplinary insights and discoveries; the discussions between scientists during this project have increased the appreciation of interconnections and interdependencies in the natural world and of common human interests among scientists. Ongoing international cooperation on interdisciplinary global environmental problems, as demonstrated by this study, can accelerate learning and broaden the base of science, and should be encouraged.

References

Ackerman, T. P., Stenback, J. M., Turco, R. P., and Grover, H. (1985b) Land use analysis and fuel estimates for U.S. missile fields. Manuscript in preparation.

Ackerman, T. P., and Toon, O. B. (1981) Absorption of visible radiation in atmosphere containing mixtures of absorbing and non-absorbing particles. *Appl. Optics* **20**, 3661–3668.

Ackerman, T. P., Toon, O. B., and Pollack, J. B. (1985a) The climatic impact of optically-thick aerosol clouds. Submitted to J. Geophys. Res.

Ackerman, T. P., and Valero, F.-P. J. (1984) The vertical structure of Arctic haze as determined from airborne net flux radiometer measurements. *Geophysical Res. Let.* **11**, 469–472.

Aleksandrov, V. (1984) Update of climatic impacts of nuclear exchange. *Presented at the International Seminar on Nuclear War, Fourth Session: The Nuclear Winter and the New Defense Systems: Problems and Perspectives,* Erice, Italy, August 19–24, 1984.

Aleksandrov. V., and Stenchikov, G. L. (1983) On the modelling of the climatic consequences of the nuclear war. *The Proceedings on Applied Mathematics (1983).* The Computing Centre of the USSR Academy of Sciences, Moscow.

Alvarez, L. W., Alvarez, W., Asaro, F., and Michael, H. W. (1980) Extraterrestrial cause for the Cretaceous–Tertiary extinction. *Science* **208**, 1095–1108.

Alvarez, W., Kauffman, E. G., Surlyk, F., Alvarez, L. W., Acaro, F., and Michel, H. V. (1984) Impact theory of mass extinctions and the invertebrate fossil record. *Science* **223**, 1135–1141.

Ambio Advisors (1982) Reference scenario: How a nuclear war might be fought. *Ambio* **11**, 94–99.

Aoki, Y. (1978) Studies on probabilistic spread of fire. Building Research Institute, 80 pp.

Arkin, W. M., Cochran, T. B., and Hoenig, M. M. (1984) Resource paper on the U.S. nuclear arsenal. *Bull. Atomic Scientists.* August/September.

Arkin, W. M., and Fieldhouse, R. W. (1985) *Nuclear battlefields: Global Links in the Arms Race.* Ballinger Publ. Co., Cambridge, MA.

Asano, S., and Sato, M. (1980) Light scattering by randomly oriented spheroidal particles. *Appl. Optics* **19**, 962–974.

Ball, D. (1981) Can nuclear war be controlled? Adelphi Paper No. 169, International Institute for Strategic Studies, London.

Ball, D. (1982) U.S. strategic forces; How would they be used? *Int. Security* **7**, 31–60.

Ball, D. (1983) Soviet strategic planning and the control of nuclear war. *Soviet-Union* **10**, 201–217.

Bankston, C. P., Zinn, B. T., Browner, R. F., and Powell, E. A. (1981) Aspects of the mechanisms of smoke generation by burning materials. *Combustion Flame* **41**, 273–292.

Banta, R. M. (1982) An observational and numerical study of mountain boundary-layer flow. Colorado State University Paper No. 350, Ft. Collins, CO.

Banta, R. M. (1985) Late-morning jump in TKE in the mixed layer over a mountain basin. *J. Atmos. Sci.* **42**, 407–411.

Baum, H. R., and Mulholland, G. (1984) Smoke dynamics in a mass fire environment. *Proceedings of the Conference on Large Scale Fire Phenomenology*, U.S. National Bureau of Standards, September 10–13, 1984.

Bennett, J. O., Johnson, P. S. C., Key, J. R., Pattie, D. C., and Taylor, A. H. (1984) Foreseeable effects of nuclear detonations on a local environment: Boulder County, Colorado. *Environmental Conservation* **11**, 155–165.

Bergstrom, R. W. (1973) Extinction and absorption coefficients of the atmospheric aerosol as a function of particle size. *Contributions to Physics of the Atmosphere* **46**, 223–234.

Berry, M. V., and Percival, I. C. (1986) Optics of fractal clusters such as smoke. *Optica Acta* **33**, 557–591.

Bigg, E. K. (1985) Carbon particles emitted by intense fires and their possible climatic impact. Paper submitted to SCOPE/ENUWAR Committee.

Birks, J. S., and Staehelin, J. (1985) Changes in tropospheric composition and chemistry resulting from a nuclear war. University of Colorado, Boulder. Draft manuscript.

Bolin, B., and Cook, R. B. (eds.) (1983) *The Major Biogeochemical Cycles and Their Interactions, SCOPE 21*. Wiley, New York.

Borghesi, A., Bussoletti, E., Colangeli, L., Minafra, A., and Rubini, F. (1983) The absorption efficiency of sub-micron amorphous carbon particles between 2.5 and 40 μm. *Infrared Physics* **23**, 85–92.

Boubnov, B. M., and Golitsyn, G. S. (1985) Theoretical and laboratory modelling of the influence of the static stability on the atmospheric general circulation regime. *Proc. USSR Acad. Sci.* In press.

Bracken, P., and Shubik, M. (1982) Strategic war: What are the questions and who should ask them? *Tech. Soc.* **4**, 155–179.

Brinkmann, A. W., and McGregor, J. (1983) Solar radiation in dense Saharan aerosol in Northern Nigeria. *Quart. J. Royal Meteorol. Soc.* **109**, 831–847.

Bruce, C. W., and Richardson, N. M. (1983) Propagation at 10 μm through smoke produced by atmospheric combustion of diesel fuel. *Appl. Optics* **22**, 1051–1055.

Budyko, M. I. (1974) *Climate and Life*. Academic Press, New York and London.

Bull, G. A. (1951) Blue sun and moon. *Meteorol. Mag.* **80**, 1–4.

Carrier, G. F., Fendell, F. E., and Feldman, P. S. (1983) Criteria for onset of firestorms. In: (1983) 17th Asilomar Conference on Fire and Blast Effects of Nuclear Weapons, University of California. Lawrence Livermore National Laboratory, Livermore, CA, 60–65.

Carter, A. B. (1985) The command and control of nuclear war. *Sci. Amer.* **252**, 32–29.

Cess, R. D. (1985) Nuclear war: Illustrative effects of atmospheric smoke and dust upon solar radiation. *Climatic Change* **7**, 237–251.

Cess, R. D., Potter, G. L., Ghan, S. J., and Gates, W. L. (1985) The climatic effects of large injections of atmospheric smoke and dust: A study of climate feedback mechanisms with one- and three-dimensional climate models. *J. Geophys. Res.* **90**, 12937–12950

CGB, see Crutzen et al., 1984.

Chandler, C. C., Storey, T. G., and Tangren, C. D. (1963) Prediction of fire spread following nuclear explosions. Research paper PSW-5, Berkeley, California, 110 pp.

Chang, J. S., and Duewer, W. H. (1973) On the possible effect of NO_x injection in the stratosphere due to past atmospheric nuclear weapons tests, Lawrence Livermore National Laboratory Report UCRL-74480, Livermore, CA.

Chang, J. S., and Wuebbles, D. J. (1983) Effects of hypothetical exchanges of strategic nuclear weapons based on current SALT II stockpile guidelines. In: *The Consequences of Nuclear War on the Global Environment.* Washington, D.C., U.S. Government Printing Office.

Charney, J. G., Stone, P. H., and Quirk, W. J. (1975) Drought in the Sahara: A biogeophysical feedback mechanism. *Science* **187**, 434–435.

Chervin, R. M., and Druyan, L. M. (1984) The influence of ocean surface temperature gradient and continentality on the Walker circulation. Part I: Prescribed tropical changes. *Mon. Weath. Rev.* **112**, 1510–1523.

Chester, C. V., and Chester, R. O. (1976) Civil defense implications of the U.S. nuclear power industry during a large nuclear war in the year 2000. *Nuclear Technology* **31**, 326–338.

Chung, Y.-S., and Le, H. V. (1984) Detection of forest fire smoke plumes by satellite imagery. *Atmos. Environ.* **18**, 2143–2151.

Chýlek, P., Ramaswamy, V., and Cheng, R. J. (1984) Effect of graphitic carbon on the albedo of clouds. *J. Atmos. Sci.* **41**, 3076–3084.

Chýlek, P., Ramaswamy, V., and Srivastava, V. (1983) Albedo of soot-contaminated snow. *J. Geophys. Res.* **88**, 10837–10843.

Clark, T. L. (1979) Numerical simulations with a three-dimensional cloud model: Lateral boundary condition experiments and multicellular storm simulations. *J. Atmos. Sci.* **36**, 2191–2215.

Clements, H. B., and McMahon, C. K. (1980) Nitrogen oxides from burning forest fuels examined by thermogravimetry and evolved gas analysis. *Thermochimica Acta* **35**, 133–139.

Cochran, T. B., Arkin, W. M., and Hoenig, M. M. (1984) Nuclear weapons databook. In: *U.S. Nuclear Forces and Capabilities,* Vol. 1, Ballinger Publ. Co., Cambridge, MA, 340 pp.

Connell, P. S., and Wuebbles, D. J. (1986) Ozone perturbations in the LLNL one-dimensional model – calculated effects of projected trends in CFC's, CH_4, CO_2, NO_2 and Halons over the next 90 years. Lawrence Livermore National Laboratory Report, UCRL-95548. pp.113.

Cooper, J. A., and Watson, J. G. (1979) Portland aerosol characterization study. Final report PACS, Oregon Graduate Center, Beaverton, Oregon, 67 pp.

Cotton, W. R. (1985) Atmospheric convection and nuclear winter. *Amer. Scientist* **73**, 275–280.

Cotton, W. R., Stephens, M. A., Nehrkorn, T., and Tripoli, G. J. (1982) The Colorado State University three-dimensional cloud/mesoscale model-1982. Part II: An ice phase parameterization. *J. Rech. Atmos.* **16**, 295–320.

Cotton, W. R., and Tripoli, C. J. (1978) Cumulus convection in shear flow-three dimensional numerical experiments. *J. Atmos. Sci.* **35**, 1503–1521.

Cotton, W. R., Tripoli, C. J., Rauber, R. M., and Mulvihill, E.A. (1985) Numerical simulation of the effects of varying ice crystal nucleation rates and aggregation processes on orographic snowfall. *J. Clim. Appl. Meteor.* **25**, no.11, 11658–80.

Covey, C., Schneider, S. H., and Thompson, S. L. (1984) Global atmospheric effects of massive smoke injections from a nuclear war: Results from general circulation model simulations. *Nature* **308**, 21–25.

Covey, C., Thompson, S. L., and Schneider, S. H. (1985) Nuclear winter: A diagnosis of atmospheric general circulation model simulations. *J. Geophys. Res.* **90**, 5615–5628.

Crutzen, P. J. (1970) The influence of nitrogen oxides on the atmospheric ozone content. *Quart. J Royal Meteorol. Soc.* **96**, 320–325.

Crutzen, P. J. (1971) Ozone production rates in an oxygen, hydrogen, nitrogen oxide atmosphere. *J. Geophys. Res.* **76**, 7311–7327.

Crutzen, P. J., and Birks, J. W. (1982) The atmosphere after a nuclear war: Twilight at noon. *Ambio* **11**, 114–125.

Crutzen, P. J., Galbally, I. E., and Brühl, C. (1984) Atmospheric effects from post-nuclear fires. *Climatic Change* **6**, 323–364.

Crutzen, P. J., Delany, A. C., Greenberg, J., Haagenson, P., Heidt, L., Lueb, R., Pollock, W., Seiler, W., Wartburg, A., and Zimmerman, P. (1985) Tropospheric chemical composition measurements in Brazil during the dry season. *J. Atmos. Chem.* **2**, 233–256.

Davies, R. W. (1959) Large-scale diffusion from an oil fire. In: Landsberg, H. E., and van Mieghem, J. (eds.), *Advances in Geophysics,* Academic Press, New York, 413–419.

Day, T., Mackay, D., Nudeau, S., and Thurier, R. (1979) Emissions from *in-situ* burning of crude oil in the Arctic. *Water, Air, and Soil Pollut.* **11**, 139–152.

DCPA, see Defense Civil Preparedness Agency.

DeCesar, R. T., and Cooper, J. A. (1983) The quantitative impact of residential wood combustion and other vegetative burning sources on the air quality in Medford, Oregon. Report of Oregon Graduate Center, Beaverton, Oregon.

Deeming, J. E., Burgan, R. E., and Cohen, J. D. (1977) The national fire-danger rating system—1978. U.S. Department of Agriculture, Washington, DC, INT-39.

Defant, F. (1951) Local winds. In: *Compendium of Meteorology.* American Meteorological Society, Boston, MA, 655–672.

Defense Civil Preparedness Agency (DCPA) (1973) Response to DCPA questions on fallout. DCPA Research Report No. 20, November 1973, U.S. Defense Civil Preparedness Agency, Washington, DC.

Duewer, W. H., Wuebbles, D. J., and Chang, J. S. (1978) The effects of a massive pulse injection of NO_x into the stratosphere. Lawrence Livermore National Laboratory Report UCRL-80397, Livermore, CA, 6 pp.

Eagan, R. C., Hobbs, P. V., and Radke, L. F. (1974) Measurements of cloud condensation nuclei and cloud droplet size distributions in the vicinity of forest fires. *J. Appl. Meteorol.* **13**, 553–557.

Eaton, F., and Wendler, G. (1983) Some environmental effects of forest fires in interior Alaska. *Atmos. Environ.* **17**, 1331–1337.

Ebert, H. V. (1963) The meteorological factor in the Hamburg fire storm. *Weatherwise* **16**, 72–75.

Edwards, L. L., Harvey, T. F., and Peterson, K. R. (1984) GLODEP2: A computer model for estimating gamma dose due to worldwide fallout of radioactive debris. Lawrence Livermore National Laboratory Report UCID-20033, Livermore, CA.

Ehrlich, P. R., Harte, J., Harwell, M. A., Raven, P. H., Sagan, C., Woodwell, G. M., Berry, J., Ayensu, E. S., Ehrlich, A. H., Eisner, T., Gould, S. J., Grover, H. D., Herrera, R., May, R. M., Mayr, E., McKay, C. P., Mooney, H. A., Myers, N., Pimentel, D., and Teal, J. M. (1983) Long-term biological consequences of nuclear war. *Science* **222**, 1293–1300.

Environmental Protection Agency (EPA) (1972) *Compilation of air pollutant emission factors.* Office of Air Programs, Research Triangle Park, NC.

Environmental Protection Agency (EPA) (1978) *Compilation of air pollutant emis-*

sion factors. Third edition, Supplement No. 8, PB-288905, Office of Air Programs, Research Triangle Park, NC, 2.4–1.

EPA, see Environmental Protection Agency.

FAO, see Food and Agriculture Organization.

Faxvog, F. R., and Roessler, D. M. (1978) Carbon aerosol visibility vs particle size distribution. *Appl. Optics* **17**, 2612–2616.

Federal Emergency Management Agency (FEMA) (1982) What the planner needs to know about fire ignition and spread. Chapter 3, *FEMA Attack Environment Manual*, Federal Energy Management Agency, CPG 2–1 A3.

FEMA, see Federal Emergency Management Agency.

Fetter, S. A., and Tsipis, K. (1981) Catastrophic releases of radioactivity. *Sci. Amer.* **244** (4), 41.

Filipelli, K. J. (1980) Theoretical calculations of mushroom clouds from multi-burst spike attacks. Report AFWL-TR-80-32, Air Force Weapons Laboratory, Kirtland Air Force Base, Albuquerque, NM.

Fitzjarrald, D. R. (1984) Katabatic wind in opposing flow. *J. Atmos. Sci.* **41**, 1143–1158.

Foley, H. M., and Ruderman, M. A. (1973) Stratospheric NO production from past nuclear explosions. *J. Geophys. Res.* **78**, 4441–4450.

Food and Agricultural Organization (FAO) (1976) *Yearbook of Forest Products 1963–1974*. United Nations, Rome, Italy.

Foote, G. B., and Fankhauser, J. C. (1973) Airflow and moisture budget beneath a northeast Colorado hailstorm. *J. Appl. Meteorol.* **12**, 1330–1352.

Ford, D. (1985) *The-Button: The Pentagon's Strategic Command and Control System*. Simon and Schuster, New York.

Fritsch, J. M., and Chappell, C. F. (1980) Numerical prediction of convectively driven mesoscale pressure systems. Part I: Convective parameterization. *J. of the Atmospheric Sciences* **37**, 1722–1733.

Fritsch, J. M., and Maddox, R. A. (1981) Convectively driven mesoscale systems aloft. Part 1: Observations. *J. Appl. Meteorol.* **20**, 9–19.

Galbally, I. E., Crutzen, P. J., and Rodhe, H. (1983) Some changes in the atmosphere over Australia that may occur due to a nuclear war. 161–185 in *Australia and Nuclear War*, Denborough, M. A. (ed.), Croom Helm, Ltd., Canberra, Australia.

Galbally, I. E., Manins, P. C., and Crutzen, P. J. (1985) Injection of elemental carbon into the stratosphere by thermonuclear fireballs. Manuscript in preparation.

Gates, W. L., and Schlesinger, M. E. (1977) Numerical simulation of the January and July global climate with a two-level atmospheric model. *J. Atmos. Sci.* **34**, 36–76.

Gaydon, A. G., and Wolfhard, H. G. (1970) *Flames, Their Structure, Radiation and Temperature*. Third Edition, Chapman and Hall, London.

Gerber, H. E., and Hindman, E. E. (eds.) (1982) *Light Absorption by Aerosol Particles*. Publ. Spectrum Press, Hampton, VA.

Ghan, S. J., Lingaas, J. W., Schlesinger, M. E., Mobley, R. L., and Gates, W. L. (1982) A documentation of the OSU two-level atmospheric general circulation model. Climatic Research Institute Report No. 61, Oregon State University, Corvallis, OR, 391 pp.

Ghan, S. J., MacCracken, M. C., and Walton, J. J. (1985) The climate response to large summer-time injections of smoke into the atmosphere: Changes in precipitation and the Hadley circulation. IAMAP/IAPSO Joint Assembly, Honolulu, Hawaii. Lawrence Livermore National Laboratory Report UCRL-92324, Livermore, CA.

Gilmore, F. R. (1975) The production of nitrogen oxides by low-altitude nuclear explosions. *J. Geophys. Res.* **80**, 4553–4554.

Ginsberg, A. S. (1973) On the radiative regime of the surface and dusty atmosphere of Mars. *Soc. Phys. Dok.* **208**, 295–298.

Ginsberg, A. S., and Feigelson, Em. M. (1971) Some peculiarities of the radiative heat exchange in atmospheres of terrestrial planets. *Atmos. Ocean Phys.* **7**, 377–384.

Glasstone, S., and Dolan, P. (1977) *The Effects of Nuclear Weapons.* U.S. Department of Defense and U.S. Energy Research and Development Administration, 653 pp.

Golitsyn, G. S., and Ginsburg, A. S. (1985) Comparative estimates of climatic consequences of Martian dust storms and possible nuclear war. *Tellus Ser. B.* **37b**, no. 3, 173–181.

Greenberg, J. P., Zimmerman, P. R., Heidt, L., and Pollock, W. (1984) Hydrocarbon emissions from biomass burning in Brazil. *J. Geophys. Res.* **89**, 1350–1354.

Greenfield, S. M. (1957) Rain scavenging of radioactive particulate matter from the atmosphere. *J. Meteoro.* **14**, 115.

Grigoriev, A. A., and Lipatov, V. B. (1978) The smoke pollution of the atmosphere as it is seen from space. *Gidrometeorisdat,* Leningrad, 36 pp (in Russian).

Güsfeldt, K.-H. (1974) Bitumen. In: *Ullmanns Encyklopädie der Technischen Chemie, Band 8,* W. Foerst, Berlin, 527–547.

Gutmacher, R. G., Higgins, G. H., and Tewes, H. A. (1983) Total mass and concentration of particles in dust clouds. Lawrence Livermore National Laboratory Report UCRL-14397, Livermore, CA, 22 pp.

Haberle, R. M., Leovy, C. B., and Pollack, J. B. (1983) Some effects of global dust storms on the atmospheric circulation of Mars. *Icarus* **50**, 322–367.

Haberle, R. M., Ackerman, T. P., Toon, O. B., and Hollingsworth, J. L. (1985) Global transport of atmospheric smoke following a major nuclear exchange. *Geophys. Res. Let.* **12**, 405–408.

Hall, R. E., and DeAngelis, D. G. (1980) EPA's research program for controlling residential wood combustion emissions. *J. Air Pollut. Contr. Assoc.* **30**, 862–867.

Hammitt, J.K., Camm, F., Connell, P.S., Mooz, W.E., Wolf, K.A., Wuebbles, D.J. and Bemazai, A. (1987) Future emission scenarios for chemicals that may deplete stratospheric ozone. *Nature* **330**, 711–716.

Hampson, J. (1974) Photochemical war in the atmosphere. *Nature* **250**, 189–191.

Harvey, T. F. (1982) Influence of civil defense on strategic countervalue fatalities, Lawrence Livermore National Laboratory Report UCID-19370, Livermore, CA.

Harvey, T. F., and Serduke, F. J. D. (1979) Fallout model for system studies, Lawrence Livermore National Laboratory Report UCRL-52858, Livermore, CA.

Harwell, M. A., and Hutchinson, T. C. (1985) *Environmental Consequences of Nuclear War. II. Ecological and Agricultural Effects.* SCOPE 28, John Wiley & Sons, Chichester.

Haselman, L. C. (1980) TDC—A computer code for calculating chemically reacting hydrodynamic flows in two dimensions. Lawrence Livermore National Laboratory Report UCID-18539, Livermore, CA.

Heft, R. E. (1970) The characterization of radioactive particles from nuclear weapons tests. In: *Radionuclides in the Environment, Adv. in Chemistry Series* **93**, 254–281.

Henderson-Sellers, A., and Gornitz, V. (1984) Possible climatic impacts of land cover transformations, with particular emphasis on tropical deforestation. *Climatic Change* **6**, 231–257.

Hidy, G. M. (1973) Removal processes of gaseous and particulate pollutants. In:

Chemistry of the Lower Atmosphere. Rasool, S. I. (ed.), Plenum Press, New York, 121–176.

Hilado, C. J., and Machado, A. M. (1978) Smoke studies with the Arapahoe chamber. *J. Fire Flammability #9*, 240–244.

Hobbs, P. V., and Matejka, T. J. (1980) Precipitation efficiencies and the potential for artificially modifying extratropical cyclones. *Proceedings of Third WMO Conference on Weather Modification,* Clermont-Ferrand, France, July 1980, 9–15.

Hobbs, P. V., Radke, L. F., and Hegg, D. A. (1984) Some aerosol and cloud physics aspects of the "Nuclear Winter" scenario. *ICSU-SCOPE Workshop on "Nuclear Winter", Supplemental Volume, Proceedings 9th International Cloud Physics Conference.* Tallinn, Estonia, USSR, August 231–228.

Hofmann, P., and Krauch, C. H. (1982) Carbon source in the future chemical industries. *Die Naturwissenschaften* **69**, 509–519.

Horiuchi, S. (1972) The speed of fire spreading in a section of a town consisting of wooden buildings. In: *Architectural Fire Protection.* Asakura Book Company, Tokyo, 175–209.

Humphreys, W. J. (1940) *Physics of the air.* New York, McGraw-Hill.

Hunt, B. G. (1976) On the death of the atmosphere. *J. Geophys. Res.* **81**, 3677–3687.

Huschke, R. E. (1966) The simultaneous flammability of wildland fuels in the United States. Memorandum RM-5073-TAB, The RAND Corporation, Santa Monica, CA, 158 pp.

ICRP Publication 30 (1980) *Limits for Intakes of Radionuclides by Workers.* Pergamon Press, Oxford and New York.

Ishikawa, E., and Swain, D. L. (1981) *Hiroshima and Nagasaki, The Physical, Medical and Social Effects of the Atomic Bombings.* Iwanami Shoten Publishers. Translation by Basic Books, New York, 706 pp.

Issen, L. A. (1980) *Single-family residential fire and live loads survey.* Report NBSIR 80-2155, Gaithersburg, MD, National Bureau of Standards, 176 pp.

Ives, J. M., Hughes, E. E., and Taylor, J. K. (1972) *Toxic atmospheres associated with real fire situations.* National Bureau of Standards, U.S. Department of Commerce, NBS Report 10807, Division of Analytical Chemistry, 77 pp.

Izrael, Y. A., Petrov, V. N., and Severov, D. N. (1983) On the impact of the atmospheric nuclear explosions on the ozone content of the atmosphere. *Meteorologia i Gidrologia* **9**, 5–13.

Jaeger, L. (1976) Monatskarten des neiderschlags für die ganze Erde. *Berichte des Deutschen Wetterdienstes, Nr 139,* Offenbach, Germany 38 pp.

Jane's Weapon Systems, 1984–1985 (1984) Pretty, R. T. (ed.), Jane's Publ. Co., London.

Janzen, J. (1980) The extinction of light by highly non-spherical strongly-absorbing colloidal particles: Spectro-photometric determination of volume distributions for carbon blacks. *Appl. Optics* **19**, 2977–2985.

Jennings, S. G., and Pinnick, R. G. (1980) Relationship between visible extinction, absorption and mass concentration of carbonaceous smokes. *Atmos. Environ.* **14**, 1123–1129.

Jernelöv, A., and Lindén, O. (1981) Ixtoc I: A case study of the world's largest oil spill. *Ambio* **10**, 299–306.

Johnston, H. (1971) Reduction of stratospheric ozone by nitrogen oxide catalysts from supersonic transport exhaust. *Science* **173**, 517–522.

Johnston, H. S., Whitten, G. Z., and Birks, J. W. (1973) Effects of nuclear explosions on stratospheric nitric oxide and ozone. *J. Geophys. Res.* **78**, 6107–6135.

Jones, A. R. (1979) Scattering efficiency for agglomerates of small spheres. *J. Phys. D: Appl. Phys.* **12**, 1661–1672.

Jurik, T. W., and Gates, D. M. (1983) Albedo following fire in a northern hardwood forest. *J. Clim. App. Meteoro.* **22**, 1733–1737.

Kang, S.-W., Reitter, T. A., and Takata, A. N. (1985) Analysis of large urban fires. Paper 85-0457 presented at AIAA 23rd Aerospace Sciences Meeting, Reno, Nevada, January 14–17, 1985, 11 pp.

Katz, A. (1982) *Life After Nuclear War*. Ballinger Publ. Co., Cambridge, MA, 411 pp.

Katz, J. (1984) Atmospheric humidity in the nuclear winter. *Nature* **311**, 417.

Kelly, P. M., and Sear, C. B. (1984) Climatic impact of explosive volcanic eruptions. *Nature* **311**, 740–743.

Kemp, G. (1974) Nuclear forces for medium powers: Part I. Targets and weapons systems. Adelphi Paper No. 106, International Institute for Strategic Studies, London, 41 pp.

Kerker, M. (1969) *The Scattering of Light and Other Electromagnetic Radiation*. Academic Press, New York.

Kerr, J. W., Buck, C. C., Cline, W. E., Martin, S., and Nelson, W. D. (1971) Nuclear weapons effects in a forest environment—thermal and fire. Report N2:TR2-70, Washington, DC, Defense Nuclear Agency, 257 pp.

Kerr, R. A. (1984) Computer models gaining on El Niño. *Science* **225**, 37–38.

Khalil, M. A. K., and Rasmussen, R. (1983) Sources, sinks and seasonal cycles of atmospheric methane. *J. Geophys. Res.* **88** (C9), 5131–5144.

Kiuchi, W. (1953) Investigation of disasters by the atomic bombs on the relation between topography and development of city. In: *Collection of the Reports on the Investigation of the Atomic Bomb Casualties*. Science Council of Japan **1**, 158.

Klemp, J. B., and Wilhelmson, R. B. (1978a) The simulation of three-dimensional convective storm dynamics. *J. Atmos. Sci.* **35**, 1070–1096.

Klemp, J. B., and Wilhelmson, R. B. (1978b) Simulation of right- and left-moving storms produced through storm splitting. *J. Atmos. Sci.* **35**, 1097–1110.

Knox, J. B. (1983) Global scale deposition of radioactivity from a large scale exchange. *Proceedings of the International Conference on Nuclear War, 3rd Session: The Technical Basis for Peace*, Erice, Sicily, Italy, August 19–24, 1983 Servizio Documentazione dei Laboratori Frascati dell'INFN, July, 1984, pp. 29–46. Also, Lawrence Livermore National Laboratory Report UCRL-89907, Livermore, CA.

Kocher, D. C. (1979) Dose-rate conversion factors for external exposure to photon and electron radiation from radionuclides occurring in routine releases from nuclear fuel cycle facilities. *Health Physics* **38**, 543–621.

Labitzke, K., Naujokat, B., McCormick, M. P. (1983) Temperature effects on the stratosphere of the April 4, 1982 eruption of El Chichón, Mexico. *Geophys. Res. Lett.* **10**, 24–26.

Larson, D. A., and Small, R. D. (1982) Analysis of the large urban fire environment: Part II: Parametric analysis and model city simulations. Pacific Sierra Research Report 1210, Los Angeles, CA.

Lee, H., and Strope, W. E. (1974) Assessment and control of the transoceanic fallout threat. Report EGU 2981, Stanford Research Institute, Menlo Park, CA, 117 pp.

Lee, K. T. (1983) Generation of soot particles and studies of factors controlling soot light absorption. Ph.D. thesis, Department of Civil Engineering, University of Washington, Seattle, WA.

Lenoble, J., and Brogniez, C. (1984) A comparative review of radiation aerosol models. *Contributions to Physics of the Atmosphere* **57**, 1–20.

Lilly, D. K., and Gal-Chen, T. (1983) *Mesoscale Meteorology—Theories, Observations and Models.* Reidel Publishing Company.

Liou, K.-N. (1980) *An introduction to atmospheric radiation.* Academic Press, New York.

London, J., and White, G. F. (eds.) (1984) *The Environmental Effects of Nuclear War.* Westview Press, Inc., Boulder, CO, 203 pp.

Luther, F. M. (1983) Nuclear war: Short-term chemical and radiative effects of stratospheric injections. *Proceedings of the International Seminar on Nuclear War 3rd Session: The Technical Basis for Peace.* Erice, Italy 19–24 August, 1985 Servizio Documentazione dei Laboratori Frascati dell'INFN, July, 1984, 108–128, also available as Lawrence Livermore National Laboratory Report UCRL-89957.

Luti, F. (1981) Some Characteristics of a Two-Dimensional Starting Mass Fire with Cross Flow. *Comb. Science and Tech.*, **26**, 25–33.

MacCracken, M. C., Ellis, J. S., Ellsaesser, H. W., Luther, F. M., and Potter, G. L. (1981) The Livermore Statistical Dynamical Climate Model. Lawrence Livermore National Laboratory Report UCID-19060, Livermore, CA.

MacCracken, M. C. (1983) Nuclear War: Preliminary estimates of the climatic effects of a nuclear exchange, *Proceedings of the International Seminar on Nuclear War 3rd Session: The Technical Basis for Peace.* Erice, Italy, August 19–24, 1983 Servizio Documentazione dei Laboratori Frascati dell'INFN, July, 1984, pp. 161–183, also available as Lawrence Livermore National Laboratory Report UCRL-89770, Livermore, CA.

MacCracken, M. C., and Walton, J. J. (1984) The effects of interactive transport and scavenging of smoke on the calculated temperature change resulting from large amounts of smoke. *Proceedings of the International Seminar on Nuclear War 4th Session: The Nuclear Winter and the New Defense Systems: Problems and Perspectives.* Erice, Italy, August 19–24, 1984. In preparation. Also, Lawrence Livermore National Laboratory Report UCRL-91446, Livermore, CA.

Machta, L., and Harris, D. L. (1955) Effects of atomic explosions on weather. *Science* **121**, 75–81.

Maddox, R. A. (1980) Mesoscale convective clusters. *Bull. Amer. Meteoro. Soc.* **61**, 1374–1387.

Maddox, R. A., Perkey, D. J., and Fritsch, J. M. (1981) Evolution of an upper tropospheric feature during the development of a mesoscale convective complex. *J. Atmos. Sci.* **38**, 1664–1674.

Malone, R. L., Aver, L. H., Glatzmaier, G. A., Wood, M. C., and Toon, O. B. (1986) Nuclear winter: Three-dimensional simulations including interactive transport, scavenging, and solar heating of smoke. *J. Geophys. Res.* **91**, 1039–1053.

Maraval, L. (1972) Characteristics of soot collected in industrial diffusion flames. *Combust. Sci. Technol.* **5**, 207–212.

Marland, G., and Rotty, R. M., (1983) Carbon dioxide emissions from fossil fuels: A procedure for estimation and results for 1950–1981, Oak Ridge Associated Universities, Institute for Energy Analysis, 75 pp.

Marwitz, J. D. (1974) An airflow case study over the San Juan mountains of Colorado. *J. Appl. Meteorol.* **13**, 450–458.

Mass, C., and Robock, A. (1982) The short-term influence of the Mount St. Helens volcanic eruption on surface temperature in the northwest United States. *Mon. Wea. Rev.* **110**, 614–622.

Matson, M., Schneider, S. R., Aldridge, B., and Satchwell, B. (1984) Fire detection using the NOAA-series satellites. NOAA Tech. Report NESDIS 7, U.S. Department of Commerce, Washington, DC.

McMahon, C. K. (1983) Characterization of forest fuels, fires, and emissions, presented at the 76th Annual Meeting, Air Pollution Control Association, Atlanta, Georgia.

McMahon, C. K., and Tsoukalas, S. N. (1978) Polynuclear aromatic hydrocarbons in forest fire smoke, in Jones P. W., and Freudenthal R. I. (eds.) *Polynuclear Aromatic Hydrocarbons*. Raven Press, New York.

Meyer, S. M. (1984) Soviet theater nuclear forces: Part I. Development of doctrine and objectives. Adelphi Paper No. 187, International Institute for Strategic Studies, London.

Middleton, W. E. K. (1952) *Vision through the atmosphere*. University of Toronto Press, Toronto, Canada.

Military Balance 1983–1984 (1984) The International Institute for Strategic Studies, 23 Tavistock Street, London, England.

Military Encyclopedic Dictionary (1983) Military Publishing House, Moscow, USSR, p. 842.

Miller, M. J., and Pearce, R. P. (1974) A three-dimensional primitive equation model of cumulonimbus convection. *Quart. J. Royal Meteoro. Soc.* **100**, 133–154.

Mintz, Y. (1984) Chapter 6: The sensitivity of numerically simulated climates to land-surface boundary conditions. In: Houghton, J. T. (ed.), *The global climate*. Cambridge University Press.

Molenkamp, C. R. (1980) Numerical simulation of self-induced rainout using a dynamic convective cloud model. *Proceedings VIII International Conference on Cloud Physics*. Cloud Physics, Clermont-Ferrand, France, July 15–19, 503–506.

Morikawa, T. (1980) Evolution of soot and polycyclic aromatic hydrocarbons in combustion. *Combustion Technology* **5**, 349–352.

Muhlbaier, J. L. (1981) A characterization of emissions from wood-burning fireplaces. In: Cooper, J. A. and Malek, D. (eds.). *Residential Solid Fuels, Environmental Impacts and Solutions* **6**, Oregon Graduate Center, Beaverton, Oregon.

Muhlbaier-Dasch, J. (1982) Particulate and gaseous emissions from wood-burning fireplaces. *Environ. Sci. Technol.* **16**, 639–645.

Muhlbaier-Dasch, J., and Williams, R. L. (1982) Fireplaces, furnaces and vehicles as emission sources of particulate carbon. In: *Particulate Carbon: Atmospheric Life Cycle* Wolff, G. T. and Klimisch, R. L. (eds.), Plenum Pres, New York, 185–205.

Nachtwey, D. S., and Rundel, R. R. (1982) Ozone change: Biological effects. In *Stratospheric ozone and man: II.* Bower, F. A. and Ward, R. B. (eds.), Boca Raton, Florida, CRC press, 81–122.

Naidu, J. R. (1984) Impact on water supplies—II. SCOPE/ENUWAR meeting, New Delhi, India, February 1984 Draft manuscript.

NAS, see National Academy of Sciences.

Nathans, M. W., Thews, K., Holland, W. D., and Benson, P. A. (1970a) Particle size distribution in clouds from nuclear airbursts. *J. Geophys. Res.* **75**, 7559–7572.

Nathans, M. W., Thews, R., and Russell, I. J. (1970b) The particle size distribution of nuclear cloud samples. In: *Radionuclides in the Environment, Adv. in Chemistry Series* **93**, 360–380.

National Academy of Sciences (NAS) (1975) *Long Term World-wide Effects of Multiple Nuclear-weapon Detonations*. National Academy Press, Washington, DC.

National Research Council (NRC) (1983) El Niño and the Southern Oscillation: A Scientific Plan. National Academy Press, Washington, DC.

National Research Council (NRC) (1984) *Causes and Effects, of Changes in Stratospheric Ozone: Update 1983*. National Academy Press, Washington, DC., 213 pp.

National Research Council (NRC) (1985) *The Effects on the Atmosphere of a Major Nuclear Exchange*. National Academy Press, Washington, DC.

Newton, C. W., and Newton, H. R. (1959) Dynamical interactions between large convective clouds and environment with vertical shear. *J. Meteoro.* **16**, 483–496.

Ng, Y. C., Colsher, C. S., Quinn, D. J., and Thompson, S. E. (1977) Transfer coefficients for the prediction of the dose-to-man via the forage-cow-milk pathway from radionuclides released to the biosphere. Lawrence Livermore National Laboratory Report UCRL-51939, Livermore, CA.

Nolan, J. L. (1979) Measurement of light absorbing aerosols from combustion sources. In: *Carbonaceous Particles in the Atmosphere*, Novakov, T. (ed.), Lawrence Berkeley Laboratory, LBL-9037, 265–269.

NRC, see National Research Council.

Oboukhov, A. M., and Golitsyn, G. S. (1983) Atmospheric consequences of a nuclear conflict. *Zemlya i Vselennaya (Earth and Universe)* **5**, 4–7.

Office of Technology Assessment (OTA) (1979) *The Effects of Nuclear War*. Washington, DC, 151 pp.

Ogren, J. A. (1982) Deposition of particulate elemental carbon from the atmosphere. In: *Particulate Carbon: Atmospheric Life Cycle*, Wolff, G. T., and Klimisch, R. L. (eds.), Plenum Press, New York.

Ogren, J. A., and Charlson, R. J. (1984) Wet deposition of elemental carbon and sulfate in Sweden. *Tellus* **36B**, 262–271.

Ohkita, T. (1985) *Report of the Third U.S.-Japan Joint Workshop for Reassessment of Atomic Bomb Radiation Dosimetry in Hiroshima and Nagasaki*, Pasadena, CA, March 12–14, 1984.

Openshaw, S., Steadman, P., and Greene, O. (1983) *Doomsday, Britain After Nuclear Attack*. Basil Blackwell, Oxford, England.

Orlanski, I. (1975) A rational sub-division of scales for atmospheric process. *Bull. Amer. Meteoro. Soc.* **56**, 527–530.

O'Sullivan, E. F., and Ghosh, B. K. (1973) The spectral transmission, 0.5–2.2 μm, of fire smokes, *Combustion Institute European Symposium 1973*, Weinberg, F. J. (ed.), Academic Press, New York.

OTA, see Office of Technology Assessment.

Otterman, J., Chou, M. D., and Arking, A. (1984) Effects of non-tropical forest cover on climate. *J. Clim. Appl. Meteoro.* **23**, 762–767.

Parish, T. R. (1984) A numerical study of strong katabatic winds over Antarctica. *Mon. Wea. Rev.* **112**, 545–554.

Patterson, E. M., and McMahon, C. K. (1984) Absorption characteristics of forest fire particulate matter. *Atmos. Environ.* **18**, 2541–2551.

Patterson, E. M., and McMahon, C. K. (1985) Studies of the physical and optical properties of forest fire aerosols. *Proceedings of Conf. on Large Scale Fire Phenomenology*, Frankel, M. (ed.), National Bureau of Standards, Gaithersburg, MD, Sept. 10–13, 1984.

Patterson, E. M., and McMahon, C. K. (1986) Absorption properties and graphitic carbon emission factors of forest fire aerosols. *Geophysical Research Letters* **13**, 129–132.

Pearman, G. I., Hyson, P., and Fraser, P. J. (1983) The global distribution of atmospheric carbon dioxide: 1. Aspects of observations and modelling. *J. Geophys. Res.* **88C**, 3581–3590.

Penner, J. E. (1983) Tropospheric response to a nuclear exchange. *Proceedings of the International Seminar on Nuclear War 3rd Session: The Technical Basis for Peace.* Erice, Italy, August 19–24, 1983 Servizio Documentazione dei Laboratori Frascati dell'INFN, July, 1984, pp. 94–106, also available as Lawrence Livermore National Laboratory Report UCRL-89956, Livermore, CA.

Penner, J. E., and Haselman, L. C. (1985) Smoke inputs to climate models: Optical properties and height distribution for nuclear winter studies. *Proceedings of the International Seminar on Nuclear War, 4th Session: The Nuclear Winter and the New Defense Systems: Problems and Perspectives,* Erice, Italy, August 19–24, 1984, also available as Lawrence Livermore National Laboratory Report UCRL–92523.

Penner, J. E., Haselman, L. C., and Edwards, L. L. (1985) Buoyant plume calculations. Paper No. 84-0459 for AIAA 23rd Aerospace Sciences Meeting, Reno, Nevada, January 14–17, 1985.

Peterson, K. R. (1970) An empirical model for estimating world-wide deposition from atmospheric nuclear detonations. *Health Phys.* 18, 357–378.

Pielke, R. A. (1974) A three-dimensional numerical model of the sea breezes over south Florida. *Mon. Wea. Rev.* 102, 115–139.

Pittock, A. B. (1984) On the reality, stability, and usefulness of Southern Hemisphere teleconnections. *Aust. Meteoro. Mag.* 32, 75–82.

Plummer, F. G. (1912) Forest fires. U.S. Department of Agriculture Forest Service Bulletin, 39 pp.

Pollack, J. B., and Cuzzi, J. N. (1980) Scattering by non-spherical particles of size comparable to a wavelength: A new semi-empirical theory and its application to tropospheric aerosols. *J. Atmos. Sci.* 37, 868–881.

Porch, W. M., Ensor, D. S., Charlson, R. J., and Heintzenberg, J. (1973) Blue moon: Is this a property of background aerosol. *J. Appl. Optics* 12, 34–36.

Porch, W. M., Penner, J., and Gillette, D. (1986) Parametric study of wind generated supermicron particle effects in large fires. *Atmos. Environ.* 20, 919–929.

Proctor, F., and Bacon, D. (1984) Numerical investigations of large area fires. Conference on Large Scale Fire Phenomenology, Gaithersburg, MD.

Prodi, F. (1983) The scavenging of submicron particles in mixed clouds: Physical mechanisms-laboratory measurements. In: *Precipitation Scavenging, Dry Deposition and Resuspension.* Pruppacher, H. R., Semonin, R. G., and Slinn, W. G. N. (eds.), Elsevier, New York, 505–516.

Pruppacher, H. R., and Klett, J. D. (1980) *Microphysics of clouds and precipitation.* Reidel Publishing Company, 714 pp.

Quintiére, J. G., Birky, M., MacDonald, F., and Smith, G. (1982) An analysis of smoldering fires in closed compartments and their hazard due to carbon monoxide. *Fire and Materials* 6, 99–110.

Radke, L. F., Benech, B., Dessens, J., Eltgroth, M. W., Henrion, X., Hobbs, P. V., and Ribon, M. (1980a) Modifications of cloud microphysics by a 1000 MW source of heat and aerosol (the Meteotron project). *Proceedings of the Third WMO Scientific Conference on Weather Modification,* World Meteorological Organization, Geneva.

Radke, L. F., Hobbs, P. V., and Eltgroth, M. W. (1980b) Scavenging of aerosol particles by precipitation. *J. Atmos. Sci.* 19, 715–722.

Radke, L. F., Stith, J. L., Hegg, D. A., and Hobbs, P. V. (1978) Airborne studies of particles and gases from forest fires. *J. Air Poll. Control Assoc.* 28, 30–34.

Ramage, C. S. (1971) *Monsoon Meteorology.* Academic Press.

Ramanathan, V., and Coakley, J. A., Jr. (1978) Climate modeling through radiative-convective models. *Reviews of Geophysics and Space Physics* 16, 465–487.

Ramaswamy, V., and Kiehl, J. T. (1985) Sensitivities of the radiative forcing due to large loadings of smoke and dust aerosols. *J. Geophys. Res.* **90**, 5597–5613.

Randhawa, J. S., and van der Laan, J. E. (1980) Lidar observations during dusty infrared Test-1. *Appl. Optics* **19**, 2291–2297.

Rasbash, D. J., and Pratt, B. T. (1979/80) Estimation of the smoke produced in fires. *Fire Safety Journal* **2**, 23–37.

Rasmusson, E. M., and Wallace, J. M. (1983) Meteorological aspects of the El Niño/Southern Oscillation. *Science* **222**, 1195–1202.

Reitter, T., McCallen, D., and Kang, S. W. (1982) Literature survey of blast and fire effects of nuclear weapons on urban areas. Lawrence Livermore National Laboratory Report UCRL-53340, Livermore, CA.

Rickover, H. G. (1980) *Naval nuclear propulsion program—1980.* Statement before the Procurement and Military Nuclear Systems Subcommittee, 96th Congress, U.S. Government Printing Office, Washington, DC.

Robock, A. (1984) Snow and ice feedbacks for prolonged effects of nuclear winter. *Nature* **310**, 667–670.

Robock, A., and Matson, M. (1983) Circumglobal transport of the El Chichón volcanic dust cloud. *Science* **221**, 195–197.

Roessler, D. M., and Faxvog, F. R. (1980) Optical properties of agglomerated acetylene smoke particles at 0.5145 μm wavelength. *J. Opt. Soc. Am.* **70**, 230–235.

Rosen, H., and Hansen, A. D. A. (1984) The role of combustion generated carbon particles in the absorption of solar radiation in the Arctic haze. *Geophys. Res. Lett.* **11**, 461–464.

Rosenblatt, M., Carpenter, G., and Eggum, G. (1978) Lofted mass characteristics and uncertainties for nuclear surface bursts, Defense Nuclear Agency Report 4760F, Washington, DC.

Rosinski, J., and Langer, G. (1974) Extraneous particles shed from large soil particles. *J. Aerosol Science* **5**, 373–378.

Rotblat, J. (1981) *Nuclear radiation in warfare.* Stockholm International Peace Research Institute SIPRI. Taylor and Francis, London.

Royal Society of Canada (1985) *Nuclear Winter and Associated Effects: A Canadian Appraisal of the Environmental Impact of Nuclear War.* Royal Society of Canada, 344 Wellington St., Ottawa, Canada.

Royal Society of New Zealand (1985). *The Threat of Nuclear War,* Royal Society of New Zealand, Science Centre and Library, Private Bag, Wellington, New Zealand.

Royal Swedish Academy of Sciences (1982) Nuclear War: The Aftermath. *Ambio* **11** (2–3), 76–176.

Russell, P. A. (1979) Carbonaceous particulates in the atmosphere: Illumination by electron microscopy. In: *Carbonaceous Particles in the Atmosphere,* Novakov, T. (ed.), Lawrence Berkeley Laboratory, LBL-9037, Berkeley CA, 133–140.

Ryan, J. A., and Henry, R. M. (1979) Mars atmospheric phenomena during major dust storms, as measured at the surface. *J. Geophys. Res.* **84**, 2821–2829.

Safronov, M., and Vakurov, A. (1981) *Fire in Forests.* Nauka Publishing House, Novosibirsk (in Russian).

Sagan, C., and Pollack, J. B. (1967) Anisotropic nonconservative scattering and the clouds of Venus. *J. Geophys. Res.* **72**, 469–477.

Sandberg, D. V., Pickford, S. G., and Daley, E. F. (1975) Emissions from slash burning and the influence of flame retardant chemicals. *J. Air Pollut. Control Assoc.* **25**, 278–281.

Sasaki, H., and Jin, T. (1979) Probability of fire spread in urban fires and their simulations. Fire Research Institute, 47.

Schlesinger, R. E. (1978) A three-dimensional numerical model of an isolated thunderstorm: Part I. Comparative experiments for variable ambient wind shear. *J. Atmos. Sci.* **35**, 690–713.

Schroeder, M. J., and Chandler, C. C. (1966) Monthly fire behavior patterns. U.S. Forest Service Research Note PSW-112, Pacific Southwest Forest and Range Experiment Station, Berkeley, CA, 15 pp.

Seader, J. D., and Einhorn, I. N. (1976) Some physical chemical, toxological and physiological aspects of fire smokes, 16th Symposium (International) on Combustion. The Combustion Institute, Pittsburgh, Pennsylvania, 1423–1445.

Seader, J. D., and Ou, S. S. (1977) Correlation of the smoking tendency of materials. *Fire Research* **1**, 3–9.

Seiler, W., and Fishman, J. (1981) The distribution of carbon monoxide and ozone in the free troposphere. *J. Geophys. Res.* **86**, 7255–7265.

Shapiro, C. S. (1974) The effects on humans of world-wide stratospheric fallout from a nuclear war and from nuclear tests. International Peace Research Institute, Oslo.

Shapiro, C. S., and Edwards, L. L. (1984) Scenario and parameter studies on global deposition of radioactivity using the computer model GLODEP2. Lawrence Livermore National Laboratory report UCID-20548, Livermore, CA.

Shimazu, Y. (ed.) (1985) Lessons from Hiroshima and Nagasaki, SCOPE-ENUWAR Report HI.01.85, p. 7.

Shostakovitch, V. B. (1925) Forest conflagrations in Siberia, with special reference to the fires of 1915. *J. For.* **23**, 365.

Shukla, J., and Mintz, Y. (1982) Influence of land-surface evapotranspiration on the Earth's climate. *Science* **215**, 1498–1501.

Singer, S. F. (1984) Is the "nuclear winter" real? *Nature* **310**, 625.

SIPRI, see Stockholm International Peace Research Institute.

Sitarski, M., and Kerker, M. (1984) Monte Carlo simulations of photophoresis of submicron aerosol particles. *J. Atmos. Sci.* **41**, 2250–2262.

Slingo, A., and Goldsmith, P. (1985) Nuclear winter: Short-wave radiative effects of soot aerosols. *Dyn. Climatol. Tech. Note* **24**, Met. Office, Bracknell, U.K.

Slinn, W. G. N. (1977) Some approximations for the wet and dry removal of particles and gases from the atmosphere. *Water, Air, Soil Pollut.* **7**, 513–543.

Slinn, W. G., and Hales, J. M. (1971) A reevaluation of the role of thermophoresis as a mechanism of in- and below-cloud scavenging. *J. Atmos. Sci.* **28**, 1465–1471.

Small, R. D., and Bush, B. W. (1985) Smoke production from multiple nuclear explosions in non-urban areas. *Science* **229**, 465–469.

Small, R. D., Larson, D. A., and Brode, H. L. (1984a) Asymptotically large area fires, *J. Heat Trans.*, **106**, 318–324.

Small, R. D., Larson, D. A., and Brode, H. L. (1984b) The physics of large fires, Conference on Large Scale Fire Phenomenology, Gaithersburg, MD.

Smith, C. D., Jr. (1950) The widespread smoke layer from Canadian forest fires during late September 1950. *Mon. Wea. Rev.* **78**, 180–184.

Stenchikov, G. L. (1985) Mathematical modeling of the influence of the atmospheric pollution on climate and nature. *SCOPE-ENUWAR* manuscript HI.07.85.

Stephens, S. L., and Birks, J. W. (1985) After nuclear war: perturbations in atmospheric chemistry. *Bio-Science.* **35***(9)* 557–562.

Stockholm International Peace Research Institute (SIPRI) Yearbook (1984) *World Armaments and Disarmament (1984)*. Taylor and Francis Ltd., London.

Stommel, H., and Stommel, E. (1983) *Volcano Weather.* Seven Seas Press, Newport, Rhode Island.

Stone, P. H., and Chervin, R. M. (1984) The influence of ocean surface temperature gradient and continentality on the Walker circulation. Part II: Prescribed global changes. *Mon. Wea. Rev.* **112**, 1524–1534.

Sud, Y. C., and Fennessy, M. (1982) A study of the influence of surface albedo on July circulation in semi-arid regions using the GLAS GCM. *J. Climatol.* **2**, 105–125.

Svirezhev, Y. M. (1985) Long-term ecological consequences of a nuclear war: Global ecological disaster. Computer Centre of USSR Academy of Sciences, Moscow. Draft manuscript.

Takata, A. (1969) Power density rating for fire in urban areas. Defense Technical Information Center, AD–695636, Arlington, VA.

Takata, A. (1972) Fire spread model adaptation. Defense Technical Information Center, AD–753989, Arlington, VA.

Takata, A., and Salzberg, F. (1968) Development and application of a complete fire-spread model: Volume 1 Development Phase. Defense Technical Information Center, AD–684874, Arlington, VA.

Takayama, H. (ed.) (1982) Studies of predictive methods for urban fire initiation and spread. In: *Ministry of Construction (1982) Development of Fire-fighting Techniques in Urban Areas.* Ministry of Construction, p. 7–136.

Tangren, C. D. (1982) Scattering coefficient and particulate matter concentration in forest fire smoke. *J. Air Poll. Control. Assoc.* **32**, 729–732.

Teller, E. (1984) Widespread after-effects of nuclear war. *Nature* **310**, 621–624.

Terrill, J. B., Montgomery, R. R., and Reinhardt, C. F. (1978) Toxic gases from fires. *Science* **200**, 1343–1347.

Tewarson, A. (1982) Experimental evaluation of flammability parameters of polymeric materials, In: *Flame Retardant Polymeric Material 3*, Plenum, New York, 97–153.

Tewarson, A. (1984) Scale effects on fire properties of materials, Technical Report for National Bureau of Standards, Center for Fire Research, *Washington, DC 20234, 42 pp.*

Tewarson, A., Lee, J. L., and Pion, R. F. (1981) The influence of oxygen concentration on fuel parameters, 18th Symposium (International) on Combustion, The Combustion Institute, Pittsburgh, Pennsylvania, 563–580.

Thompson, S. L., Aleksandrov, V. V., Stenchikov, G. L., Schneider, S. H., Covey, C., and Chervin, R. M. (1984) Global climatic consequences of nuclear war: Simulations with three-dimensional models. *Ambio* **13**, 236–243.

Thompson, S. L. (1985) Global interactive transport simulations of nuclear war smoke. *Nature* **317**, 35–39.

Tomaselli, V. P., Rivera, R., Edewaard, D. C., and Möller, K. D. (1981) Infrared optical constants of black powders determined from reflection measurements. *Appl. Optics* **20**, 3961–3967.

Toon, O. B., Pollack, J., Ackerman, T., Turco, R., McKay, C., and Liu, M. (1982) Evolution of an impact-generated dust cloud and its effects on the atmosphere. *Geological Society of America* **SP 190**, 13.

Toon, O. B., Pollack, J. B. (1982) Stratospheric aerosols and climate. In: *The stratospheric aerosol layer.* Whitten, R. C. (ed.), Springer-Verlag, Berlin, 121–147.

Treitman, R. D., Burgess, W. A., and Gold, A. (1980) Air contaminants encountered by firefighters. *Am. Ind. Hyg. Assoc. J.* **41**, 796–802.

Tripoli, G. J., and Cotton, W. R. (1980) A numerical investigation of several factors contributing to the observed variable intensity of deep convection over south Florida. *J. Appl. Meteoro.* **19**, 1037–1063.

Tripoli, G. J., and Cotton, W. R. (1982) The Colorado State University three-dimensional cloud/mesoscale model—1982. Part I: General theoretical framework and sensitivity experiments. *J. Rech. Atmos.* **16**, 185–220.

Tripoli, G. J., and Cotton, W. R. (1986) An intense quasi-steady thunderstorm over mountainous terrain. Part IV: Three-dimensional numerical simulation. *J. Atmos. Sci.* **43***(9)*, 894–912.

TTAPS, see Turco et al., 1983a.

Turco, R. P., Toon, O. B., Ackerman, T. P., Pollack, J. B. and Sagan, C. (1983a) Nuclear winter: Global consequences of multiple nuclear explosions. *Science* **222**, 1283–1292.

Turco, R. P., Toon, O. B., Ackerman, T. P., Pollack, J. B. and Sagan, C. (1983b) Global Atmospheric Consequences of Nuclear War. Interim Report, R & D Associates, Marina del Rey, CA, 144 pp.

Twomey, S. (1977) The influence of pollution on the short-wave albedo of clouds. *J. Atmos. Sci.* **34**, 1149–1152.

Twomey, S. A., Pietgrass, M., and Wolfe, T. L. (1984) An assessment of the impact of pollution on global cloud albedo. *Tellus* **36B**, 356–366.

United Nations (1979) The Effects of Weapons on Ecosystems. United Nations Environmental Programme, New York, 70 pp.

United Nations (1980a) *Nuclear Weapons,* Autumn Press, Brookline, MA 223 pp.

United Nations, (1980b) *Patterns of Urban and Rural Population Growth: Population Studies.* 68, United Nations, New York.

United Nations (1981) *Statistical Yearbook 1979/1980.* United Nations, New York.

United Nations Scientific Committee on the Effects of Atomic Radiation (UNSCEAR) (1982) United Nations, New York.

United States Senate Foreign Relations Committee (1975) *Analyses of Effects of Limited Nuclear Warfare,* Committee Print, Subcommittee on Arms Control, International Organization and Security Agreements of the Senate Committee on Foreign Relations, Washington, DC.

UNSCEAR, see United Nations Scientific Committee on the Effects of Atmospheric Radiation.

van de Hulst, H. C. (1957) *Light Scattering by Small Particles.* John Wiley & Sons, New York.

van der Heijde, P. K. M. (1985) *Groundwater contamination following a nuclear exchange.* SCOPE/ENUWAR Workshop Report. Delft, The Netherlands, October 3–5, 1984.

Vines, R. G., Gibson, L., Hatch, A. B., King, N. K., MacArthur, D. A., Packham, D. R., and Taylor, R. J. (1971) On the nature, properties and behavior of bush-fire smoke, Technical Paper No. 1, CSIRO, Australia, 32 pp.

Vittori, O. (1984) Transient Stephan flow and thermophoresis around an evaporating droplet, *Il Nuovo Cimento* **7**, 254–269.

Voice, M. E., and Gauntlett, F. J. (1984) The 1983 Ash Wednesday fires in Australia. *Mon. Wea. Rev.* **112**, 584–590.

Wade, D. D. (1980) An attempt to correlate smoke from Everglades fires with urban south Florida air quality. *Proceedings 6th Conference on Fire and Forest Meteorology.* Seattle, Washington, April 22–24, 156–162.

Waggoner, A. P., Weis, R. E., Ahlquist, N. C., Covert, D. S., Will, S., and Charlson, R. J. (1981) Optical characteristics of atmospheric aerosols. *Atmos. Environ.* **15**, 1891–1909.

Wagner, H. (1980) Soot formation—An overview, In: *Particulate Carbon Formation During Combustion*, Siegla, D. and Smith, G. W. (eds.), Plenum, New York, 1–29.

Walton, J. J. (1973) Scale-dependent diffusion. *J. Appl. Meteoro.* **12**, 547–549.

Walton, J. J., MacCracken, M. C., and Ellsaesser, H. W. (1983) Preliminary report on the LSDM transport sub-model TRANZAM. Lawrence Livermore National Laboratory Report UCID-19829, Livermore, CA.

Walton, J. J., and MacCracken, M. C. (1984) Preliminary report on the global transport model GRANTOUR. Lawrence Livermore National Laboratory Report UCID-19985, Livermore, CA.

Ward, D. E., McMahon, E. K., and Johansen, R. W. (1976) An update on particulate emissions from forest fires, Paper 76-2.2 at 69th Annual Meeting Air Pollution Control Association, Portland, Oregon, 15 pp.

Warren, S. G., and Wiscombe, W. J. (1985) Dirty snow after nuclear war. *Nature* **313**, 467–470.

Washington, W. M. (ed.) (1982) Documentation for the community climate model (CCM) version 0. NTIS PB82-194192, National Center for Atmospheric Research, Boulder, CO.

Webster, P. J. (1981) Monsoons. *Sci. Amer.* **245** (8), 108–118.

Weissermel, K., and Müller, W. H., (1981) Primärchemikalienent-wicklung—Bedarf und Deckung. *Chemische Industrie* **12**, 1–6.

Wexler, H. (1950) The Great Smoke Pall, September 24–30, 1950. *Weatherwise* **3**, 6–11.

WHO, see World Health Organization.

Wiersma, S. J., and S. B. Martin (1975) The nuclear fire threat to urban areas, April, 1975, AD-A018342, Available from Defense Technical Information Center, Arlington, VA.

Wik, M. et al. (1985) URSI factual statement on nuclear electromagnetic pulse (EMP) and associated effects. *International Union of Radio Science* **232**, March, 1985, 4–12.

Williamson, D. L. (1983) Description of NCAR community climate model (CCMOB). TN-210-+STR. National Center for Atmospheric Research, Boulder, CO.

Wilson, R. (1951) The blue sun of 1950 September. *Royal Astronomical Society Monthly Notices* **111**, 477–489.

WMO, see World Meteorological Organization.

Wohlstetter, A. (1983) Bishops, statesmen, and other strategists on the bombing of innocents. *Commentary*, June 15, p. 35.

Wohlstetter, A. (1985) Between an unfree world and none: Increasing our choices. *Foreign Affairs* **63**, 962–994.

Wolff, G. T., Countess, R. J., Groblicki, P. J., Ferman, M. A., Cadle, S. H., and Muhlbaier, J. L. (1981) Visibility-reducing species in the Denver "Brown Cloud"—II. Sources and temporal patterns. *Atmos. Environ.* **15**, 2485–2502.

Wolff, G. T., and Klimisch, R. L. (eds.) (1982) *Particulate Carbon: Atmospheric Life Cycle*. Plenum, 411 pp.

Woolley, W. D. and Fardell, P. J. (1982) Basic aspects of combustion toxicology. *Fire Safety J.* **5**, 29–48.

World Bank (1978) *Forestry*. (Sector Policy Paper), G. Donaldsen et al., 65 pp., Washington, DC.

World Health Organization, (WHO) (1983) *Effects of Nuclear War on Health and Health Services*. Report A36/12, Geneva, Switzerland, 152 pp.

World Meteorological Organization (WMO) (1982) *The stratosphere 1981: Theory and measurement,* Hudson, R. D. et al. (eds.), WMO Global Ozone Research and Monitoring Project Report No. 11, Geneva, World Meteorological Organization.

Wyrtki, K. (1975) El Niño—The dynamic response of the equatorial Pacific Ocean to atmospheric forcing. J. Phys. Oceanogr. **5**, 572–584.

Yoon, B. L., Wilensky, G. D., Yoon, D. C., and Grover, M. K. (1985) Nuclear dust and radiation cloud environments for aircraft and optical sensors, R & D Associates, Los Angeles, CA. Manuscript in preparation.

Zurek, R. W. (1982) Martian great dust storms: An update. *Icarus* **50**, 288–310.

APPENDIX 1

Executive Summary of Volume II:
Ecological and Agricultural Effects

by Mark A. Harwell and Thomas C. Hutchinson

The potential consequences to the global environment of a nuclear war have been the focus of several studies in the four decades since the first detonations of nuclear weapons in Japan. During this time, the *potential consequences* that would ensue from a modern nuclear war have increased dramatically, and the combination of much larger yields and much greater numbers of nuclear warheads could now result in a large-scale nuclear war having little in common with the relatively limited experiences of Hiroshima and Nagasaki. Simultaneously, the projections of the magnitude of impacts from a nuclear war have also increased steadily; however, the *perception of the consequences* of a large-scale nuclear war consistently have lagged behind the reality. New global-scale phenomena continue to be identified, even up to the present, and there remains a concern that decision-makers are operating with obsolete analyses and basing their policies on a foundation of misunderstanding of the total consequences of nuclear war.

The SCOPE-ENUWAR project had as one of its objectives the development of a comprehensive understanding of the nature of a post-nuclear war world, based on the full range of available information and models. Volume 1 of the ENUWAR report presented the bases for estimating potential effects on the physical environment, including possible climatic disturbances as well as fallout, UV-B, air pollutants, and other effects. The present volume takes up where the first left off, by specifically considering the potential consequences of such physical and chemical stresses on biological systems and on the ultimate endpoint of concern, i.e., effects on the global human population.

The approach taken in the biological analyses was to synthesize current understanding of the responses of ecological and agricultural systems to perturbations, relying on the expertise of over 200 scientists from over 30 countries around the Earth. Much of the synthesis took place in the context of a series of workshops that addressed specific issues; other work included conducting simulation modeling and performing detailed calculations of potential effects on the human populations of representative countries. We do not present the evaluation of a single nuclear war scenario as estimated by a single methodology; rather, a suite of methodologies were drawn upon collectively to develop an image of the aftermath of a large-scale nuclear war. The range of possible nuclear war scenarios is great; the estimates from the physical scientists of potential climatic consequences are not yet certain and continue to evolve with time. Those estimates are complex in their spatial and temporal distribution over the Earth, and the global landscape is covered by extremely complex ecological, agricultural, and human systems that react to perturbations in complex manners. For these reasons, the present volume investigates the *vulnerability* of these sytems to the types of perturbations possible after a nuclear war, offering readers the opportunity to form their own specific projections of biological and human consequences by providing calculations of vulnerabilities to benchmark assumptions.

Nevertheless, many conclusions are evident from considering these vulnerabilities to nuclear war perturbations. These include:

- Natural ecosystems are vulnerable to extreme climatic disturbances, with differential vulnerability depending on the ecosystem type, location, and season of effects. Temperature effects would be dominant for terrestrial ecosystems in the Northern Hemisphere and in the tropics and sub-tropics; light reductions would be most important for oceanic ecosystems; precipitation effects would be more important to grasslands and many Southern Hemisphere ecosystems.
- The potential for synergistic responses and propagation of effects through ecosystems implies much greater impacts than can be understood by addressing perturbations in isolation. For example, increased exposure to UV-B and to mixtures of air pollutants and radiation, while not crucially harmful for any one stress, might collectively be very detrimental or lethal to sensitive systems because of synergistic interactions.
- Fires as a direct consequence of a major nuclear exchange could consume large areas of natural ecosystems, but fire-vulnerable ecosystems are generally adapted to survive or regenerate via a post-fire succession. Other direct effects of nuclear detonations on ecological systems would be limited in extent or effect.
- The recovery of natural ecosystems from the climatic stresses postulated for an acute phase following a major nuclear war would depend on normal adaptations to disturbance, such as through presence of spores, seed banks, seedling banks, vegetative growth, and coppicing. For some systems, the initial damage could be very great and recovery very slow, with

full recovery to the pre-disturbed state being unlikely. Human-ecosystem interactions could act to retard ecological recovery.

- Because of limitations in the amounts of utilizable energy, natural ecosystems cannot replace agricultural systems in supporting the majority of humans on Earth, even if those natural ecosystems were not to suffer any impacts from nuclear war.

- Consequently, human populations are highly vulnerable to disruptions in agricultural systems.

- Agricultural systems are very sensitive to climatic and societal disturbances occurring on regional to global scales, with reductions in or even total loss of crop yields possible in response to many of the potential stresses. These conclusions consistently follow from a suite of approaches to evaluating vulnerabilities, including historical precedents, statistical analyses, physiological and mechanistic relationships, simulation modeling, and reliance on expert judgment.

- The vulnerabilities of agricultural productivity to climatic perturbations are a function of a number of different factors, any one of which could be limiting. These factors include: insufficient integrated thermal time for crops over the growing season; shortening of the growing season by reduction in a frost-free period in response to average temperature reductions; increasing of the time required for crop maturation in response to reduced temperatures; the combination of the latter two factors to result in insufficient time for crops to mature prior to onset of killing cold temperatures; insufficient integrated time of sunlight over the growing season for crop maturation; insufficient precipitation for crop yields to remain at high levels; and the occurrence of brief episodic events of chilling or freezing temperatures at critical times during the growing season.

- Potential disruptions in agricultural productivity and/or in exchange of food across national boundaries in the aftermath of a large-scale nuclear war are factors to which the human population is highly vulnerable. Vulnerability is manifested in the quantities and duration of food stores existing at any point in time, such that loss of the continued agricultural productivity or imports that maintain food levels would lead to depletion of food stores for much of the world's human population in a time period before it is likely that agricultural productivity could be resumed.

- Under such a situation, the majority of the world's population is at risk of starvation in the aftermath of a nuclear war. Risk is therefore exported from combatant countries to noncombatant countries, especially those dependent on others for food and energy subsidies and those whose food stores are small relative to the population.

- The high sensitivity of agricultural systems to even relatively small alterations in climatic conditions indicates that many of the unresolved issues among the physical scientists are less important, since even their lower estimates of many effects could be devastating to agricultural production and thereby to human populations on regional or wider scales.

• Longer-term climatic disturbances, if they were to occur, would be at least as important to human survival as the acute, early extremes of temperature and light reductions, suggesting that much greater attention should be given to those issues. Similarly, much greater attention is needed to resolve uncertainties in precipitation reduction estimates, since many of the agricultural systems are water-limited, and reduced precipitation can significantly reduce total production.

• Factors related to the possibility and rates of redevelopment of an agricultural base for the human population would have much influence on the long-term consequences to the human population. Interactions with societal factors would be very important.

• Global fallout is not likely to result in major ecological, agricultural, or human effects, as compared to effects of other global disturbances. Local fallout, on the other hand, could be highly consequential to natural and agricultural systems and to humans; however, the extent of coverage of lethal levels of local fallout and the levels of internal doses to humans from such fallout are inadequately characterized.

• Human populations are highly vulnerable to possible societal disruptions within combatant and noncombatant countries after a large-scale nuclear war, such as in the consequent problems of distribution of food and other limited resources among the immediate survivors. This is an area requiring a level of serious scientific investigation that has not yet been brought to bear on these issues.

As a part of the SCOPE-ENUWAR project, a workshop was held in Hiroshima, Japan in order for the scientists to gain a fuller appreciation of the human consequences of nuclear detonations. The considerations listed above indicate that as devastating as the Japanese atomic bombings were, as consequential to their victims even to the present day, and as important to the development of the 20th Century, they cannot provide a sense of what the global aftermath of a modern nuclear war could be like. Hiroshima today is a thriving, dynamic city reborn from complete devastation by interactions and support from the outside world; after a large-scale nuclear war, there would be essentially no outside world, and qualitatively new global-scale effects would occur that could devastate not just an urban population but the entirety of humanity. Although issues remain to be resolved, the information in this volume demonstrates some of the great vulnerabilities of agricultural, ecological, and societal support systems to the potential direct and indirect consequences of nuclear war. This demonstration of global frailties mandates the formulation of new global perspectives on avoiding the aftermath of nuclear war.

RECOMMENDATIONS FOR FURTHER RESEARCH

One of the more important outcomes of the process of holding workshops around the world to investigate the environmental consequences of nuclear war has been the recognition of the broad subject areas that have not yet received adequate research treatment. It is instructive to realize that many of the following listed topic areas involve the field of stress ecology; thus, improved understanding of the environmental consequences of nuclear war will progress in concert with advances in stress ecology.

During the synthesis workshop held at the Wivenhoe Conference Center, University of Essex in June 1985, the review committee for this volume compiled their recommendations for further research. (See Appendix A of Volume II for a listing of review committee members.) These were supplemented by a more detailed list compiled by M. Harwell, T. Hutchinson, W. Cropper, Jr., and C. Harwell. (See Appendix B of Volume II for both sets of recommendations.)

In brief summary:

1. Progress in this field will be strongly dependent on research efforts in stress ecology.
2. There needs to be an enhanced cooperation between physical and biological scientists in identifying the research priorities of the physical scientists. Examples of biologically-based suggestions include:
   - Climatologists need to do research on short-term variability in climate and the relationships between average climatic conditions and variances in climatic conditions in a post-nuclear-war framework.
   - Better resolution is needed of the potential levels of air pollution likely to result from a nuclear war.
   - Better resolution is needed of the potential levels of precipitation reductions likely to result from a nuclear war.
   - The potential for long-term climatic changes needs investigation, particularly involving feedback mechanisms, such as albedo changes, ice pack dynamics, and greenhouse-effect gases.
3. There is a need for models of environmental and ecosystem responses extending into the chronic post-nuclear-war phase. These should include better estimates of chronic phase parameters of temperature, light, and precipitation. These should also include much more experimental work on the effects of beta-radiation on plants and crops. Microcosm or enclosure experiments would be appropriate.
4. Explicit experimentation is needed to investigate synergisms including, for example, the interactive effects on biota of radiation, UV-B, and air pollution.
5. Synthesis studies addressing the specific conditions at local, regional, and

national levels are the next logical step in the process of understanding the effects of global nuclear war.

6. There is a need for further analysis of food stores, the likelihood of their destruction in a major nuclear war, their location, and other data, on a country-by-country basis.

7. Experiments need to be conducted using microcosms and enclosed whole ecosystems and agricultural systems to examine systems-level responses to climatic disturbances; particular attention can be given to recovery processes by using this approach.

8. New model development is needed to determine responses of ecosystems to climatic perturbations. Such models will have general applicability to other important issues in addition to nuclear war.

9. Existing sophisticated models of local fallout patterns need to be used to evaluate the range of dose levels that would be experienced after a nuclear war, based on a variety of nuclear war scenarios and weather conditions. Similarly, existing dose models need to be used to evaluate the range of internal radiation doses to be expected in the aftermath of a large-scale nuclear war.

10. There is a critical need for comprehensive and concerted study of the potential societal responses to nuclear war.

APPENDIX 2

*Glossary**

Absorption: The process whereby *visible* or *infrared radiation* passing through a medium is transformed via interactions into another form of energy, often heat.

See also *Optical depth, Extinction, Scattering.*

Absorption coefficient: A physical parameter expressing the efficiency of a medium in absorbing radiation. The absorption coefficient of homogeneous materials depends on the nature of the material and on the wavelength of the incident radiation.

See also *Extinction coefficient.*

Absorption optical depth: The part of the *optical depth* due to the *absorption* process.

See also *Optical depth.*

Activity: The number of nuclear disintegrations per second in a radioactive substance. The *SI* unit of activity is the Becquerel.

See also *Radioactivity.*

Advection: The process of transport of a fluid property as a result of the motion of the fluid itself. Advection refers conventionally to transport mainly by large scale horizontal and vertical winds. Vertical mixing by small scale processes (e.g., clouds) is usually referred to as *convection.*

Aerosol: A suspension of liquid or solid particles in a gas, usually the atmosphere. Most of the particles constituting the aerosols are 10 micrometers or less in diameter. Haze, most *smoke*, and some fogs are aerosols.

See also *Cloud condensation nuclei, Dust, Residence time, Soot.*

Air burst: A nuclear explosion that takes place high enough in the atmosphere that the *fireball* does not reach the surface. As a consequence, air bursts do not raise large quantities of surface materials (e.g., *dust*) in the atmosphere, nor do they give rise to *local fallout.*

See also *Blast wave, Electromagnetic pulse.*

---

* Any word printed in *italics*, whether in the text or in the 'See also' sections, refers to a concept defined elsewhere in the glossary. When such a cross-reference is indicated in the text, it is *not* repeated in the 'See also' section.

Albedo: The ratio of the reflected to incident radiation flux, usually for *visible radiation*, or *near-infrared radiation*. The *planetary albedo* of the Earth's surface-atmosphere system in the absence of smoke is about 30%.

Amorphous carbon: Small spheres of randomly-arranged crystallites of graphite formed in flames.

See also *Elemental carbon, Soot.*

Anvil: Elevated layer of ice particles produced by outflow at the top of a cumulonimbus cloud.

Atmospheric pressure: The static pressure that results from the weight per unit surface area of the mass of air lying above a location. The atmospheric pressure at sea level is about 100 kilopascals (kPa), equal to 1000 millibars (mb) or 14.7 pounds per square inch (psi).

Atomic bomb: A weapon that derives its destructive power from the energy released by the splitting or *fission* of nuclei such as *uranium* and *plutonium*.

Attenuation: Used either as a synonym of *extinction* of a directed flux of energy resulting from both *absorption* and *scattering*, or as the reduction in a directed flux of energy due to absorption and backscattering only.

Back-scattering: An instance of *scattering* where the change in direction of the *photon* or particle relative to its direction before the interaction is greater than 90°.

See also *Albedo, Forward-scattering.*

Beta particle: A beta particle is an electron emitted in a nuclear disintegration.

Biomass: The amount of carbon stored in the living tissues of an ecosystem. Sometimes also the dry weight of living materials, typically expressed in kg (carbon, or C), or in kg (dry matter), respectively, or kg/m^2 for areal loading.

Black rain: Name given to the dark, smoke-stained rain that fell on some neighborhoods of Hiroshima and Nagasaki shortly after the atomic bombings.

See also *Rainout, Washout.*

Blast wave: An intense *shock wave* created by a nuclear explosion in which the pressure, temperature, and fluid velocity discontinuities are very large.

See also *Peak overpressure.*

Boundary layer: The layer of the atmosphere that is affected by the nature and characteristics of the surface. The atmospheric or planetary boundary layer is typically several hundred meters deep at night, and up to a few km during the day, in the undisturbed atmosphere.

C^3: Acronym for Command, Control, and Communications, three components essential for the management and operation of military forces.

See also C^3I, *Electromagnetic pulse.*

C^3I: Acronym for Command, Control, Communications, and Intelligence.

See also C^3, *Electromagnetic pulse.*

CCN: Acronym for *Cloud Condensation Nuclei.*

Climate: The word climate is used in a number of ways, but usually refers to the state of the atmosphere, oceans and surface averaged over some area and period of time, usually greater than several days.

Cloud Condensation Nuclei [CCN]: Small, usually *hygroscopic, aerosols* on which water vapor condenses in clouds. These nuclei are essential for the formation of clouds and precipitation.

See also *Aerosol, Washout.*

CO: The chemical symbol for carbon monoxide, a gas produced in combustion processes, particularly when the availability of oxygen is limited.

Coagulation: Any process which agglomerates small *particles* or aerosols into larger ones.

See also *Residence time, Washout.*

Combustible burden: The amount of fuel available for combustion per unit area of building floor or ground area, expressed in kg/m^2.

See also *Fuel loading.*

Conflagration: A spreading fire driven by winds and *thermal radiation.*

Convection: The process of transport and mixing of a fluid as a result of localized vertical air motions.

Counterforce attack: A strategic attack that aims at destroying the military potential of the adversary, such as *missile silos,* air and naval bases, military depots, etc.

See also *Countervalue attack.*

Countervalue attack: A strategic attack that aims at destroying the industrial and economic base of the adversary.

See also C^3, *Counterforce attack.*

Delayed fallout: See *Radioactive fallout.*

Dose: The energy of *ionizing radiation* absorbed in a medium, per unit mass, integrated over time.

See also *Gray, Rad.*

Dry deposition: The return to the Earth's surface of *particles* in a precipitation-free environment, principally by impaction brought on by mixing of the air in the *boundary layer.*

See also *Radioactive fallout, Precipitation scavenging.*

Dust: Here, mainly small soil particles picked up by winds.

See also *Aerosol, Ground burst*

Dynamic pressure: The *pressure* on a surface produced by a wind.

See also *Overpressure, Peak overpressure.*

Early fallout: See *Radioactive fallout.*

Electromagnetic pulse [EMP]: The sharp and intense pulse of *electromagnetic radiation* within the radio frequency spectrum that is produced by a high altitude nuclear explosion. EMP can damage electrical power distribution systems, and unprotected electrical and electronic equipment, including telecommunications networks and computer systems.

Electromagnetic radiation: A form of energy contained in oscillating electrical and magnetic fields, traveling at the speed of light.

Elemental carbon: An approximately pure form of carbon, composed of either *amorphous carbon* or graphitic carbon, as distinguished from organic carbon or carbonates.

See also *Soot.*

EMP: Abbreviation for *Electromagnetic pulse.*

Entrainment: The process of mixing air from the environment into a cloud, plume or *fireball*, contributing to the dilution of the cloud.

ENUWAR: Acronym for Environmental Consequences of Nuclear War, the name of the *SCOPE* project studying this subject.

Epicenter: The location of the *nuclear weapon* at the time of the explosion.

See also *Ground zero, Hypocenter.*

Extinction: The depletion or reduction of a direct flux of radiation as it penetrates a substance (e.g., aerosols in the atmosphere), due both to *absorption* and *scattering.*

See also *Extinction coefficient.*

Extinction coefficient: A physical parameter expressing the efficiency of a substance in causing extinction. The extinction coefficient depends on the nature of the material and on the wavelength of the incident *radiation*.

See also *Absorption coefficient.*

Extinction cross-section: A measure, in units of area, of the probability of an interaction occurring between a particle and a photon, either by scattering or absorption.

Extinction efficiency: For a particle or molecule, the ratio of the extinction cross-section to the geometric cross-section or projected area (e.g., πr^2 for a spherical particle of radius r) of the particle or molecule.

Extinction optical depth: See *optical depth.*

Fallout: See *Radioactive fallout.*

Fireball: The volume of hot air that forms immediately after the explosion of a *nuclear weapon* in a *surface* or *air burst.* The fireball is luminous due to its high temperature.

See also *Mushroom cloud.*

Firestorm: The severe meteorological conditions resulting from an intense surface heating as a result of extensive and vigorous stationary fires. Firestorms develop a swirling cyclonic inflow of air that limits entrainment and lofts smoke to great heights. They have been observed to occur after intense incendiary bombing in Hamburg and Dresden and may have occurred after the nuclear explosion in Hiroshima.

Fissile material: Material capable of undergoing *fission* by interaction with slow (thermal) neutrons.

See also *Black rain, Conflagration.*

Fission: The process of splitting the nucleus of heavy chemical elements, usually *radionuclides,* into two or more smaller nuclei. This process can be accompanied by the emission of beta particles, gamma rays, and neutrons.

See also *Fusion.*

Fluence: The radiative energy flux per unit surface integrated over a finite period of time, and generally measured in cal/cm^2 (1 cal/cm^2 = 4.2×10^4 J/m^2).

See also *Ignition threshold.*

Forward-scattering: An instance of *scattering* where the change in direction of the incident *photon* or *particle* is less than $90°$.

See also *Back-scattering.*

Fuel loading: The mass amount of combustible fuel per unit area.

See also *Combustible burden.*

Fusion: The process of combining the nuclei of light chemical elements, such as deuterium or tritium, into larger nuclei, accompanied by the release of large amounts of energy. *Thermonuclear weapons* rely mainly on the fusion process.

See also *Fission.*

Gamma radiation or gamma ray: Energetic *photons* of *electromagnetic radiation*, with wavelengths nominally shorter than 0.1 nanometers. Gamma rays are emitted by nuclear processes, including explosions.

See also *X-ray.*

GCM: Acronym for *General Circulation Model.*

General Circulation Model [GCM]: Three-dimensional computer model based on the physical equations for fluid flow used to study the evolution of the large-scale features of the atmospheric circulation.

Gray [Gy]: The *SI* unit of *ionizing radiation* absorbed by a medium. It is equivalent to the *absorption* of 1 Joule per kilogram of material.

See also *Rad, Rem.*

Greenhouse effect: The common name for designating several radiative processes where specific atmospheric molecules (such as water vapor and carbon dioxide) allow most of the *solar radiation* to pass through the atmosphere to the surface, while most of the *infrared radiation* emitted by the surface is absorbed by the atmosphere and re-emitted both downward and upward. The downward re-radiation warms the surface, while the upward component is lost to space or absorbed higher up in the atmosphere.

Ground burst: A nuclear explosion on a land surface. Ground bursts raise large quantities of radioactive *dust* in the atmosphere, and produce *local fallout.*

Ground shock: An intense shock or pressure wave travelling in the ground as a result of a nuclear explosion.

See also *Blast wave.*

Ground zero: Synonym of *hypocenter* for an *air* or *ground burst.*

Half-life: The time required for the *radioactivity* of a given material to decrease to half of its initial value due to nuclear disintegrations.

See also *Plutonium, Uranium.*

H-bomb: Synonym of hydrogen or *thermonuclear weapon.*

Height of burst [HOB]: The altitude of a *nuclear weapon* at the time it explodes. The height of burst affects the type of damage generated.

Hertz [Hz]: Unit of frequency equal to 1 cycle per second.

Hydrophobic: The tendency of particles not to attract water vapor or become dissolved in water droplets.
See also *Hygroscopic.*

Hygroscopic: The tendency of particles to be susceptible to condensation of water vapor on them and to be soluble in water droplets.
See also *Hydrophobic.*

Hypocenter: The location at the ground surface immediately below a nuclear explosion, or the site of a surface burst (also called *ground zero*).

ICSU: Acronym for International Council of Scientific Unions.

Ignition threshold: The *fluence* from a nuclear fireball needed to ignite a given material. It is dependent on the type, size, dryness, and orientation of the material as well as on the yield of the *nuclear weapon.*

Incendiary efficiency: The area on the ground subject to fire ignition by a nuclear explosion per *kiloton* of *yield.*

Infrared radiation: The band of *electromagnetic radiation* in the wavelength interval 0.8 to 100 micrometers. About half the *solar energy* available at the Earth's surface is contained in the radiation band from 0.8 to 5 micrometers: this is often called the *near-infrared*. In contrast, *radiation* at longer wavelengths is called *thermal radiation*, or the thermal infrared.
See also *Greenhouse effect, Visible radiation.*

Ionizing Radiation: A high energy *photon* (such as an *X* or *gamma ray*) or particle (such as electrons or alpha particles) capable of ionizing an atom or molecule by stripping off electrons.

Isotope: Isotopes of a given chemical element differ in the number of neutrons present in the nucleus of the atom. Isotopes have similar, but not identical, physical and chemical properties.

Kiloton [kt]: An amount of energy equal to 4.2×10^{12} J. This is approximately the amount of energy that would be released by the simultaneous explosion of roughly a thousand tons of *TNT.*
See also *Hiroshima, Megaton, Yield.*

Local fallout: See *Radioactive fallout.*

Mass fire: A moving or stationary fire involving a large area.
See also *Conflagration, Firestorm.*

Megaton [Mt]: An amount of energy equal to 4.2×10^{15} J. This is approximately the amount of energy that would be released by the simultaneous explosion of roughly a million tons of *TNT*.

See also *Kiloton, Yield.*

Mesoscale: Term applied to atmospheric features on horizontal scales ranging from a one to several hundred kilometers and temporal scales from about an hour to a day.

Mesosphere: The atmospheric layer between the stratopause and the mesopause, i.e. roughly from 50 to 80 km above the surface.

Microwave radiation: The band of the *radio frequency spectrum* with a wavelength in the interval 0.1 to 500 millimeters, used in telecommunication and radar systems.

See also *Infrared radiation.*

Missile silo: A hardened concrete structure, usually totally buried, that contains a nuclear missile.

Mushroom cloud: The rising cloud formed just after a nuclear *air* or *ground burst* by convective winds that entrain and carry upward dust debris, and condensed water.

Near-infrared radiation: See *Infrared radiation.*

NO$_x$: A generic chemical symbol to designate a nonspecific mixture of the oxides of nitrogen, mainly NO and NO$_2$.

Nuclear fuel cycle facilities: All facilities involving nuclear fuel, from mining through to waste storage facilities, including nuclear reactors, spent fuel storage facilities, reprocessing plants, and waste depositories.

Nuclear radiation: A general term designating all of the ionizing radiation of a nuclear explosion; e.g., gamma rays, neutrons, electrons (or beta particles) and alpha particles.

Nuclear weapon: A generic name for a device that derives its explosive energy from *fission, fusion,* or both.

See also *Atomic bomb, Thermonuclear weapon.*

Nuclear winter: A phrase used to refer mainly to the sharp and widespread cooling and near-darkness that could result from the emission of massive amounts of *smoke* and other materials as a result of widespread fires induced by extensive nuclear attacks on urban areas, oil and gas storage facilities, and other developed and wildland areas. The extent of the cooling and associated reduction in precipitation would vary strongly with latitude, season, and proximity to coastlines. A wide range of perturbations is possible because of necessary assumptions and remaining uncertainties.

In its broadest usage, the term "nuclear winter" is sometimes considered to include the entire set of adverse environmental consequences following a nuclear war, including the additional effects of lofted *dust* (which could prolong or augment the cooling) and of *radioactive fallout.*

By extension, the term is sometimes used to refer to the consequent effects on ecosystems, agriculture, and human health and welfare.

Optical depth: A non-dimensional number used to describe the cumulative depletion, or extinction (due to both *absorption* and *scattering*) that a direct beam of *visible* or *infrared radiation* experiences as it travels through a medium. An extinction optical depth of 1 reduces the direct beam to 37% of its original value.

Overpressure: The excess of the local static pressure above the normal atmospheric pressure.

See also *Peak overpressure, Dynamic pressure.*

Ozone [O$_3$]: A molecule composed of three atoms of oxygen. In the unperturbed atmosphere, ozone is concentrated mainly in the lower *stratosphere* (20–50 km).

Peak overpressure: The maximum value, above the normal atmospheric pressure, of the static pressure attained during the passage of a *shock* or *blast wave.*

See also *Dynamic pressure, Overpressure.*

Photon: The quantized unit of electromagnetic energy. The energy carried by a photon is proportional to the electromagnetic wave frequency.

See also *Electromagnetic radiation.*

Planetary albedo: The average *albedo* of the planet Earth, as seen from space, about 0.3 or 30% for solar radiation under present conditions.

Plutonium [Pu]: A fissionable chemical element produced by nuclear reactions and used in *nuclear weapons.*

Precipitation scavenging: A term to designate the removal of gases, *aerosols* or nuclear debris from the atmosphere by precipitation processes. Precipitation scavenging includes both *rainout* and *washout* processes.

See also *Fallout, Scavenging.*

Primary ignition: The fires ignited as a direct result of the *thermal radiation* or *thermal flash* of a nuclear explosion.

Prompt Radiation: The ionizing *radiation* emitted during, and in the first minute or so after, a nuclear explosion, as opposed to that emitted later by the *radioactive fallout*. The prompt radiation contains about 5% of the total energy liberated by the explosion.

Protection factor: The attenuation factor of ionizing radiation due to the shielding provided by a structure or material.

Psi [psi]: A non-*SI* unit of pressure equal to one pound per square inch. The *SI* unit of pressure is the Pascal (1 psi = 6894 Pa).

Pyrotoxin: A term to designate toxic chemicals released during combustion, particularly from plastics and industrial chemicals.

Rad [rad]: A non-*SI* unit of absorbed energy from *ionizing radiation*, equivalent to the absorption of 100 ergs per gram, or 10^{-2} Joule per kilogram. The corresponding *SI* unit is the *Gray*, and 1 Gy = 100 rads. The name of this unit derives from the abbreviation for Radiation Absorbed Dose.

See also *Rem*.

Radioactive fallout: The radioactive debris (particles) generated by a nuclear explosion that is deposited on the surface at various times by dry deposition or in precipitation. *Early* or *local fallout* refers to the surface deposition within one day in the vicinity of the explosion. Global or *delayed fallout* refers to the settling or washout of radioactivity after the first day. Intermediate timescale fallout is deposited generally within the first month after the explosion, while long term fallout is deposited over times of months to years.

See also *Air burst, Black rain, Ground burst, Residence time, Scavenging, Washout*.

Radioactivity: The property of unstable chemical elements or isotopes to decay by emitting *nuclear radiation*.

See also *Activity, Fission, Plutonium, Radionuclide, Uranium*.

Radio frequency spectrum: The band of *electromagnetic radiation* with wavelengths between 1 millimeter (300 GHz) and 10^6 m (300 Hz).

See also *Infrared radiation*.

Radionuclide: A radioactive element or isotope.

See also *Plutonium, Radioactivity, Uranium*.

Rainout: The in-cloud removal of *aerosols* by incorporation into cloud water and subsequent precipitation.

See also *Black rain, Fallout, Precipitation scavenging, Scavenging, Washout*.

Rayleigh scattering or absorption: A particular type of scattering or absorption that occurs in cases where the wavelength of the incident radiation is much larger than a typical dimension of the particle.

Rem: A non-*SI* unit for absorbed radiation *dose*, defined as the *dose* that will produce the same biological effect as the *absorption* of 1 roentgen of X-rays or *gamma rays*. For *gamma rays*, 1 *rad* is equivalent to 1 rem. The name of this unit derives from the abbreviation for Roentgen Equivalent Man.

Residence time: The average length of time that material is expected to remain within a given system before being removed (e.g., the atmosphere).

See also *Scavenging, Washout.*

Scattering: The interaction of electromagnetic radiation with a particle, resulting in the deflection of the radiation.

See also *Absorption, Forward-scattering, Back-scattering, Rayleigh scattering.*

Scattering optical depth: The fraction of the *optical depth* due to the *scattering* process.

See also *Absorption optical depth.*

Scavenging: A general term referring to the processes of particle or gas collection, agglomeration and/or removal from the atmosphere, particularly by clouds and precipitation.

See also *Black rain, Rainout, Washout.*

Scenario: A description of the hypothetical development of a nuclear war, with specific assumptions about the number, yield, and space and time distribution of nuclear explosions.

SCOPE: Acronym for Scientific Committee on Problems of the Environment. SCOPE was established in 1969 by *ICSU* to identify interdisciplinary environmental issues requiring the most urgent attention.

Shock wave: A supersonic pressure wave in a medium, such as air, associated with nuclear detonations.

See also *Blast wave*

SI: Acronym for Système International of units. SI is a coherent system of units based on the meter, the kilogram, the second, the ampere, the kelvin, the mole, and the candela. A large number of derived units can be created by combining these 7 base units.

The SI system defines standard prefixes to designate multiples and fractions of the standard units, including:

Multiplication Factors			Prefix	SI Symbol
1 000 000 000 000	=	10^{12}	tera	T
1 000 000 000	=	10^9	giga	G
1 000 000	=	10^6	mega	M
1 000	=	10^3	kilo	k
100	=	10^2	hecto	h
10	=	10^1	deka	da
0.1	=	10^{-1}	deci	d
0.01	=	10^{-2}	centi	c
0.001	=	10^{-3}	milli	m
0.000 001	=	10^{-6}	micro	μ
0.000 000 001	=	10^{-9}	nano	n
0.000 000 000 001	=	10^{-12}	pico	p

Silo: See *missile silo*.

Single scattering albedo: The ratio of the *scattering* coefficient and the *extinction* coefficient for a given particle or medium.

Smoke: A suspension of solid and/or liquid particles and gases produced by combustion processes as well as wind blown debris and ash. Smoke is typically a heterogeneous mixture of particles of different sizes, structures, and composition, and may vary from white and oily (from smoldering combustion) to black and sooty (from flaming combustion).

See also *Black rain, Soot*.

Smoldering: A process of low temperature combustion in the absence of open flames.

Soot: A *smoke* component composed largely of *amorphous* or *elemental carbon*. Soot is particularly effective in absorbing *solar radiation*.

Strategic nuclear weapons: Weapons, generally of high yield, placed on intercontinental missiles (ICBMs) and other long range delivery systems (SLBMs, bombers with intercontinental range).

See also *Tactical nuclear weapons, Theater nuclear weapons*.

Stratosphere: The atmospheric layer between the *tropopause* and the stratopause, roughly between 12 and 50 km above the surface, characterized by a general increase in temperature with increasing altitude. The *residence time* of gases and particles in the stratosphere is much longer than in the troposphere.

See *Temperature inversion*.

Surface burst: A nuclear explosion occurring at or very close to the surface, either over water or land, so that the *fireball* intercepts the surface.

Synoptic: Organized atmospheric motions on spatial scales of about one to ten thousand kilometers and lasting one to several days.

Tactical nuclear weapons: Generally speaking, low yield nuclear weapons mounted on short range (less than 200 km) delivery systems, and built to be used on the battlefield.

See also *Strategic nuclear weapons, Theater nuclear weapons.*

Temperature inversion: A layer of the atmosphere in which the temperature is constant or increases with height. Such layers are usually stable against convection and mixing from below.

Theater nuclear weapons: Generally speaking, medium or large *yield nuclear weapons* carried by intermediate range missiles and aircraft for use in continental-size geographical regions.

See also *Strategic nuclear weapons, Tactical nuclear weapons.*

Thermal radiation: The radiation emitted by any body or substance as a result of its heat content (as measured by its temperature). For substances at normal temperatures (250 to 320 K), the thermal radiation is in the thermal-*infrared.* For the Sun and a nuclear fireball ($\sim$6000 K), the radiation is in the visible and near-infrared parts of the spectrum.

Thermal pulse or thermal flash: The intense but brief emission of heat and light in a nuclear explosion.

Thermonuclear weapons: A *nuclear weapon* that derives a substantial part of its explosive energy from nuclear *fusion.* Such weapons are sometimes called *H-bombs,* because they use hydrogen *isotopes* (typically a few kg of deuterium and tritium) as the fuel for the *fusion* process. Thermonuclear weapons use *fission* devices as a trigger. The energy *yield* of a thermonuclear weapon can be much larger than that of an *atomic bomb* of comparable size.

TNT: Trinitrotoluene, a conventional high intensity explosive, also known as dynamite. The explosive power of *nuclear weapons* is expressed in equivalents of *kilotons* or *megatons* of TNT.

Troposphere: The lowest layer of the *atmosphere,* from the surface to about 12 kilometers, characterized by a general decrease of temperature with increasing altitude. The troposphere contains 90% of the total mass of air, and most of the water vapor in the atmosphere; thus, most precipitation originates in this layer.

See also *Stratosphere, Tropopause.*

Tropopause: The boundary between the *troposphere* and the *stratosphere.*

Ultraviolet radiation: The band of *electromagnetic radiation* with wavelengths in the interval 100 to 400 nanometers. This *radiation* band is further divided into three other regions: UV-A (320–400 nanometers), UV-B (290–320 nanometers), and UV-C (200–290 nanometers). The UV-B radiation has the greatest biological significance.

Underground burst: A nuclear explosion that takes place below the surface of the Earth.

Underwater burst: A nuclear explosion that takes place below the surface of the water, in a sea or ocean.

Uranium [U]: A heavy fissionable chemical element found in nature. Its *isotopes* ^{235}U and ^{238}U are used to build *nuclear weapons*, as well as to operate nuclear power stations.

UV-B: See *Ultraviolet radiation.*

Visible radiation: The band of *electromagnetic radiation* with wavelength in the interval 0.4 to 0.8 micrometers. About half of the solar energy received at the surface is in the visible radiation band.

See also *Infrared radiation, Ultraviolet radiation.*

Washout: The process of *aerosol* removal from the atmosphere through capture by precipitation, particularly below the cloud base.

See also *Fallout, Rainout, Scavenging.*

Wind shear: A situation in which the winds at a location vary in speed and direction at different heights.

X-ray: The band of *electromagnetic radiation* with wavelengths nominally in the interval 0.1 to 10 nanometers. X-rays are emitted by nuclear explosions, but are relatively ineffective at penetrating air.

Yield: The amount of energy released by a nuclear explosion. The yield of a *nuclear weapon* is often expressed in terms of *kilotons (kt)* or *megatons (Mt)*. The energy released by *delayed radioactive fallout* is not usually counted in the yield of a nuclear weapon.

See also *Atomic bomb, Thermonuclear weapon.*

APPENDIX 3

Contributors, Reviewers and Workshop Participants*

ACKERMAN, T.
NASA Ames Research Center
Moffett Field
California 94035
U.S.A.

ADAMS, M.
Marine Sciences Information
Department of Fisheries and
   Oceans
Ontario
Canada

ADDISCOTT, T. M.
Rothamsted Experimental
   Research Station
Harpenden, Herts AL5 2JQ
U.K.

ALBINI, F.
U.S. Forest Service
Missoula, Montana
U.S.A.

ALEKSANDROV, V. V.
Computing Centre
USSR Academy of Sciences
Vavilov St. 40
Moscow 117333
U.S.S.R.

ANTONI, F.
Institute of Biochemistry
Semmelweiss University Medical
   School
1444 Budapest
Hungary

ANTOSHECHKIN, A. G.
Institute of Molecular Genetics
USSR Academy of Sciences
Moscow
U.S.S.R.

ARIAS, L.
CENIAP-FONAIAP
Maracay 2101
Estado Aragua
Venezuela

ATHAVALE, R. N.
National Geophysical Research
   Institute
Hyderbad
India

AUCLAIR, A. N.
Canadian Forest Service
Agriculture Canada
Hull
Quebec
Canada

---

* This list includes participants at workshops and those involved in preparation and review
 of various parts of Volumes I and II. Inclusion in this list should not necessarily be
 construed as an endorsement of the contents of the reports, responsibility for which
 rests solely with the principal authors. Because of the open nature of the study, some
 of those who participated or attended only some parts of a workshop may have been
 unintentionally missed.

AYYAD, M. A.
Botany Department
University of Alexandria
Moharren Bay
Alexandria
Egypt

BAKER, F. W. G.
International Council
  of Scientific Unions
51 Boulevard de Montmorency
75016 Paris
France

BANTA, R. M.
Air Force Geophysics Laboratory
Hanscom AFB
Massachusetts 01731
U.S.A.

BARBER, S. A.
Agricultural Department
Purdue University
West Lafayette
Indiana 47907
U.S.A.

BARDACH, J.
Resources Systems Institute
East-West Center
Honolulu, Hawaii
U.S.A.

BARRACLOUGH, D.
I.C.I.
Jeallot's Hill Research Station
Bracknell
Berks RG12 6EY
U.K.

BARRETT, S. C. H.
Department of Botany
University of Toronto
Toronto, Ontario
Canada

BARRIOS, E.
Instituto Venezolana de
  Investigaciones Cientificas
Caracas
Venezuela

BAZZAZ, F. A.
Biological Laboratories
Harvard University
Cambridge
Massachusetts 02138
U.S.A.

BÉNARD, J.
La Maison de la Chimie
28 rue St. Dominique
75007 Paris
France

BERGSTROM, S. K. D.
Karolinska Institute
P.O. Box 60400
104 01 Stockholm
Sweden

BERRY, J.
Department of Plant Biology
Carnegie Institute of
  Washington
Stanford
California
U.S.A.

BEZUNEH, T.
SAFGRAD
OAU/S TRC
P.O. Box 173
Ougadougou
Burkina-Faso

BIGG, K.
C.S.I.R.O.
Atmospheric Research
Mordialloc
Victoria
Australia

BING, G.
Lawrence Livermore National
  Laboratory
P.O. Box 808
Livermore
California 94550
U.S.A.

BLISS, L. C.
Department of Botany
University of Washington
Seattle
Washington 98195
U.S.A.

BOLIN, B.
Department of Meteorology
University of Stockholm
Arrhenius Laboratory
Stockholm
Sweden

BOLLING, L.
2025 Massachusetts Avenue
Washington, DC 20036
U.S.A.

BORISENKOV, E. P.
Main Geophysical Observatory
Leningrad 194918
U.S.S.R.

BOULDING, E.
Department of Sociology
Dartmouth College
Hanover, New Hampshire 03755
U.S.A.

BRACE, A.
Australia

BRICKNER, Rabbi
Balfour Synagogue
30 West 68th Street
New York, New York 10023
U.S.A.

BROMLEY, J.
Atomic Energy Research
  Establishment
Harwell, Oxon OX11 ORA
U.K.

BROWN-WEISS, E.
Georgetown University Law
  Center
Washington, DC
U.S.A.

BROYLES, A.
Department of Physics
University of Florida
Tallahassee, Florida
U.S.A.

BUDD, W.
University of Melbourne
Parkville, Victoria
Australia

BUDOWSKI, G.
C.A.T.I.E.
Apartado Postal 59
Turrialba
Costa Rica

BUETNER, E. K.
State Hydrological Institute
Leningrad
U.S.S.R.

CAIRNIE, A. B.
Office of the Science Advisor
Department of the Environment
Ottawa, Ontario K1A OH3
Canada

CAMPBELL, G. S.
Nottingham University Agriculture
  School
Loughborough
Leics LE12 5RD
U.K.

CANNELL, M. G. R.
Institute Terrestrial Ecology
Edinburgh Research Station
Scotland
U.K.

CAPUT, C.
Service de l'Environnement
BP 6, 92260 Fontenay aux Roses
France

CARLETON, T. J.
Department of Botany
University of Toronto
Toronto, Ontario
Canada

CASANOVA, J.
Facultad de Ciencias
Prada de Magdalena s/n
Spain

CESS, R.
Laboratory for Planetary
   Atmosphere Research
State University of New York
Stony Brook
New York 11794
U.S.A.

CHAMBERLAIN, A. C.
A.E.R.E., Harwell
Didcot, Oxon OX1 OSH
U.K.

CHERRY, N.
Lincoln College
Canterbury
New Zealand

CHERRY, R. J.
Department of Chemistry
University of Essex
Colchester
CO4 3SQ, Essex
U.K.

CHIOZZA, E.
Asociacion Argentina de
   Ambientalistas
Buenos Aires
Argentina

CHOPRA, V. L.
Genetics Division
Indian Institute of Agricultural
   Research
New Delhi
India

CHURCHILL, D.
Royal Botanic Gardens
Melbourne
Victoria
Australia

CLARK, J. A.
Nottingham University Agriculture
   School
Loughborough
Leics. LE12 5RD
U.K.

CLARKE, M.
Social Sciences
Deakin University
Geelong
Victoria
Australia

CLEMENT, C. F.
Theoretical Physics Division
Atomic Energy Research
   Establishment
Harwell, Oxon OX11 ORA
U.K.

COLBECK, I.
Chemistry Department
Essex University
Colchester
Essex
U.K.

COLEMAN, J. R.
Department of Botany
University of Toronto
Toronto, Ontario
Canada

CONNELL, P. S.
Atmospheric and Geophysical
  Sciences Division
Lawrence Livermore National
  Laboratory
Livermore
California 94550
U.S.A.

COOPER, C. F.
Department of Biology
San Diego State University
San Diego
California 92182
U.S.A.

COTTON, W. C.
Department of Atmospheric
  Science
Colorado State University
Ft. Collins, Colorado
U.S.A.

COVEY, C.
University of Miami
Miami, Florida
U.S.A.

CROPPER, W. P., Jr.
Center for Environmental
  Research
Cornell University
Ithaca, New York 14853
U.S.A.

CRUTZEN, P. J.
Max Planck Institut für Chemie
P.O. Box 3060
D 6500 Mainz
F.R.G.

CUTSHALL, N.
Oak Ridge National Laboratory
P.O. Box X
Oak Ridge Tennessee 37831
U.S.A.

CWYNAR, L.
Department of Botany
University of Toronto
Toronto, Ontario
Canada

DAS, S. K.
Meteorology Department
Lodi Road
New Delhi 110003
India

DAVY, D.
Environmental Science Division
Australian Atomic Energy
  Commission
Sutherland
Australia

DAY, W.
Physics Department
Rothamsted Experimental Station
Harpenden, Herts
U.K.

DENBIGH, K. G.
3/4 St. Andrews Hill
London EC4V 5BY
U.K.

DENG, F.
4040 Glasco Turnpike
Woodstock, New York 12498
U.S.A.

DENNETT, M. D.
Department Agriculture and
Botany
Whiteknights
Reading, Berks RG6 2AH
U.K.

DETLING, J. K.
NRE Laboratory
Colorado State University
Ft. Collins, Colorado 80523
U.S.A.

DILWORTH, J.
Department of Chemistry
University of Essex
Colchester CO4 3SQ, Essex
U.K.

DITCHBURN, R. W.
9 Summerfield Rise
Goring, Reading RG8 ODS
U.K.

DOBROSELSKI, V. K.
Academy of Sciences Computing
Centre
Vavilov St. 40
Moscow 117333
U.S.S.R.

DOLAN, P. J.
Lockheed Missiles and
Space Co., Inc.
Sunnyvale, California 94086
U.S.A.

DÖÖS, B. R.
International Meteorological
Institute
S10691, Stockholm
Sweden

DORODNICYN, A. A.
Computing Centre
USSR Academy of Sciences
Moscow
U.S.S.R.

DOTTO, L.
52 Three Valleys Drive, #9
Don Mills, Ontario M3A 3B5
Canada

DREW, E.
Australian Institute of
Marine Science
Townsville
Queensland
Australia

DRYSDALE, D.
University of Edinburgh
Department of Fire Safety
Engineering
The King's Buildings
Edinburgh, Scotland
U.K.

DUTRILLAUX, B.
Institut Curie
26 rue d'Ulm
75005 Paris
France

EDWARDS, J. H.
Genetics Laboratory
University of Oxford
South Park Road
Oxford OX1 3OY
U.K.

EHRENBERG, L.
Department of Radiobiology
University of Stockholm
Stockholm
Sweden

ELLINGSON, R.
Department of Meteorology
University of Maryland
College Park
Maryland 20742
U.S.A.

ERNSTER, L.
Arrhenius Laboratory
University of Stockholm
S-106 91 Stockholm
Sweden

FARRINGTON, J. W.
Woods Hole Oceanography
  Institute
Woods Hole, Massachusetts 02543
U.S.A.

FEDDES, R.
Institute for Land and Water
  Management Research
P.O. Box 35
6700 AA Wageningen
The Netherlands

FEDOROV, I. V.
D.V.N.C.
USSR Academy of Sciences
Nauka
U.S.S.R.

FEINENDEGEN, L.
Institute of Medicine
Kernforschungsanlage
Jülich
F.R.G.

FIOCCO, G.
Citta Universita
Rome I-00185
Italy

FONTAN, J.
University Paul Sabatier
118 Route de Narbonne
31062 Toulouse, Cedex
France

FORESTER, A.
Innis College
University of Toronto
Toronto, Ontario
Canada

FOUQUART, F.
University de Lille
Lab. d'Opt. Atm.
59650 Villeneuve d'Ascq, Cedex
France

FREEMAN, A. C.
SCOPE Unit
Chemistry Department
University of Essex
Colchester
CO4 3SQ, Essex
U.K.

FRIEND, J. P.
Drexel University
Philadelphia
Pennsylvania 19104
U.S.A.

FROGGATT, P.
Department of Geology
Scarborough College
University of Toronto
Toronto
Ontario
Canada

FUKUSHIMA, Y.
The Science Council of Japan
Minato-ku
Tokyo 106
Japan

GALBALLY, I.
C.S.I.R.O.
Atmospheric Research
Mordialloc
Victoria
Australia

GELEYN, M.
C.M.R.D.
2 Avenue Rapp
75007
Paris
France

GEYER, A.
4500 Massachusetts Avenue, N.W.
Washington, DC 20016
U.S.A.

GHAN, S.
Atmospheric and Geophysical
  Sciences Division
Lawrence Livermore National
  Laboratory
Livermore
California 94550
U.S.A.

GILMORE, F. R.
R & D Associates
Marina del Rey
California 90295
U.S.A.

GINZBERG, A. S.
Institute of Physics of the
  Atmosphere
USSR Academy of Sciences
Moscow
U.S.S.R.

GLANTZ, M.
National Center for Atmospheric
  Research
Boulder, CO 80303
U.S.A.

GOBBI, G. P.
IFA/CNR
CP. 27
0004 Frascati
Italy

GODFREY, J. S.
C.S.I.R.O.
Division of Oceanography
Hobart
Australia

GOLDSMITH, P.
Meteorological Office
Bracknell
Berks
U.K.

GOLDSTEIN, G.
Department de Biology
Universidad de los Andes
Merida
Venezuela

GOLITSYN, G. S.
Institute of Physics of the
  Atmosphere
USSR Academy of Sciences
Moscow
U.S.S.R.

GOLUBOVA, N. A.
Council on Problems of the
  Biosphere
USSR Academy of Sciences
Moscow
U.S.S.R.

GONTAREV, B. A.
Systems Research Institute
USSR Academy of Sciences
Moscow
U.S.S.R.

GOPALASIRAM, R.
Scientific Adviser to MOD
South Bolck
New Delhi
India

GREEN, O.
Faculty of Science
Open University
Milton Keynes
MK7 6AA
U.K.

GREENWOOD, D.
Natural Vegetation Research
  Station
Wellesbourne
Warwick
CV35 9EF
U.K.

GREGORY, P. J.
Department of Soil Science
University of Reading
Whiteknights, Reading RG6 2AY
U.K.

GRIGORIEV, A. A.
L.G.P.I.
Moiki nab., 48
Leningrad 191186
U.S.S.R.

GRIME, J. P.
Unit of Comp. Plant Ecology
Department of Botany
University of Sheffield
U.K.

GROVER, H.
Department of Biology
University of New Mexico
Albuquerque, New Mexico 87131
U.S.A.

HALLETT, J.
Desert Research Institute
Reno, Nevada
U.S.A.

HAHN, J.
Max Planck Institut für Chemie
P.O. Box 3060
D 6500 Mainz
F.R.G.

HANSON, J.
U.S.D.A. Agricultural
  Research Station
Ft. Collins, CO
U.S.A.

HARRISON, R. M.
Chemistry Department
Essex University
Colchester, Essex
U.K.

HARVEY, T. F.
Atmospheric and Geophysical
  Sciences Division
Lawrence Livermore National
  Laboratory
Livermore, California 94550
U.S.A.

HARWELL, C. C.
Center for Environmental
  Research
303 Corson Hall
Cornell University
Ithaca, New York 14853
U.S.A.

HARWELL, M. A.
Center for Environmental
  Research
303 Corson Hall
Cornell University
Ithaca, New York 14853
U.S.A.

HAVAS, M.
Institute for Environmental
  Studies
University of Toronto
Toronto
Ontario
Canada

HERRERA, R.
Instituto Venezolana de
  Investigaciones Cientificas
Centro de Ecologia
APDO 1827, Caracas
Venezuela

HESBURGH, Rev. T.
President
University of Notre Dame
Notre Dame
Indiana 46556
U.S.A.

HOBBS, P.
Atmospheric Sciences Department
University of Washington
Seattle, Washington
U.S.A.

HOLLAND, G.
Bureau of Meteorology Research
  Centre
Melbourne
Australia

HUGHES, H.
New Zealand

HUNT, B.
C.S.I.R.O.
Atmospheric Research
Private Bag 1
Mordialloc
Victoria 3195
Australia

HUNT, E.
The Blackett Laboratory
Imperial College
London SW7 2BZ
U.K.

HUTCHINSON, T. C.
Institute for Environmental
  Sciences
University of Toronto
Toronto, Ontario M5S 1A1
Canada

HUXLEY, A. Sir
The Royal Society
6 Carlton House Terrace
London SWY 5AG
U.K.

IIJIMA, Dr.
Nagoya University
Nagoya-shi
Japan

IMAHORI, S.
Hiroshima Womens University
Hiroshima
Japan

ISARD, W.
Peace Science
Cornell University
ITHACA, NY 14853
U.S.A.

IVANITCHSHEV, V. V.
Leningrad Research Computing
  Centre
USSR Academy of Sciences
Leningrad
USSR

IZRAEL, Ya A.
USSR State Committee for
  Hydrometeorology and Control
  of the Natural Environment
Moscow
U.S.S.R.

JACOBSEN, J.
Dept. of Geography
University of Colorado
Boulder, CO
U.S.A.

JASPER, D.
Bureau of Meteorology
Melbourne, Victoria
Australia

JEFFERIES, G. W.
C.S.I.R.O.
Marine Laboratories
Hobart
Australia

JEFFERIES, R. L.
Department of Botany
University of Toronto
Toronto, Ontario
Canada

JONES, A.
Department of Chemical
  Engineering
Imperial College
London
U.K.

JONES, G.
Pan Heuristics
Los Angeles, California
U.S.A.

JOUSMA, G.
I.G.W.M.C.
TNO-DGV Institute
P.O. Box 285
2600 AG Delft
The Netherlands

JOVANOVIC, P.
8 Omladinskih Brigada
11070, Beograd
Yugoslavia

KAHN, I.
World Muslim Congress
P.O. Box 5050
Karachi
Pakistan

KANG, S. W.
Nuclear Test Engineering Division
Lawrence Livermore National
  Laboratory
Livermore, California 94550
U.S.A.

KAROL, I.
Main Geophysical Observatory
Leningrad
U.S.S.R.

KATZ, A. M.
7 Stevenage Circle
Rockville, MD
U.S.A.

KEAST, A.
Department of Biology
Queens University
Kingston
Ontario
Canada

KELLY, G. J.
C.S.I.R.O.
Marine Laboratories
Hobart
Australia

KELLY, J. R.
Ecosystems Research Center
Cornell University
Ithaca
New York 14853
U.S.A.

KELLY, P. M.
Climatic Research Unit
School of Environmental Sciences
University of East Anglia
Norwich NR4 7TJ
U.K.

KIEHL, J. T.
National Center for Atmospheric
  Research
P.O. Box 3000
Boulder, Colorado 80309
U.S.A.

KIRILL, Archbishop
Bishop of Vyborg
Leningrad Theological Academy
Leningrad
U.S.S.R.

KIRCHMANN, R.
Department of Radiology
CEN-SCK
Boertang 200 2400 Mol
Belgium

KLEIN, J.
Institute Francais Relations
  Internationales
6 rue Ferrus
75014 Paris
France

KNOX, J. B.
Atmospheric and Geophysical
  Sciences Division
Lawrence Livermore National
  Laboratory
Livermore
California 94550
U.S.A.

KOKOSHIN, A.
USSR Academy of Sciences
Leninsky Prospect 14
Moscow V-17
U.S.S.R.

KONDRATYEV, K. Ya.
Institute for Lake Research
USSR Academy of Sciences
Leningrad
U.S.S.R.

KOROTKEVICH, O. E.
Institute of Ozerovendenya
USSR Academy of Sciences
Leningrad
U.S.S.R.

KOSTROWICKI, M. J.
Institute of Geography
Polish Academy of Sciences
Warsaw
Poland

KRAPIVIN, V. F.
Applied Mathematics Laboratory
I.R.E.
USSR Academy of Sciences
Moscow
U.S.S.R.

KRISHNASWAMY, S.
School Biological Sciences
Kamaraj University
Madurai 65021
India

KUDERSKII, L. A.
N.P.O. Promribvod
Leningrad
U.S.S.R.

KUZIN, A. M.
Computing Centre
USSR Academy of Sciences
Vavilov St. 40
Moscow 117333
U.S.S.R.

KYBAL, D.
Federal Emergency Management
  Agency
Washington, DC
U.S.A.

LABOURIAU, M. S.
Instituto Venezolana de
  Investigacione Cientificas
Caracas
Venezuela

LAFUMA, Dr.
C.E.N., BP 6
92260 Fontenay aux Roses
France

LAKSHMIPATHI, N.
Institute of Nuclear Medicine and
  Applied Sciences
Delhi 110007
India

LA RIVIÈRE, J. W. M.
International Institute for
  Hydraulic Engineering
P.O. Box 3015
2601, DA Delft
The Netherlands

LATERJET, R.
Institute Curie
26 rue d'Ulm
75005 Paris
France

LAVAL, K.
Laboratoire de Météorologie
  Dynamique
24 rue d'Ulm
75005 Paris
France

LEAF, A.
Harvard Medical School
Cambridge
Massachusetts
U.S.A.

LEITH, C.
Lawrence Livermore National
  Laboratory
P.O. Box 808
Livermore
California 94550
U.S.A.

LENOBLE, J.
UER Physique Fondamentale
59655 Villeneuve d'Ascq
Cedex
France

LEVITT, J.
Carnegie Institute Washington
Stanford
California 94305
U.S.A.

LIFTON, R.
Dept. of Psychiatry and
  Psychology
City University of New York
New York
NY 10019
U.S.A.

LIMBURG, K.
Ecosystems Research Center
Cornell University
Ithaca, New York
U.S.A.

LIVERMAN, D.
Dept. of Geography
University of Wisconsin
Madison, WI 53706
U.S.A.

LLOYD, J. W.
Department of Geological Sciences
University of Birmingham
P.O. Box 363
Birmingham B15 2TT
U.K.

LOEWE, W.
Physics Department
Lawrence Livermore National
  Laboratory
Livermore
California 94550
U.S.A.

LONDON, J.
University of Colorado
Department of Astrophysical,
  Planetary and
  Atmospheric Sciences
P.O. Box 391
Boulder, Colorado 80309
U.S.A.

LOPES, E. S.
Instituto Agronomico
Campinas, S.P.
Brazil

LOWE, D.
Institute of Nuclear Sciences
D.S.I.R.
Lower Hutt
New Zealand

LUKYANOV, N. K.
Computing Centre
USSR Academy of Sciences
Vavilov St. 40
Moscow 117333
U.S.S.R.

LUNDBERG, H.
Royal Swedish Academy of
  Sciences
P.O. Box 50005-104 05
Stockholm
Sweden

LUNDBOM, P. O.
National Institute for Materials
  Testing
Stockholm
Sweden

LUTHER, F. M.
Atmospheric and Geophysical
  Sciences Division
Lawrence Livermore National
  Laboratory
Livermore, California 94550
U.S.A.

LYSTAD, M.
National Institute of Mental Health
Rockville, MD 20857
U.S.A.

MacCRACKEN, M. C.
Atmospheric and Geophysical
  Sciences Division
Lawrence Livermore National
  Laboratory
Livermore, California 94550
U.S.A.

MAHONY, R.
The Chancery
1105 N. Lincoln Street
Stockton, California 95204
U.S.A.

MALONE, R.
Los Alamos National Laboratory
Los Alamos, New Mexico
U.S.A.

MALONE, T.
5 Bishop Road
West Hartford
Connecticut 06119
U.S.A.

MANINS, P.
C.S.I.R.O.
Atmospheric Research
Mordialloc, Victoria
Australia

MARIANA, J. C.
Centre Recherche Agronomique
  de Tours-Nouzilly
37380 Monnaie
France

MARTELL, D.
Faculty of Forestry
University of Toronto
Toronto, Ontario
Canada

MARTON-LEFEVRE, J.
51 Boulevard de Montmorency
75016 Paris
France

McEWAN, A. C.
National Radiation Laboratory
Christchurch
New Zealand

McGREGOR, J.
C.S.I.R.O.
Atmospheric Research
Private Bag 1
Mordialloc
Victoria
Australia

McILROY, I. C.
Mount Eliza, Victoria
Australia

McLAREN, A. L.
MRC Mammalian Development
Unit
Wolfson
London NW1 2HE
U.K.

McNAUGHTON, S. J.
Biological Research Laboratory
Syracuse University
Syracuse, New York 13210
U.S.A.

MEDINA, E.
Instituto Venezolana de Investiga-
ciones Cientificas
Caracas
Venezuela

MEEMA, K.
Department of Botany
University of Toronto
Toronto, Ontario M5S 1A1
Canada

MENON, M. G. K.
Member of Planning Commission
Yojana Bhavan
New Delhi
India

MENSHUTKIN, V. V.
Institute of Evolutionary Physics
and Biochemistry
USSR Academy of Sciences
Leningrad
U.S.S.R.

METALNIKOV, A. P.
GKNT USSR
Gorki Street
Moscow
U.S.S.R.

MIKHAILOV, V. V.
Leningrad Computer Centre
USSR Academy of Sciences
Leningrad
U.S.S.R.

MILETI, D.
Dept. of Sociology
Colorado State University
Fort Collins, CO 80523
U.S.A.

MITCHELL, J. K.
Dept. of Geography
Rutgers University
New Brunswick
NJ 08903
U.S.A.

MITRA, A. P.
National Physical Laboratory
Hillside Road
Pusa
New Delhi 110012
India

MOISEEV, N. N.
USSR Academy of Sciences
Vavilov St. 40
Moscow 117333
U.S.S.R.

MOLENKAMP, C. R.
Atmospheric and Geophysical
  Sciences Division
Lawrence Livermore National
  Laboratory
Livermore
California 94550
U.S.A.

MOOK, W. G.
University of Groningen
Westersingel 34
9718 CM Groningen
The Netherlands

MOORE, D. J.
C.E.G.B.
Kelvin Avenue
Leatherhead, Surrey
U.K.

MULHOLLAND, G.
B356
National Bureau of Standards
Washington, DC 20234
U.S.A.

MUNN, R. E.
Institute of Environmental Studies
University of Toronto
Toronto, Ontario M5S 1A4
Canada

MYERS, R. L.
Archbold Biological Station
P.O. Box 2057
Lake Placid
Florida 33852
U.S.A.

NAIDU, J. R.
Brookhaven National Laboratory
Associated Universities, Inc.
Upton, New York 11973
U.S.A.

NAGAI, M.
United Nations University
Toho Seimi Building
Tokyo 150
Japan

NECHAYEV, L. G.
USSR Academy of Sciences
Moscow
U.S.S.R.

NICHOLLS, N.
Bureau of Meteorology Research
Mordialloc
Australia

NIKOLSKI, G. A.
Leningrad State University
Leningrad
U.S.S.R.

NIX, H.
C.S.I.R.O.
Water and Land Resources
Canberra
Australia

O'CONNOR, D. J.
Phillips Institute of Technology
Bundoora, Victoria
Australia

OHASTI, J.
Universidad Central de Venezuela
Caracas 1041A
Venuzuela

OHKITA, T.
Nat Nagoya Hospital
Nagoya-shi
Japan

OLDS, G.
Canadian Pacific University
Anchorage
Alaska 99508
U.S.A.

OSAEV, S.
USSR Academy of Sciences
Akademia Nauk
Leninsky Prospekt 14
Moscow V-17
U.S.S.R.

PACENKA, S.
Environmental Research
468 Hollister
Cornell University
Ithaca
New York
U.S.A.

PACKHAM, D.
Rural Fire Research
South Melbourne
Victoria
Australia

PALAU, J.
Instituto de Biologia
CSIC c/Jorge Girona
Barcelona 34
Spain

PALTRIDGE, G.
C.S.I.R.O.
Atmospheric Research
Mordialloc
Victoria
Australia

PATTERSON, E.
Georgia Institute of Technology
Atlanta
Georgia
U.S.A.

PEARSON, R.
AREC, Ltd.
Whiteshell Nuclear Research
  Establishment
Pinawa, Manitoba ROE ILO
Canada

PENNER, J. E.
Atmospheric and Geophysical
  Sciences Division
Lawrence Livermore National
  Laboratory
Livermore, California 94550
U.S.A.

PERCIVAL, I. C.
Queen Mary College
University of London
Mile End Road
London
U.K.

PETERSON, K. R.
Atmospheric and Geophysical
  Sciences Division
Lawrence Livermore National
  Laboratory
Livermore, California 94550
U.S.A.

PETTY, R.
Australia

PIMENTEL, D.
Department of Entomology
Comstock Hall
Cornell University
Ithaca
New York 14853
U.S.A.

PITTOCK, A. B.
C.S.I.R.O.
Atmospheric Research
Private Bag 1
Mordialloc, Victoria 3195
Australia

PLOCQ, V.
SCOPE
51 Boulevard de Montmorency
75016, Paris
France

PORTER, J.
Long Ashton Research Station
Bristol BS18 9AS
U.K.

POSADAS, R.
College of Science
University of the Philippines
Philippines

POSTNOV, A.
USSR Academy of Sciences
Moscow V-71
U.S.S.R.

POTTER, D. J.
Sydney University
N.S.W.
Australia

POZDNYAKOV, D. V.
Institute of Ozerovedenia
USSR Academy of Sciences
Leningrad
U.S.S.R.

PRUPPACHER, H.
University of Mainz
Mainz
F.R.G.

RADKE, L.
University of Washington
Seattle
Washington
U.S.A.

RADKIN, A.
Australia

RAINEY, R. C.
Elmslea
Old Risborough Road
Stoke Mandeville
Bucks HP225XJ
U.K.

RAMANA, R.
Tata Institute of Fundamental
  Research
Bombay
India

RASBASH, D. J.
University of Edinburgh
Department of Fire Safety
  Engineering
The King's Buildings
Edinburgh
Scotland
U.K.

RASCHKE, E.
Institut für Geophysik zu Koln
D-5000 Koln 41
F.R.G.

RATHJENS, G.
Massachusetts Institute of
  Technology
Cambridge, Massachusetts 02138
U.S.A.

RAWSON, G.
P.O. Box 873
Del Mar, California 92014
U.S.A.

REITTER, T.
Nuclear Test Engineering Division
Lawrence Livermore National
  Laboratory
Livermore, California 94550
U.S.A.

RENOUX, P.
Faculte des Science
12 Ave. Charles de Gaulle
94000 Creteil
France

RIEBSAME, W.
Dept. of Geography
University of Colorado
Boulder, CO 80309
U.S.A.

RISCH, S. J.
University of California
1050 San Pablo Avenue
Berkeley, California
U.S.A.

RITCHIE, J. C.
Department of Biology
University of Toronto
Toronto, Ontario
Canada

ROBOCK, A.
Department of Meteorology
University of Maryland
College Park, Maryland
U.S.A.

ROBSON, M. H.
The Grassland Research Institute
Hurley
Maidenhead, Berks SL6 5LR
U.K.

RODHE, H.
Department of Meteorology
Arrhenuis Laboratory
University of Stockholm
Stockholm
Sweden

ROSENBLITH, W.
National Research Council
2101 Constitution Avenue
Washington, DC 20148
U.S.A.

ROSER, D. J.
Hawkesbury Agricultural College
N.S.W.
Australia

ROSSWALL, T.
Department of Water in
   Environment and Society
University of Linkoping
Sweden

ROTBLAT, J.
8 Asmara Road
London, NW2 3ST
U.K.

ROYER, P.
Centre Meteor.
42 Avenue Coriolis
31057 Toulouse Cedex
France

RUNCIMAN, A.
Australian National University
Canberra
Australia

SAGAN, C.
Laboratory for Planetary
   Studies
Cornell University
Ithaca
New York 14853
U.S.A.

SAGDEEV, R. Z.
USSR Academy of Science
Akad. Nauk
Leninsky Prospect 14
Moscow V-71
U.S.S.R.

SALAM, A.
Third World Academy of Sciences
   International Center for
   Theological Studies
P.O. Box 586
I-34100 Trieste
Italy

SALINGER, M. J.
New Zealand Meteorological
   Service
Wellington
New Zealand

SANHUEZA, E.
Ingenieria
Instituto Venezolana de
   Investigaciones Cientificas
Caracus
Venezuela

SCHLICHTER, T.
Department de Ecologia
1417 Buenos Aires
Argentina

SCHNEIDER, S.
National Center for Atmospheric
  Research
P.O. Box 8000
Boulder, Colorado 80307
U.S.A.

SCOTT, D.
Drew University
Madison
New Jersey
U.S.A.

SESHAGIRI, N.
I.P.A.G.
Electronics Commission
Madangir Road
New Delhi
India

SEYMOUR, A.
5855 Oberlin N.E.
Seattle
Washington 98105
U.S.A.

SGRILLO, R.
CENA/USP
C.P. 96
Piracicaba
Brazil

SHAPIRO, C. S.
San Francisco State University
1600 Holloway Avenue
San Francisco
California 94132
U.S.A.

SHIMAZU, Y.
Faculty of Science
Nagoya University
Furo-Cho, Chikusa
Nagoya-shi,
Aichi 464
Japan

SHINE, K. P.
University of Oxford
Atmospheric Physics
South Parks Road
Oxford OX1 3PU
U.K.

SHINN, R.
Union Theological Seminary
3041 Broadway
New York
New York 10027
U.S.A.

SHIRATORI, R.
Department of Government
University of Essex
Colchester
CO4 3SQ, Essex
U.K.

SIEGEL, R. H.
c/o International Institute
  for Strategic Studies
London WC2E 7NQ
U.K.

SIMPSON, R. W.
Griffith University
Nathan
Queensland
Australia

SINCLAIR, T.
Agricultural Research Service
Agronomy Department
University Florida
Gainsville, Florida
U.S.A.

SINCLAIR, W.
7910 Woodmont Avenue
Suite 1016
Bethesda
Maryland
U.S.A.

SISLER, D.
Warren Hall
Cornell University
Ithaca, New York 14853
U.S.A.

SKRYABIN, G. K.
USSR Academy of Sciences
Akademia Nauk
Leninsky Prospect 14
Moscow V-71
U.S.S.R.

SLATYER, R. O.
Australian National University
P.O. Box 474
Canberra, ACT 2601
Australia

SLEPYAN, E. I.
Institute of Zoology
USSR Academy of Sciences
Leningrad
U.S.S.R.

SLINGO, A.
British Meteorological Office
London Road
Bracknell, Berkshire
U.K.

SLINN, G.
Battelle Pacific Northwest
    Laboratory
Richland, Washington
U.S.A

SLOVIC, P.
Decision Research Inc.
Eugene, OR 97401
U.S.A.

SMALL, R.
Pacific Sierra Research
    Corporation
Los Angeles, California 90025
U.S.A.

SMIRNOV, N. N.
Institute of Evolutionary
    Morphology and Ecololgy
    of Animals
USSR Academy of Sciences
Moscow
U.S.S.R.

SMIRNYAGIN, V. A.
Presidium of Academy of Sciences
Moscow
U.S.S.R.

SMITH, H.
N.R.P.B.
Chilton Didcot
Oxon OX11 ORQ
U.K.

SMOKTII, O. I.
Leningrad Research Computing
    Centre
USSR Academy of Sciences
Leningrad
U.S.S.R.

SPEED, R.
R & D Associates
Marina del Rey, California 90295
U.S.A.

SPICER, B.
University of Melbourne
Parkville, Victoria
Australia

STASSEN, G. S.
Baptist Theological Seminary
2825 Lexington Road
Louisville, Kentucky 40280
U.S.A.

STEADMAN, P.
Faculty of Science
Open University
Milton Keynes MK7 6AA
U.K.

STENCHIKOV, G. L.
Computing Centre
USSR Academy of Sciences
40 Vavilov St.
Moscow
U.S.S.R.

STEPHENS, G. L.
C.S.I.R.O.
Division of Atmospheric Research
Private Bag 1
Mordialloc
Victoria 3195
Australia

STETSENKO, G. A.
USSR Academy of Sciences
Moscow V-71
U.S.S.R.

STEWART, R. B.
Crop Production Division
Agriculture Canada
Ottawa
Ontario KIA 0G6
Canada

STOCKS, B.
Great Lakes Forest Research
   Centre
Sault Ste. Marie, Ontario
Canada

SU, J. C.
Institute of Agricultural
   Chemistry
National Taiwan University
Taipei 6
Taiwan

SUBRAHMANYAM, K.
I.D.S.A.
Sapru House
Barakhamba Road
New Delhi 110001
India

SUMMERFIELD, A.
Birkbeck College
University of London
London
U.K.

SVIREZHEV, Y. M.
Laboratory of Mathematical
   Ecology
USSR Academy of Sciences
Moscow 117333
U.S.S.R.

SVOBODA, J.
Department of Biology
University of Toronto
Mississauga
Canada

TAMM, C. O.
Department of Ecology
Swedish University of Agricultural
   Science
Uppsala
Sweden

TARKO, A. M.
Computing Centre
USSR Academy of Sciences
Vavilov St. 40
Moscow 117333
U.S.S.R.

TASSIOS, F.
Rural Fire Research
South Melbourne
   Victoria
Australia

TAO, S.
c/o Ms. Wu Ganmei
China Association of Science and
   Technology
Beijing
Peoples Republic of China

TAYLOR, H. W.
Department of Biology
University of Toronto
Mississauga
Canada

TEAL, J. M.
Woods Hole Oceanographic
  Institute
Woods Hole
Massachusetts 02543
U.S.A.

TEWARSON, A.
Factory Mutual Research Corp.
1151 Boston Providence Turnpike
Norwood
Massachusetts 02061
U.S.A.

THOMAS, P. H.
Fire Research Station
Borehamwood, Herts
U.K.

THOMPSON, S.
National Center for Atmospheric
  Research
P.O. Box 3000
Boulder, Colorado 80307
U.S.A.

THRUSH, B. A.
Department of Physical Chemistry
Cambridge
U.K.

TINKER, P. B.
Rothamsted Exp. Stat.
Harpenden, Herts AL5 2JQ
U.K.

TOON, O. B.
NASA Ames Research Center
Moffett Field, California
U.S.A.

TOWNES, C. H.
Department of Physics
LeConte Hall
University of California
Berkeley
California 94720
U.S.A.

TRAVESI, A.
Junta Energia Nuclear
Medio Cliubreule
Madrid 3
Spain

TRIPOLI, G.
Department of Atmospheric
  Science
Colorado State University
Ft Collins, Colorado 80523
U.S.A.

TSUSHIMA, K.
United Nations University
Tokyo
Japan

TUCKER, B.
C.S.I.R.O.
Atmospheric Research
Private Bag 1
Mordialloc, Victoria 3195
Australia

TURCO, R. P.
R & D Associates
P.O. Box 9695
4640 Admiralty Way
Marina Del Rey
California 90295
U.S.A.

UCHIJIMA, Z.
National Institute of
  Agri-Environmental Sciences
Ibaraki-ken
Japan

UNSWORTH, M. H.
I.T.E.
Bush Estate
Penicuik
Midlothian EH26 OQB
Scotland
U.K.

URABE, T.
Computer Centre
Nagoya University
Furocho
Nagoya-shi, 464
Japan

VAN DER HEIJDE, P. K. M.
Butler University
4600 Sunset Avenue
Indianapolis
Indiana 46208
U.S.A.

VAN DER MEER, J.
Division of Biological Science
University of Michigan
Ann Arbor
Michigan 48109
U.S.A.

VERSTRAETE, M.
National Center for Atmospheric
  Research
P.O. Box 3000
Boulder, Colorado 80307
U.S.A.

VILLEVIEILLE, A.
73-77 rue de Se vres.
92100 Boulogne
France

VOICE, M.
Bureau of Meteorology Research
Centre
Melbourne
Australia

WALKER, B.
University of the Witwatersrand
1 Jan Smuts Ave.
Johannesburg 2001
South Africa

WALTON, D. W. H.
British Antarctic Survey
High Cross
Cambridge
U.K.

WALTON, J. J.
Atmospheric and Geophysical
  Sciences Division
Lawrence Livermore National
  Laboratory
Livermore
California 94550
U.S.A.

WARNER, Sir F.
SCOPE Unit
Department of Chemistry
University of Essex
Colchester
CO4 3SQ
Essex
U.K.

WARREN, P.
The Royal Society
6 Carlton House Terrace
London SW1Y 5AG
U.K.

WARRICK, R.
School of Environmental Sciences
University of East Anglia
Norwich
NR4 7TJ
U.K.

WEERAMANTRY, C.
Moash University
Australia

WEIN, R. W.
Fire Research Centre
University of New Brunswick
Canada

WESTGATE, J.
Department of Geology
University of Toronto
Mississauga
Ontario
Canada

WHITE, G. F.
Institute of Behavioral Science
University of Colorado
Boulder
CO 80309
U.S.A.

WIK, M.
Defense Mat. Admin.
Electronics Directorate
S115 88 Stockholm
Sweden

WILLIAMS, M.
Department of Nuclear
  Engineering
Queen Mary College
London
U.K.

WILLIAMS, P. G.
New South Wales Department of
  Agriculture
Australia

WINTER, H.
Commission for Environment
Wellington
New Zealand

WISE, K.
Australian Radiation Laboratory
Yallambie
Australia

WOLFENDALE, A. W.
Department of Physics
University of Durham
South Road
Durham
DH1 3LE
U.K.

WOODWELL, G. M.
Marine Biological
  Laboratory
Woods Hole
Massachusetts 02543
U.S.A.

WORREST, R. C.
Corvallis Environmental Research
  Laboratory
U.S. Environmental Protection
  Agency
Corvallis Oregon 97330
U.S.A.

YANASE, M.
Rev. Pont. University Greg
Piazza della Pillotta 4
00187 Rome
Italy

YOON, B. L.
R & D Associates
Marina del Rey
California 90295
U.S.A.

ZAVARZIN, G. A.
USSR Academy of Sciences
Profsojusnaja 7a
117312 Moscow
U.S.S.R.

ZHOU XIUJI, Dr.
Academy of Meteorological
  Sciences
Peoples Republic of China

ZUBROW, E.
877 Building 4 Spalding
　Quadrangle
University of New York
Buffalo, New York 14621
U.S.A.

ZUCKERMAN, Lord S.
The Shooting Box
Burnham Thorpe
King's Lynn
Norfolk
U.K.

Index